Ecological Studies, Vol. 137

Analysis and Synthesis

Edited by

M. M. Caldwell, Logan, USA
G. Heldmaier, Marburg, Germany
O. L. Lange, Würzburg, Germany
H. A. Mooney, Stanford, USA
E.-D. Schulze, Jena, Germany
U. Sommer, Kiel, Germany

Ecological Studies

Volumes published since 1992 are listed at the end of this book.

Springer

Berlin
Heidelberg
New York
Barcelona
Hong Kong
London
Milan
Paris
Singapore
Tokyo

F. Rodà   J. Retana   C. A. Gracia   J. Bellot   (Eds.)

Ecology of Mediterranean Evergreen Oak Forests

With 84 Figures and 60 Tables

 Springer

Ferran Rodà
Javier Retana
Carlos A. Gracia
Center for Ecological Research and Forestry Applications (CREAF)
Autonomous University of Barcelona
08193 Barcelona
Spain

Juan Bellot
Department of Ecology
Faculty of Sciences
University of Alicante
PO Box 99
03080 Alicante
Spain

ISSN 0070-8356
ISBN 3-540-65019-9 Springer-Verlag Berlin Heidelberg New York

Library of Congress Cataloging-in-Publication Data

Ecology of Mediterranean evergreen oak forests / F. Roda ... [et al.], (eds.) p. cm. (Ecological studies / Analysis and synthesis, 0070-8356; vol. 137) Includes bibliographical references and index. ISBN 3-540-65019-9 (hardcover: alk. paper) 1. Forest ecology – Mediterranean Region. 2. Holm oak – Ecology – Mediterranean Region. 3. Forst ecology – Spain – Catalonia. 4. Holm oak – Ecology – Spain - Catalonia. 5. Plant ecophysiology – Mediterranean Region. 6. Biogeochemistry – Mediterranean Region. 7. Plant ecophysiology – Spain – Catalonia. 8. Biogeochemistry – Spain – Catalonia. I. Roda, F. (Ferran), 1953 – II. Series: Ecological studies; v. 137.
Q314.5.E36 1999; 577.3'09182'2–dc21; 98-44830

Cover Design: Design & Production GmbH, Heidelberg
Camera ready by Ulrich Kunkel, Reichartshausen

SPIN 10566121   31/3137   5 4 3 2 1 0 – Printed on acid-free paper

Preface

Broadleaved evergreen forests where holm oak (*Quercus ilex* L.) is almost the only canopy tree are a distinctive ecosystem of the Mediterranean Basin. Biogeographically, these forests lie between cool-temperate deciduous forests to the north and drier shrublands to the south. In a more general view, they are ecologically intermediate between these deciduous forests and humid warm-temperate evergreen forests, such as the laurisilva of the Macaronesian islands or the broadleaved evergreen forests of eastern Asia. Holm oak forests are characterized by small-stature trees (usually 5 to 12 m tall), which are slow growing, cast a deep shade, and have small, evergreen, sclerophyllous leaves. Summer drought is the major environmental constraint.

Mediterranean broadleaved evergreen forests have been underrepresented in the open literature. For example, The Woodlands Data Set of the International Biological Program (IBP) contained data from just one holm oak forest plot, Le Rouquet in southern France (Reichle 1981). Also, some models of European forests consider just two major forest types, broadleaved deciduous and evergreen conifers, forgeting the extensive forests and woodlands of holm oak and cork oak (*Quercus suber* L.), which are the major evergreen oak tree species in the Mediterranean Basin. The extensive literature on Mediterranean-type ecosystems has also largely neglected these forests, being centred mostly on Mediterranean-type shrublands.

To bridge this gap in knowledge, a research programme on the ecology of holm oak forests was started in 1978 under the direction of Ramon Margalef, Jaume Terradas, and Antoni Escarré. This programme brought together researchers from three Spanish universities: the University of Barcelona, the Autonomous University of Barcelona, and the University of Alicante.

The programme was originally intended to obtain a functional understanding of holm oak forests, combining three approaches: catchment budgets, plot-level analyses of the distribution and cycling of matter, and ecophysiological studies. Its inception was inspired by the IBP plot studies, particularly those conducted by Maurice Rapp in Montpellier (France) and by P. Duvigneaud and Simone Denaeyer-De Smet in Belgium, and by the catchment ecosystem studies pioneered by F. Herbert Bormann and Gene E. Likens at Hubbard Brook (USA). The programme later evolved to include also demographic approaches to forest dynamics.

In the 1970s, closed-canopy holm oak forests were still considered the paradigm of the natural, undisturbed ("climacic") vegetation in those parts of the Mediterranean Basin receiving enough rainfall. Later findings pointed out that the prevalence of holm oak was closely related to it being favoured by humans from about 5000 years B.P. This raises the question of what would be the contribution of deciduous oaks and pines to the natural landscape. To some extent, our own programme reflected such changing views. We started selecting plots and catchments that appeared as homogeneous and undisturbed as possible. Very soon it became obvious that these holm oak forests bore the imprint of past cultivation or coppicing and that their present structure, often characterized by high-density stands of stool resprouts, was bound to change as a result of abandonment or shifts in forest management. In later studies we therefore placed more emphasis on the effects of disturbance and forestry practices, and on the relationships between holm oak and pines in mixed forests.

The present book brings together and summarizes what we have learnt about the ecology of Mediterranean evergreen forests dominated by holm oak in Catalonia, northeast Spain. Some results have been published before, either in international publications of scattered disciplines, or in sources or languages little available to foreign audiences. Nonetheless, most chapters of this book present new material, and attempt a new synthesis of the involved topics. Also, data have been often updated through increased temporal and spatial coverage.

Chapter 1 introduces holm oak and holm oak forests, particularly for those readers that are unfamiliar with Mediterranean forests. Chapter 2 describes our two major experimental areas (Montseny and Prades) where most of the results of this book were obtained. Chapters 3 and 4 describe the range of stand structures found in holm oak forests and analyze the patterns of above- and belowground biomass and production of our experimental sites. Chapters 5 to 7 deal with the ecology of resprouts, seeds, and seedlings of holm oak. Chapter 8 integrates the information on the above recruitment processes and formulates a demographic model of stand dynamics of mixed holm oak-Aleppo pine forests. Chapters 9 to 11 take an ecophysiological view to light and water use by holm oak leaves and canopies. Chapter 12 integrates the above knowledge into a water-driven model of ecosystem functioning. Chapters 13 to 18 focus on the distribution and cycling of matter at the stand level. Chapters 19 to 21 summarize water and element budgets at the catchment scale, and attempt to model the ecosystem functioning. The effects of fire and of experimental forestry practices are treated in Chapters 22 and 23. Finally, Chapters 24 and 25 outline the ecology of two major animal groups in the holm oak forest: soil arthropods and birds.

This book was made possible through the efforts of many people and institutions. We would like to express our gratitude to those investigators who were deeply involved in the first research phase, particularly: Lluís Ferrés who studied the biomass, primary production, and plant nutrient contents in the permanent plot at Montseny; Claret Verdú who studied litterfall and litter decay at the same site and other forest types; and Luis López-Soria who in 1978 devised the first application of dimensional analysis for our studies. We are also grateful to F.H. Bormann, George Hornberger, and Keith Beven for their early advice on catchment studies, to Carles Gené for his collaboration in fieldwork and data analysis, and to Alicia Bernabé for her contribution in data analysis.

We are grateful to the former Institute of Nature Conservation (ICONA) of the Spanish Ministry of Agriculture, and to the Department of Agriculture, Livestock and Fisheries of the Catalan government, who throughout these years have made possible the continued use of the experimental areas at Montseny and Prades for research. Our work was funded by the US-Spanish Joint Committee for Science and Technology, the European Commission DGXII, the Spanish Ministry of Education through the former CAICYT and then the CICYT, the Spanish Ministry of Agriculture through the INIA and the former ICONA, the Catalan government through the CIRIT and IRTA, and the government of the Valencian Autonomous Community through the Center of Environmental Mediterranean Studies (CEAM).

December 1998

F. Rodà
J. Retana
C.A. Gracia
J. Bellot

References

Reichle DE (ed) (1981) Dynamic properties of forest ecosystems. Cambridge University Press, Cambridge

Contents

Part 3 Light Harvesting and Gas Exchange 119

Contributors

EVA ALBEZA, Department of Ecology, University of Alicante, Ap. correus 99, 03080 Alicante, Spain. E-mail: albaus@ua.es

PILAR ANDRÉS, CREAF, Autonomous University of Barcelona, 08193 Bellaterra, Spain. E-mail: pilar.andres@blues.uab.es

CARLOS ASCASO, CREAF, Autonomous University of Barcelona, 08193 Bellaterra, Spain. (Present address: Unit of Biostatistics, Faculty of Medicine, University of Barcelona, c/ Casanovas 143, 08036 Barcelona, Spain.) E-mail: ascaso@med.ub.es

ANNA ÀVILA, CREAF, Autonomous University of Barcelona, 08193 Bellaterra, Spain. E-mail: avila@uab.es

JUAN BELLOT, Department of Ecology, University of Alicante, Ap. correus 99, 03080 Alicante, Spain. E-mail: Juan.Bellot@ua.es

DAVID BONILLA, CREAF, Autonomous University of Barcelona, 08193 Bellaterra, Spain. (Present address: EYCAM Perrier, camí vell de Viladrau s/n, 17406 Viladrau, Spain.)

PEP CANADELL, Department of Biological Sciences, Stanford University, Stanford, California 94305-5020, USA. (Present address: CSIRO Wildlife and Ecology, PO Box 284, Canberra, ACT 2601, Australia.) E-mail: pep.canadell@dwe.csiro.au

ADORACIÓN CARRATALÁ, Department of Ecology, University of Alicante, Ap. correus 99, 03080 Alicante, and CEAM, Parque Tecnológico c/4 sector oeste, 46980 Paterna, Spain. E-mail: dori@ceam.es

CARLES CASTELL, Servei de Parcs Naturals, Diputació de Barcelona, Urgell 187, 08036 Barcelona, Spain. E-mail: castellpc@diba.es

VICTORIA DIEGO, CREAF, Autonomous University of Barcelona, 08193 Bellaterra, Spain.

AREZKI DJEMA, Department of Ecology, Faculty of Biology, University of Barcelona, Diagonal 645, 08028 Barcelona, Spain.

ANTONI ESCARRÉ, Department of Ecology, University of Alicante, Ap. correus 99, 03080 Alicante, Spain. E-mail: Escarre@ua.es

JOSEP MARIA ESPELTA, CREAF, Autonomous University of Barcelona, 08193 Bellaterra, Spain. E-mail: JosepMaria.Espelta@uab.es

José A. Gil-Delgado, Department of Microbiology and Ecology, University of Valencia, c/ dr. Moliner 50, 46100 Burjassot, Spain. E-mail: gild@uv.es

Carlos A. Gracia, CREAF, Autonomous University of Barcelona, 08193 Bellaterra, and Department of Ecology, Faculty of Biology, University of Barcelona, Diagonal 645, 08028 Barcelona, Spain. E-mail: gracia@porthos.bio.ub.es

Marc Gracia, CREAF, Autonomous University of Barcelona, 08193 Bellaterra, and Escola Tècnica Superior d'Enginyeria Agrària, University of Lleida, Spain. E-mail: MGracia@pvcf.UdL.es

Agnès Hereter, Escola Superior d'Agricultura, c/ Urgell 187, 08036 Barcelona, Spain. E-mail: Hereter@esab.upc.es

J.J. Ibàñez, CREAF, Autonomous University of Barcelona, 08193 Bellaterra, Spain. E-mail: iefc@uab.es

María José Lledó, Department of Ecology, University of Alicante, Ap. correus 99, 03080 Alicante, Spain. E-mail: MJ.Lledo@ua.es

Francisco Lloret, CREAF, Autonomous University of Barcelona, 08193 Bellaterra, Spain. E-mail: lloret@uab.es

Germán López-Iborra, Department of Ecology, University of Alicante, Ap. correus 99, 03080 Alicante, Spain. E-mail: german.lopez@ua.es

Bernat López, Department of Ecology, Faculty of Biology, University of Barcelona, Diagonal 645, 08028 Barcelona, and CREAF, Autonomous University of Barcelona, 08193 Bellaterra, Spain. Email: lopez@porthos.bio.ub.es

Juan Manuel Martínez, Department of Ecology, University of Alicante, Ap. correus 99, 03080 Alicante, Spain. E-mail: JM.Martinez@ua.es

Eduardo Mateos, Department of Animal Biology, Faculty of Biology, University of Barcelona, Diagonal 645, 08028 Barcelona, Spain. E-mail: eduard@porthos.bio.ub.es

Xavier Mayor, CREAF, Autonomous University of Barcelona, 08193 Bellaterra, Spain. (Present address: Departament de Medi Ambient, Diagonal 525, 08029 Barcelona, Spain.) E-mail: wxmayor@correu.gencat.es

Núria Melià, Department of Ecology, University of Alicante, Ap. correus 99, 03080 Alicante, and Department of Plant Biology, University of Barcelona, 08028 Barcelona, Spain. (Present address: Laboratori Polivalent de la Garrotxa, ctra. de Ridaura s/n, 17800 Olot, Spain.) E-mail: lpgarro@ddgi.es

Millán Millán, CEAM, Parque Tecnológico c/4 sector oeste, 46980 Paterna, Spain. E-mail: millan@ceam.es

Josep Piñol, CREAF, Autonomous University of Barcelona, 08193 Bellaterra, Spain. E-mail: jpinol@uab.es

JAVIER RETANA, CREAF, Autonomous University of Barcelona, 08193 Bellaterra, Spain. E-mail: retana@uab.es

MIQUEL RIBA, CREAF, Autonomous University of Barcelona, 08193 Bellaterra, Spain. E-mail: ibec3@uab.es

FERRAN RODÀ, CREAF, Autonomous University of Barcelona, 08193 Bellaterra, Spain. E-mail: roda@uab.es

ANSELM RODRIGO, CREAF, Autonomous University of Barcelona, 08193 Bellaterra, Spain. E-mail: sahara@blues.uab.es

SANTIAGO SABATÉ, Department of Ecology, Faculty of Biology, University of Barcelona, Diagonal 645, 08028 Barcelona, and CREAF, Autonomous University of Barcelona, 08193 Bellaterra, Spain. E-mail: santis@porthos.bio.ub.es

ANNA SALA, Division of Biological Sciences, University of Montana, Missoula, Montana 59812, USA. Email: sala@selway.umt.edu

JUAN RAFAEL SÁNCHEZ, Department of Ecology, University of Alicante, Ap. correus 99, 03080 Alicante, Spain. E-mail: JR.Sanchez@ua.es

ROBERT SAVÉ, Departament de Tecnologia Hortícola, Institut de Recerca i Tecnologia Agroalimentària (IRTA), ctra. de Cabrils s/n, 08348 Cabrils (Barcelona), Spain. E-mail: save@cabrils.irta.es

ISABEL SERRASOLSES, Department of Plant Biology, Faculty of Biology, University of Barcelona, Diagonal 645, 08028 Barcelona, Spain. (Present address: Unit of Ecology, Autonomous University of Barcelona, 08193 Bellaterra, Spain.) E-mail: isabel.serrasolses@uab.es

DANIEL SISCART, CREAF, Autonomous University of Barcelona, 08193 Bellaterra, Spain.

ESTÍBALIZ TELLO, CREAF, Autonomous University of Barcelona, 08193 Bellaterra, Spain. E-mail: e.tello@creaf.uab.es

JAUME TERRADAS, CREAF, Autonomous University of Barcelona, 08193 Bellaterra, Spain. E-mail: jaume.terradas@uab.es

V. RAMON VALLEJO, Department of Plant Biology, Faculty of Biology, University of Barcelona, Diagonal 645, 08028 Barcelona, and CEAM, Parque Tecnológico c/4 sector oeste, 46980 Paterna, Spain. E-mail: ramonv@ceam.es

MIGUEL ANGEL ZAVALA, Department of Ecology and Evolutionary Biology, Princeton University, Princeton, New Jersey 08544, USA. E-mail: mazavala@phoenix.princeton.edu

Part 1 Introduction

1 Holm Oak and Holm Oak Forests: An Introduction

Jaume Terradas

Why should we study the Mediterranean evergreen forests of holm oak (*Quercus ilex* L.)? Besides the pursuit of knowledge, two major reasons can be put forward. First, holm oak forests are a dominant type of vegetation in a transition zone between temperate forests, mostly dominated by deciduous trees and the scrublands (maquis, chaparral, phrygana, etc.) that herald the tropical regions. In this transition zone, plants have had to cope with a selective pressure resulting from a double stress – winter cold and summer drought – that has determined their morphological and ecophysiological evolutive responses. One of our aims is to provide an insight into the features of holm oak as related to environmental factors and in comparison with other tree types (broadleaved deciduous hardwoods, needle-leafed conifers).

Second, holm oak is distributed throughout areas harbouring large human populations, massive tourism and, in places, intense industrial activity. Furthermore, these are zones where we can reasonably expect a very sensitive response to global change processes. Climate change could decrease water resources beyond the threshold of the forests' needs (Piñol et al. 1995) and the resulting changes in landscape would be dramatic. The great number of ecotones (favoured by a great heterogeneity in bedrocks and soils, and by a steep terrain) along aridity gradients in the boundaries between shrub and forest formations should be highly sensitive to change. Land use patterns are also rapidly changing, mainly in two opposite directions for the north and south rims of the Mediterranean Basin: land abandonment, with young, poorly managed and fuel-accumulating forests in the north; and a fast increase in firewood extraction and unsustainable forest abuse in the south. Wildfire frequency could increase due to higher temperature, higher evapotranspiration and increased human activities, accelerating these changes. An understanding of forest function concerning aspects such as individual recruitment and mortality, canopy regeneration, hydrology, nutrient use, biodiversity and others would be essential to coping with all these future events (Romane and Terradas 1992).

The aim of this chapter is to summarize the major ecological characteristics of holm oak and holm oak forests. Many of these characteristics will be

Ecological Studies, Vol. 137
Ferran Rodà et al. (eds) Ecology of Mediterranean
Evergreen Oak Forests
© Springer-Verlag, Berlin Heidelberg 1999

developed later in this book, but I will try here to relate them to some more general aspects of holm oak ecology. Other characteristics will be outlined now as an introduction for those readers who are unfamiliar with Mediterranean forests.

1.1 Biogeography of Holm Oak

In the Mediterranean Basin and the Middle East, holm oak is present over a large area extending 6000 km longitudinally (Debazac 1983), from Portugal to Syria, and 1500 km latitudinally from Morocco and Algeria to France. It has been frequently considered as a typical Mediterranean plant due to its evergreenness, sclerophyllous leaves, and drought resistance. Nevertheless, many authors consider that the typical Mediterranean vegetation types are not forests, but dense evergreen shrub formations, such as maquis or garrigue in the Mediterranean Basin, chaparral in California, fynbos in South Africa, matorral in Chile or mallee in Australia. Holm oak can be present in maquis and has leaves quite similar to those of evergreen shrub oaks (*Quercus coccifera, Q. calliprinos*), but its optimum lies on the geographical periphery of maquis and garrigue formations. Holm oak is mostly found from cold semi-arid to temperate humid Mediterranean bioclimates (Emberger 1955), whereas maquis and garrigues are more frequent from temperate semi-arid to warm humid bioclimate zones.

Biogeographical aspects of holm oak have been reviewed by Barbero et al. (1992). They describe the species as a circum-Mediterranean one, more abundant in the western part of the Basin. It is distributed mainly in the thermo-, meso- and upper Mediterranean altitudinal "étages" and in semi-arid to humid Mediterranean climates (Rivas Martínez 1987). It is absent from those parts of the Mediterranean Basin where summer drought is too long. In North Africa, southern Spain and southern Greece, holm oak is rarely found at altitudes below 350 m and it can be found up to 1500–1800 m, reaching even 2700 m on the Atlas range. It can compete with deciduous oaks and some conifers in the coldest and moistest limits of its distribution range, or with other sclerophyllous evergreen oaks or conifers near its driest border. Its limits of distribution have been modified by management practices favouring holm oak, or by different livestock uses. Overuse by agriculture and grazing for many centuries and frequent fires have contributed to a gradual degradation of forests and soils and increased aridity. This, in the moister parts of the holm oak range, has sometimes favoured the spread of evergreen forests over deciduous forests (Debazac 1983), whereas in the drier parts it has favoured pine forests with xerophytic shrubs or maquis. In most parts of Iberia, holm oak is largely indifferent to soil reaction, whereas in the northern areas it prefers calcareous soils and southern slopes where it competes successfully with deciduous species.

In Spain, holm oak dominates in about 3 million ha, i.e. 25% of the area where trees cover at least 10% of the ground (Primer Inventario Forestal Nacional, 1965–1974), and it accounts for 16% of total wood production. Almost 1 million ha is coppices used for firewood, and another 1.3 million ha is "dehesas", a savannah-type man-managed formation, whereas the remaining areas consist of poorly developed coppices, practically without economic use. The forest area dominated by holm oak is smaller in Catalonia (NE Spain), totalling around 100 000 ha (about 10% of forested lands). In the French Mediterranean region, holm oak covers about 250 000 ha (Barbero et al. 1992). In Italy, Corsica and Sicily it is the dominant forest tree in Mediterranean climate areas at altitudes ranging from only 500 m in Liguria to 1400 m in Calabria, the basal parts being mostly occupied by a "macchia" dominated by holm oak. In Sardinia its climacic area spreads over most of the island, but forests and "macchia" are strongly degraded. In Morocco, holm oak grows between 600 and 2700 m, covering 1.4 million ha, over 30% of the natural forest area (Braun-Blanquet and Maire 1924; Khatouri 1992), but it is declining fast through overexploitation.

1.2 History of Holm Oak Forests

Holm oak has been present in the Mediterranean Basin at least since the Miocene (Barbero et al. 1992) and its dominance has expanded from south to north: in central and southern Spain it was abundant as early as 9500 B.P. (before this, it was probably limited to a few warm refuges), whereas similar proportions of *Q. ilex* and of the deciduous *Q. humilis* (= *Q. pubescens*) appeared in NE Spain after 5000 B.P. (Riera-Mora and Esteban-Amat 1994). In the north-western Mediterranean Basin, holm oak has spread only in very recent times, replacing deciduous *Q. humilis* forests (Pons and Vernet 1971). For instance, in Low Provence (southern France) it became dominant only after the SubAtlantic period (2000 B.P.), with the increase in human activity and the spread of agriculture in Gallo-Roman times. A similar history has been described for the Eastern Mediterranean Basin, with the increase of *Q. ilex* pollen associated with an increase in human activity at the end of the Atlantic period (4700 B.P.). In northern Africa, there were probably extensive holm oak forests much before, but a human-related increase is also observed.

1.3 Species Composition Patterns in Holm Oak Forests

It is difficult to assess the natural diversity of forests that have been managed over many centuries. Holm oak forests are now, in most cases, monospecific in the tree layer (Chap. 3), perhaps as a result of human activities. The main sources of information on the floristic composition of holm oak forests are phytosociological studies, which are very detailed in the western part of the Mediterranean Basin. Broadleaved evergreen associations are grouped in the *Quercetalia ilicis* order, with the alliances *Quercion ilicis* and *Oleo-Ceratonion* corresponding respectively to the areas where holm oak forests and *Pistacia lentiscus* maquis are dominant in the western Mediterranean.

Holm oak forests are complex, rather diverse ecosystems, with a relatively high degree of specificity, at least in their floristic composition. Braun-Blanquet (1936) considers that a plot of 100 m^2 is usually needed to have a good representation of the normal specific composition of holm oak forests: in the *Viburno-Quercetum ilicis* (= *Quercetum ilicis gallo-provinciale*) association, there is an average of 34.7 vascular plant species per relevé, whereas at higher altitudes, above 400–500 m, holm oak forests of the *Asplenio-Quercetum ilicis* (= *Quercetum mediterraneo-montanum*) association become poorer in vascular plant species but richer in macrofungi.

Dense holm oak forests have usually a dominant tree layer 6–12 m high. In many cases, mixed forests of holm oak with *Pinus halepensis* or *P. pinea* occur, pines being frequently taller. A lower tree layer can be formed by *Arbutus unedo* and *Phillyrea latifolia* ssp. *media* (3–5 m). An upper shrub layer can contain other species such as *Viburnum tinus*, *Rhamnus alaternus* and *Pistacia lentiscus*. Also, vines (evergreen or deciduous) like *Hedera helix*, *Smilax aspera*, *Clematis flammula*, *C. vitalba*, *Lonicera* spp., *Rosa* spp., *Rubia peregrina*, *Asparagus acutifolius* or *Tamus communis* can be an important part of the system, mainly in coastal, relatively moist habitats, and give it a quasi-tropical appearance. The lower shrub layer is characterized mainly by *Ruscus aculeatus*. The grass layer rarely reaches a 30% cover, including sometimes *Hedera helix* as an important component.

In the relevé table of holm oak forests provided by Braun-Blanquet (1936) the dominant plant life forms are evergreen and deciduous phanerophytes (37 and 17%, respectively, of the listed species), and hemicryptophytes (24%), followed by chamaephytes (9%), geophytes (9%) and therophytes (4%). This is a very different composition from that of a temperate deciduous forest, a beech forest for instance, where hemicryptophytes and geophytes can represent over 90% of the number of species present. Nevertheless, there is a gradual transition between holm oak and deciduous oak forests. In the montane communities, the vertical structure is simpler, the lower tree layer disappears and the number of vine species decreases, whereas herbs greatly increase. These changes show a transition to non-Mediterranean deciduous forests patterns.

1.4 Biology of Holm Oak

According to Schwarz (1964) and d'Amaral Franco (1983), holm oak is a monoecious, wind-pollinated oak tree included in the subgenus *Sclerophyl-*

lodrys O. Schwarz. It can grow up to 27 m high but rarely reaches 15 m. Two main forms exist: *Q. ilex* ssp. *ilex* and *Q. ilex* ssp. *ballota* (Desf.) Sampaio (= *Q. ilex* ssp. *rotundifolia*), sometimes considered to be different species (*Q. ilex* L. and *Q. rotundifolia* Lam. Schwarz 1964, respectively). Both forms can interbreed, but their main distribution ranges are separated, with the ssp. *ilex* in coastal areas and the ssp. *ballota* mostly in interior areas. The morphological differences between the leaves of both forms are probably related to the constraints of their respective environments. In this book we will deal with the ssp. *ilex*.

Holm oak is a hardwood tree that grows slowly, forming dense canopies. Its leaves are small and coriaceous, the upper side is dark green, while the lower side is covered by white hairs. Stomata are relatively small (length is 22–27 μm) and limited to the abaxial surface of the leaf, hidden under a dense cover of stellate hairs. There are between 100 and 500 stomata mm^{-2}. A fine adjustment of holm oak stomata allows a hydrostable behaviour (Larcher 1995). The upper epidermis and cuticle are much thicker than their lower counterparts. We must consider these characteristics as a part of a conservative strategy in order to maintain a low cuticular transpiration and a high resistance to desiccation (Chap. 10).

Holm oak is a shade-tolerant tree, becoming dominant in late successional stages. Photosynthesis is conducted throughout the year (Eckardt et al. 1978). Leaf chlorophyll content is very high. A LAI of 4.5 m^2 m^{-2} could be near optimal (Sala et al. 1994). LAI seems to be poorly sensitive to water availability. In closed-canopy stands, leaves are mostly concentrated in the uppermost meter of the canopy (Chap. 9).

Wood is dense and its structure is diffuse semiporous, with a 50- to 150-μm porus diameter (Marchal 1995). Vessels are irregularly distributed and have slightly decreasing diameters towards the outer part of the growth ring. There are many relatively narrow and quite short vessels compared with the ring-porous woods found in deciduous oaks. Annual rings can be difficult to recognize because false tree-rings are frequent and pluriseriate medullar rays obscure the growth pattern (Gené et al. 1993). As a result, few data are available on radial growth rate. Wood hardness and density have made holm oak wood useful for making carts, tool handles, jack planes and other carpentry tools, gear teeth, parquets, etc. As a result of its vessel characteristics, holm oak has slow midday peak velocities of water flow, ranging from 1 to 6 m h^{-1} (0.3–1.7 mm s^{-1}). The cavitation of one vessel in that organization causes much less damage than in ring-porous trees.

Holm oak has deep roots. In only one year, the unbranched taproot of a holm oak seedling can reach 60–80 cm. In adult holm oaks, there is also a complex system of spreading lateral roots that can extend many meters, growing from a thick, sometimes massive root crown, particularly in coppices as a result of tissue accumulation during successive resprout production. There are few data on aboveground/underground biomass ratio, but a high variability can be expected under different climate conditions and management (Chap. 4).

Vegetative growth occurs in May–June in the montane north-facing slopes where most of our experimental sites are located (Chap. 2), and about one month earlier in the lowlands. After the growth of dolicoblasts and the opening of new leaves in May–June shoot growth ceases in July. A second flush of shoot growth, carrying a smaller generation of leaves, can be produced in late summer or early autumn (De Lillis and Fontanella 1992), but this is not common for adult trees in montane sites. Maximum leaf fall is in May–June, coinciding with the growth of new shoots. So, old leaves yielding low carbon returns are discarded just before the summer drought and at a time when the growth demands for nutrients and mobile carbon are highest. Old leaves, twigs and small branches, as well as stem and roots, can act as nutrient reservoirs useful for regenerating the tree crowns after disturbance. Leaf lifespan ranges from less than 1 to 4 years, being usually around 2 years since most leaves are shed at the start of their third year. Leaf turnover rates change with crown position, habitat characteristics and weather, with a high interannual variability. So, changes in leaf features within the canopy or between sites are sometimes the result of changes in the age structure of the leaf population.

Flower buds and flowers appear mostly in May–June on the newly elongated branches. Flowering can last until late June, when acorn growth begins. Probably, holm oak has genetic auto incompatibility, and it shows large differences in flowering time between trees and between years; both factors favour gene combination within populations. This is an important characteristic, because it ensures high genetic variability right from the beginning of a colonization process and onwards (Michaud et al. 1992). Genetic introgression in oaks can explain much of their intraspecific variation (Schwarz 1964). Genetic varieties of holm oak are probably distinguishable not only by their leaf morphology, but also by fruit forms.

Acorns mature and fall mostly in November–January. Acorn production has a great individual and interannual variability (Chap. 6). High acorn production has been said to occur once every (2)4–6 years, but long series of systematic records are lacking. As in other *Quercus* species, acorns remain viable for only a short time (1–2 months). The highest germination rates are obtained within 2 months after collecting the fruits. Predation on acorns can be a problem in some cases. Germination is stimulated under reduced light levels and the associated increase in soil moisture that occurs at canopy closure (Bran et al. 1990). Survival is also better under closed canopies. There, the shade-tolerant holm oak seedlings can survive even decades without significant growth (Espelta et al. 1993; Chap. 7). So, there is usually no significant recruitment into the canopy under the selection thinning management currently used in most of Catalonia, and only substantial reductions in tree density could lead to an effective recruitment. When a disturbance occurs, new genets can develop from acorns but, as noted above, most regeneration comes from resprouts (Chap. 5), which are produced abundantly from stumps or root crowns. Resprout leaves benefit from an increased root to shoot ratio, and resprout growth rates are high during the first year (Castell et al. 1994).

1.5 Responses of Holm Oak to Environmental Conditions

The genus *Quercus* contains a large variety of leaf types, from summer-deciduous (leaves withering in spring and falling in summer), to evergreen (with leaves persisting more than a year), to semi-evergreen (with leaves withering in autumn, persisting on the tree through winter and falling in spring), and to winter deciduous (withering and falling in autumn). A transition between these types seems to occur along environmental gradients of vegetative period and drought season lengths. Holm oak is one of a number of evergreen oaks distributed in a rather large area between tropical and temperate regions of the world. To the continental European species (*Q. ilex, Q. suber, Q. calliprinos, Q. coccifera*) we can add *Q. alnifolia* from Cyprus, *Q. baloot* from the Middle East, and many species from North America, Japan, etc. There is usually an inverse correlation between summer and winter stresses (Mitrakos 1980) that determines the changes in leaf type dominance. Winter-deciduous oaks correspond mostly with cold-temperate climates, while evergreen and semi-evergreen or summer-deciduous oaks extend well into subtropical regions. In evergreen species, leaf lifespan lengthens from mesic to drier sites. Resistance to desiccation and to drought injury increases from deciduous to evergreen oaks. Larcher (1960) measured very large differences in these aspects between the frequently coexisting *Q. humilis* (winter deciduous) and holm oak (evergreen). Whereas cuticular transpiration in *Q. humilis* is 30% of open-stomata transpiration, in holm oak it is only 3%. Drought injury delay – measured by the ratio (available water content at time of complete stomatal closure)/(cuticular transpiration) – for *Q. ilex* is 15 times higher than for *Q. humilis*.

Holm oak has been considered a Mediterranean paradigm, partly because of its evergreen habit. Evergreenness can be advantageous because, under appropriate conditions, it permits carbon fixation at any season. Evergreen dominance is also usually explained as a result of low nutrient loss rates in nutrient-poor environments, with slower decomposition rates. Although in Mediterranean-type ecosystems leaf chemistry and leaf decomposition may not differ consistently between evergreen and deciduous species (Aerts 1995), deciduous oaks do have leaves of higher nutrient concentrations than evergreen oaks. Therefore, in the Mediterranean Basin the evergreen habit could be more related to the double-stress resistance than to oligotrophy (Mitrakos 1980). The interrelationships between water and nutrient economies are, however, very complex, and we need to know much more about them. Water relationships are discussed in Chapters 10 and 11. Holm oak is much more resistant to drought than winter-deciduous trees, but not enough to survive in the drier areas of the Mediterranean region. Extreme droughts, such as that of 1994 in Spain, can produce high rates of crown withering and tree mortality. High losses of hydraulic conductivity can occur during summer drought and losses can be much larger than in other Mediterranean trees (*Ceratonia siliqua, Olea oleaster, Quercus suber*). Sala and Tenhunen (1994) suggest that when xylem water potential is below -3 MPa cavitation may occur and strong leaf shedding can be observed. Salleo and Lo Gullo (1990) consider that sclerophylly does not have the ecological meaning of water saving. Holm oak does not behave as a drought avoider but as a drought tolerant, because its stomata remain open at least twice as long as those of *Q. humilis* or *Q. suber*, and because it has a greater capacity to recover from water losses and to maintain a higher minimum diurnal relative water content.

Such relatively stable water status is possible because the deep roots of holm oak can tap groundwater, enabling it to resist drought. However, the Californian *Q. agrifolia* and *Q. chrysolepis* are drought avoiders, unable to tap groundwater to a large extent, with lower recovery values than the drought-deciduous *Q. douglasii* and the deciduous *Q. kelloggii* (Knops and Koenig 1994). This shows that conclusions about an evergreen oak cannot be extrapolated to other species with an apparently similar strategy nor can we explain the evergreenness of Mediterranean trees just by reasoning on water budget aspects.

Dense holm oak forests seem to use around 500–600 mm of water a year (Chap. 19). When annual rainfall is on average below 400–450 mm, holm oak forests cannot maintain closed canopies, and the species becomes scarce and restricted to the best sites. Mature vegetation is then formed by Mediterranean pines and sclerophyllous or summer-deciduous shrubs with phrygana-type malacophyllous chamaephytes or steppe-type perennial grasses.

Holm oak resists quite low temperatures, down to -15 °C approximately, and it can, in fact, survive temperatures as low as -20 to -25 °C in winter if cold periods do not last long enough to freeze the thick trunks (Larcher and Mair 1969). Apparently, holm oak suffers from water stress due to frost-drought more than from frost injury (Savé et al. 1988). Under very cold temperatures, all leaves may die, but even then trees can still survive. Holm oak is more cold-resistant than other sclerophyllous species which frequently co-exist with it, like *Q. suber, Rhamnus alaternus, Q. coccifera* or *Ceratonia siliqua* (Larcher and Mair 1969). Nevertheless, soil surface temperatures of only -4 °C can kill the seedlings. The regular occurrence of -8 to -10 °C temperatures over a number of years makes the survival of young trees impossible (Larcher 1995). Seasonal hardening for cold resistance is described by Larcher (1970).

Nutrient availability could also be a factor limiting holm oak distribution, but simultaneous variation of water- and nutrient-holding capacity of soils makes this effect difficult to test. Sclerophylly and evergreenness are advantageous when nutrients are scarce. Within the holm oak range, on oligotrophic soils, deciduous oaks (*Q. humilis, Q. faginea, Q. cerrioides*) are sometimes limited to mesic and nutrient-rich sites. Where deciduous trees become dominant, holm oak refuges are usually on calcareous, non-oligotrophic soils with relatively low water availability. Water, and water-related competition, not nutrients, seems to be the usual limiting factor. Experiments aimed at understanding the relative roles of water, nitrogen and phosphorus availability on holm oak performance are presented in Chapter 13.

Wildfires are an environmental threat to Mediterranean forests. Whereas Mediterranean pine forests are frequently devastated by fire, holm oak forests score lower in yearly burnt surfaces. For instance, in Catalonia the yearly burnt area relative to the area occupied by each species is double for *Pinus halepensis* than for *Quercus ilex* + *Q. suber*. It is well known that chemical composition and the production of volatile organic substances are related to flammability. Whereas most other oak species emit isoprene, holm oak, surprisingly, emits monoterpenes, with an emission rate that seems to be apparently higher than any other values reported in the literature for other plant species (Seufert et al. 1995). This behaviour differs from that of pines, which accumulate monoterpenes in the leaves, twigs or bark. The relevant aspect for fire behaviour is that no massive emission will occur from a holm oak canopy as a result of heating, making it less flammable than pine.

1.6 Management

As noted above, holm oak forests were probably favoured by man in early historical times. However, they have been heavily exploited for centuries, and have been progressively restricted to marginal areas, like shallow soils on steep slopes in mountain areas, due to the spread of agriculture, human settlements and, sometimes, because faster growing pines have been favoured for timber production.

There are great geographical differences in management. On very oligotrophic soils in central and western Spain and in Portugal, holm oak woodlands are managed as dehesas (*montados* in Portugal), a multi-secular manmade type of savannah-like ecosystem with large, isolated trees emerging from a grassland (Gómez Gutiérrez 1992; Huntsinger and Bartolome 1992). About 1.2 million ha is exploited in Spain in this way for grazing and acorn production, with acorn yields between 400 and 700 kg ha^{-1} (Montoya 1993). Since these acorns are sweet, in the past they were used not only as hog food but also even for human use, and holm oak in dehesas can be considered as a semi-domesticated fruit tree (Laguna 1883). Trees also provide shade for livestock, firewood from pruning and refuge and breeding sites for a large number of vertebrates, whereas the grassland is used by cows and sheep for milk and meat production. Dehesas are also excellent for hunting. The system is managed by preventing woody plants from invading grasslands through grazing, disking and hand weeding, with a high labour effort. This sustainable and ecologically diverse system is endangered by the spread of cereal-intensive monocultures using large amounts of fertilizers and pesticides.

Dehesas are very interesting ecosystems, but our subject in this book are the closed-canopy holm oak forests prevalent in eastern Spain. In contrast to the dehesas, in eastern Spain, including Catalonia, and in most of the Mediterranean regions of France, Italy, and Greece, charcoal and firewood were the main products obtained from holm oak. As a result, these forests have been managed as coppices or coppice-with-standards at least since the Neolithic period. At Montseny, cutting intervals were short for charcoal (usually 10–15 years), increasing to 20–25 years if the largest boles were used for timber (Gutiérrez 1996). Selection thinning was the most frequently used coppicing method, with about one third of the usable stems harvested at each cutting (Gutiérrez 1996). Clear-cutting or almost clear-cutting was also sometimes used. Regular and frequent coppicing of holm oak forests was also the rule in other Mediterranean countries (Ducrey and Boisserie 1992; Giovannini et al. 1992; Bacillieri et al. 1993). Coppicing has created forests with high tree densities, and the ensuing intense competition leads to a low growth rate of individual stems (Chap. 3). Charcoal production, very important in the past, has practically disappeared. Tannin from the bark of young trees was another major product, but synthetic tannins have superseded such production in the last 30 years. The main current use is for firewood. In Catalonia, selection thinning at intervals of (15) 20–35 years is mostly used, leading to uneven-aged stands where restocking is almost exclusively based on stump resprouting (Chaps. 3 and 5). Natural seedling recruitment (usual in other oak forests) is minimal in Catalan holm oak forests (Espelta et al. 1995; Chap. 7). Multiple use is in fact occurring, but not as a result of planning. For instance, coppicing has been always accompanied by sheep and goat grazing (cutting holm oak induces sprouting, with an increase in leaf production that can be used by livestock). Acorns play an important role in the diet of sheep during the winter; sheep, and sometimes cows, may also graze on the herbaceous layer of recently-coppiced stands, whereas goats feed directly on holm oak leaves and young twigs. When deciduous oaks are present, they are preferred to holm oaks by browsers, due to the higher content of cutin, lignin and tannins and the lower content of proteins in holm oak. This can give holm oak a competitive advantage over deciduous oaks in some parts of its distribution range. However, in the experimental areas on which this book is based, grazing by domestic herbivores is currently not a major ecological factor.

References

Aerts R (1995) The advantages of being evergreen. Trends Ecol Evol 10:402-407

Bacillieri R, Bouchet MA, Bran D, Grandjanny M, Maistre M, Perret P, Romane F (1993) Germination and regeneration mechanisms in Mediterranean degenerate forests. J Veg Sci 4: 241-246

Barbero M, Loisel R, Quézel P (1992) Biogeography, ecology and history of Mediterranean Quercus ilex ecosystems. Vegetatio 99/100:19-34

Bran D, Lobreaux O, Maistre M, Perret P, Romane F (1990) Germination of Quercus ilex and Quercus pubescens in a Quercus ilex coppice. Vegetatio 87:45-50

Braun-Blanquet J (1936) La chênaie d'Yeuse méditerranéenne. SIGMA 45 et Mem Soc Sci Nat Nîmes 5:1-147

Braun-Blanquet J, Maire R (1924) Études sur la végétation et la flore marocaines. Mém Soc Sci Nat Maroc VIII, 212 pp

Castell C, Terradas J, Tenhunen JD (1994) Water relations, gas exchange and growth of resprouts and mature plant shoots of Arbutus unedo L. and Quercus ilex L. Oecologia 98:201-211

d'Amaral Franco J (1983) Quercus. In: Castroviejo S, Laínz M, López G, Montserrat P, Muñoz F, Paiva J, Villar L (eds) Flora Iberica, vol II. CSIC, Madrid, pp 15-20

Debazac EF (1983) Temperate broad-leaved evergreen forests of the Mediterranean region and Middle East. In: Ovington JD (ed) Temperate broad-leaved evergreen forests. Elsevier, Amsterdam, pp 107-123

De Lillis M, Fontanella A (1992) Comparative phenology and growth in different species of the Mediterranean maquis of central Italy. Vegetatio 99/100:83-96

Ducrey M, Boisserie M (1992) Recrû naturel dans les taillis de chêne vert (Quercus ilex L.) à la suite d'exploitations partielles. Ann Sci For 49:91-109

Eckardt FE, Berger A, Méthy M, Heim G, Sauvezon R (1978) Interception de l'énergie rayonnante, échanges de CO_2, régime hydrique et production chez différents types de végétation sous climat méditerranéen. In: Moyse A (ed) Les processus de la production végétale primaire. Géobiologie, écologie, aménagement. Gauthier-Villars, Paris, pp 1-75

Emberger L (1955) Une classification biogéographique des climats. Rec Trav Lab Bot-Zool Fac Sci Univ Montpellier Bot 7:3-43

Espelta JM, Retana J, Gené C, Riba M (1993) Supervivencia de plántulas de pino carrasco (Pinus halepensis) y encina (Quercus ilex) en bosques mixtos de ambas especies. In: Silva-Pando FJ, Vega G (eds) Congreso Forestal Español. Ponencias y comunicaciones, vol II. Xunta de Galicia, Vigo, pp 393-398

Espelta JM, Riba M, Retana J (1995) Patterns of seedling recruitment in west-Mediterranean Quercus ilex L. forests influenced by canopy development. J Veg Sci 6:465-472

Gené C, Espelta JM, Gracia, M, Retana J (1993) Identificación de los anillos anuales de crecimiento de la encina (Quercus ilex L.). Orsis 8:127-139

Giovannini G, Perulli D, Piussi P, Salbitano F (1992) Ecology of vegetative regeneration after coppicing in macchia stands in central Italy. Vegetatio 99/100:332-344

Gómez Gutiérrez JM (ed) (1992) El libro de las dehesas salmantinas. Junta de Castilla y León, Salamanca

Huntsinger L, Bartolome JW (1992) Ecological dynamics of Quercus dominated woodlands in California and southern Spain. Vegetatio 99/100:299-305

Gutiérrez C (1996) El carboneig. L'exemple del Montseny. Alta Fulla, Barcelona

Khatouri M (1992) Growth and yield of young Quercus ilex coppice stands in the Tafferte forest (Morocco). Vegetatio 99/100:77-82

Knops JMH, Koenig WD (1994) Water use strategies of five sympatric species of Quercus in central coastal California. Madroño 41:290-301

Laguna M (1883) Flora forestal española. Ed Facsimile, Xunta de Galicia, Vigo, 1993

Larcher W (1960) Transpiration and photosynthesis of detached leaves and shoots of *Quercus pubescens* and *Q. ilex* during desiccation under standard conditions. Bull Res Counc Isr 8D:213–224

Larcher W (1970) Kälteresistenz und Überwinterungsvermögen mediterraner Holzpflanzen. Acta Oecol Oecol Plant 5:267–286

Larcher W (1995) Physiological plant ecology. Springer, Berlin

Larcher W, Mair B (1969) Die Temperaturresistenz als ökophysiologisches Konstitutionsmerkmal: 1. *Quercus ilex* und andere Eichenarten des Mittelmeergebietes. Acta Oecol Oecol Plant 4:347–376

Marchal R (1995) Propriétés des bois de chêne vert et chêne pubescent. Caractéristiques physiques. For Méditerr 16:425–438

Michaud H, Lumaret R, Romane F (1992) Variation in the genetic structure and reproductive biology of holm oak populations. Vegetatio 99/100:107–114

Mitrakos K (1980) A theory for Mediterranean plant life. Acta Oecol Oecol Plant 1:245–252

Montoya JM (1993) Encinas y encinares. Agroguías Mundi-Prensa, Madrid

Piñol J, Terradas J, Àvila A, Rodà F (1995) Using catchments of contrasting hydrological conditions to explore climate change effects on water and nutrient flows in Mediterranean forests. In: Moreno JM, Oechel WC (eds) Global change and Mediterranean-type ecosystems. Springer, New York, pp 371–385

Pons A, Vernet JL (1971) Une synthèse nouvelle de l'histoire du chêne vert. Bull Soc Bot Fr 118: 841–850

Riera-Mora S, Esteban-Amat A (1994) Vegetation history and human activity during the last 6000 years on the central Catalan coast (north-eastern Iberian Peninsula). Veg Hist Archaebot 3:7–23

Rivas Martínez S (1987) Memoria del mapa de series de vegetación de España. ICONA, Madrid

Romane F, Terradas J (eds) (1992) *Quercus ilex* L. ecosystems: function, dynamics and management. Kluwer, Dordrecht

Sala A, Tenhunen JD (1994) Site-specific water relations and stomatal response of *Quercus ilex* L. in a Mediterranean watershed. Tree Physiol 14:601–617

Sala A, Sabaté S, Gracia C, Tenhunen JD (1994) Canopy structure within a *Quercus ilex* forested watershed: variations due to location, phenological development and water availability. Trees 8:254–261

Salleo S, Lo Gullo MA (1990) Sclerophylly and plant water relations in three Mediterranean *Quercus* species. Ann Bot 65:259–270

Savé R, Rabella R, Terradas J (1988) Effects of low temperature on *Quercus ilex* L. ssp. *ilex* water relationships. In: Di Castri F, Floret C, Rambal S, Roy J (eds) Time scales and water stress. Proc 5th Int Conf on Mediterranean ecosystems. International Union of Biological Sciences, Paris, pp 1103–1105

Schwarz O (1964) *Quercus* L. In: Tutin TG, Heywood VH, Burges NA, Valentine DH, Walters SM, Webb DA (eds) Flora Europaea, vol 1. Cambridge University Press, Cambridge, pp 61–64

Seufert G, Kotzias D, Spartà C, Versino B (1995) Volatile organics in Mediterranean shrubs and their potential role in a changing environment. In: Moreno JM, Oechel WC (eds) Global change and Mediterranean-type ecosystems. Springer, New York, pp 343–370

2 Experimental Areas of Prades and Montseny

Agnès Hereter and Juan Rafael Sánchez

2.1 The Montseny and Prades Mountains

Most of the research on holm oak forests described in this book was conducted in two mountain locations in Catalonia, northeast Spain: the Montseny mountains and the Prades mountains (Fig. 2.1). Both massifs belong to
the Catalan precoastal range, a chain of mid-sized mountains paralleling the
Mediterranean seaboard of Catalonia, 20–30 km from the coast. Both Montseny and Prades are extensively forested. The distance between them is
120 km.

The Montseny mountains are situated 40 km NNE from Barcelona. They
cover about 400 km² and the highest altitude is 1707 m a.s.l. The inner 75%
of the massif is a natural park and biosphere reserve. To the north and west
Montseny connects with other forested mountains. To the south lie the intensively occupied plains of the precoastal depression, with urban, industrial,
and agricultural areas, and major motorways. To the northwest lies a smaller
rural plain with intensive pig husbandry. Montseny is underlain mostly by

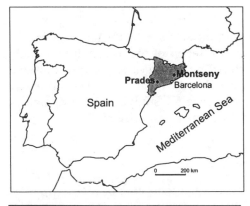

Fig. 2.1. Location of experimental areas in
Spain

Ecological Studies, Vol. 137
Ferran Rodà et al. (eds) Ecology of Mediterranean
Evergreen Oak Forests
© Springer-Verlag, Berlin Heidelberg 1999

metamorphic phyllites and schists, and by granodiorites. The relief is very steep. Climatically and biogeographically Montseny lies at a northern boundary of the Mediterranean region. At mid-altitudes, where our experimental sites are located, the climate is Mediterranean, but the characteristic summer drought is attenuated by summer storms. The massif is forested, except in its upper reaches. Altitudinal changes in vegetation are prominent, with pine forests, evergreen holm oak forests, beech forests, and heathlands and grasslands dominating in this order at increasing altitudes (Bolòs 1983). Holm oak forests cover about 50% of the natural park.

The Prades mountains are located 100 km WSW of Barcelona. They cover about 260 km², and the highest altitude is 1201 m a.s.l. The massif is designated as a site of natural interest, a conservation category that conveys a mild degree of protection. Part of the massif, the Valley of the Poblet Monastery including the Poblet Forest, benefits from stronger protection as a natural site of national interest. The Prades mountains are surrounded by lowlands devoted to dry farming. The cities of Tarragona and Reus, totalling 200 000 people and harbouring major petrochemical industries, lie 30 km to the south. The climate is typically Mediterranean with a marked summer dry period. There are two major geological units: a Palaeozoic base (metamorphic and igneous rocks) and a Mesozoic cover (limestone, dolomite, clay and sandstone) (Poblet and Pujadas 1988). The relief is flat on the Mesozoic top of the massif and very steep on the Palaeozoic slopes. The vegetation changes from Aleppo pine forests at low altitudes to holm oak forests and Scots pine forests (Mestres and Massalles 1988). Extensive shrublands occur, particularly on the driest and thinnest soils.

2.2 Experimental Areas at Montseny

The major experimental area used at Montseny was La Castanya Biological Station (41° 46' N, 2° 21' E), in the central sector of the massif. It was established in 1977 and used mainly for functional studies at the tree, plot and catchment scales. A second experimental area, Figaró, was established in 1990 in the southwestern sector of Montseny for studying stand dynamics. The major characteristics of the study sites are summarized in Table 2.1.

2.2.1 Experimental Layout

La Castanya valley lies at the heart of the Palaeozoic core of Montseny. Most studies at La Castanya Biological Station were conducted in and around a 200–ha catchment (named TM0) drained by the Torrent de la Mina stream, which flows in a NE direction (Fig. 2.2). On the north-facing slope of the Torrent de la Mina catchment, a small tributary, named TM9, draining a 5.9-ha

Table 2.1. Major characteristics of the study sites in the Montseny and Prades experimental areas

Location	Type of site	Major focus	Area[a] (ha)	Mean slope[b] (°)	Aspect[c]	Altitude (m)
Montseny						
La Castanya						
Torrent de la Mina (TM0)	Catchment	Water and nutrient budgets	200	34	NE	652–1343
TM9	Catchment	Water and nutrient budgets	5.9[d]	37	N	700–1035
Forest survey plots	Plots (*n*=81)	Tree growth; stand biomass and production	0.0154	20–38	N and E	700–1200
Closed-canopy plots	Plots (*n*=18)	Nutrient cycling	0.0154	20–35	N and E	800–1000
Permanent plot	Plot (*n*=1)	Biomass, production, nutrient cycling, hydrochemistry, ecophysiology	0.23	7–23	W to NW	660–675
Llançà	Plot (*n*=1)	Soil mesofauna	0.02	9	SE	870
Figaró	Plots (*n*=43)	Stand dynamics	0.0314	7–40	All	450–750
Prades						
Titllar						
Avic	Catchment	Water and nutrient budgets; ecophysiology; soil studies	51.6	28	NW	700–1018
Teula	Catchment	Water and nutrient budgets	38.5		NW	740–1084
Forest survey plots[e]	Plots (*n*=151)	Tree growth; stand biomass and production	0.0025		W to N	750–950
Torners						
Fertilization/irrigation experiment	Plots (*n*=24)	Resource limitation	0.0064	11–24	SE	900
Burning/clear-cutting experiment	Plots (*n*=3)	Disturbance effects	0.0400		SSE	800–900
Thinning experiment	Plots (*n*=12)	Water and carbon use, stand growth, modelling	0.5		NE	650–750

[a] Area of each individual catchment or plot.
[b] For catchments, the mean slope of sideslopes is given.
[c] For catchments, the dominant orientation of the main stream is given.
[d] The previously reported area for TM9 (4.3 ha), based on 1:10 000 topographic maps, was found to underestimate the true catchment area according to the Cl⁻ budget. Catchment area was adjusted to give a balanced Cl⁻ budget for 1984–1992, as detailed in Piñol et al. (1995).
[e] Sixty-nine 25-m² plots within the Avic catchment; eighty-two 49-m² plots within the Teula catchment.

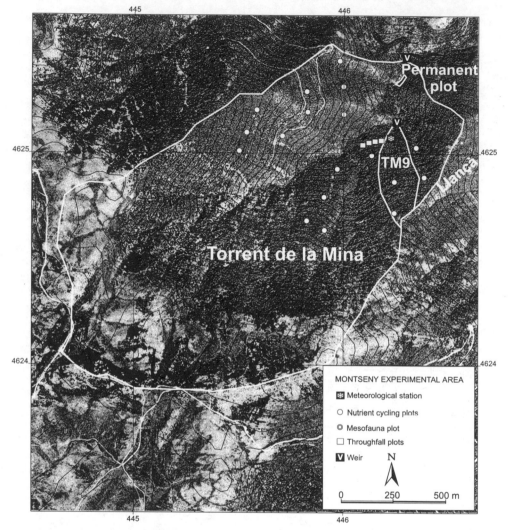

Fig. 2.2. Study sites in La Castanya (Montseny). Contour lines at 20-m intervals

catchment was gaged in 1983 for measuring catchment budgets of water and
nutrients in holm oak forest. The whole Torrent de la Mina catchment, the
lower half of which is covered by holm oak, was gaged for the same purpose
in 1990. A further catchment (TM5) in the upper reaches of the Torrent de la
Mina catchment was also gaged (Belillas and Rodà 1991), but since it is covered by heathlands and grasslands it will not be dealt with in this book.

A 0.23-ha forest plot near the outlet of the Torrent de la Mina catchment
was the site of our earlier studies on biomass, production, and nutrient cycling in the late 1970s and early 1980s. This plot, thereafter referred to as the
permanent plot, was representative of holm oak stands growing in high
quality sites. In 1985 a network of 81 154-m² plots was systematically laid out

all over the holm oak forest area of the Torrent de la Mina catchment for deriving unbiased forest surveys of the catchment and for monitoring tree growth. The same year, 18 154-m² closed-canopy plots (12 of which belonged to the previous network) were laid out at different altitudes and slope aspects within the Torrent de la Mina catchment for production and nutrient cycling studies (Fig. 2.2).

In the Figaró experimental area a network of plots varying in aspect, slope and altitude were used for studies on stand dynamics. Sixty-eight 314-m² plots were established in 1990. Nevertheless, in the present book we will deal mostly with results obtained in those plots where the canopy is dominated by holm oak.

2.2.2 Climate

At La Castanya the climate is subhumid meso-Mediterranean. Annual precipitation is relatively high, averaging 879 mm at the outlet of the TM9 catchment (annual range 627–1178 mm, period 1983–1995). Summer drought is present, though attenuated by the relatively high June rainfall and by frequent storms in August and September (Fig. 2.3). As in all our experimental areas, there is a large interannual variability both in the annual amount and the monthly distribution of precipitation. Snowfall accounts on average for only 3% of annual precipitation at the lower altitudes. Snowpacks are sporadical and short-lived in the holm oak forest area, but they persist longer at altitudes above 1200 m a.s.l. The mean annual air temperature probably varies strongly with altitude, aspect and physiographic position: for the holm oak forest in the Torrent de la Mina catchment it is estimated to range from ca. 9 °C on the north-facing slope at 1000 m a.s.l. to ca. 13 °C on the lower reaches of the east-facing slope. At Figaró the climate is drier than at La Castanya, the mean annual precipitation being 625 mm (range 423–876 mm, period 1977–1991). Climate at Figaró ranges from meso-Mediterranean in the upper slopes to thermo-Mediterranean in the lower, south-facing slopes. Topographic variations in mean annual air temperature are likely to vary strongly also here. At La Garriga, at 250 m a.s.l. and 4 km to the south of the Figaró area, the mean annual temperature is 15 °C, meaning that for our forest plots in the Figaró area it must lie around 12–14 °C.

2.2.3 Lithology and Relief

The bedrock at La Castanya is a metamorphic phyllite, with quartz, chlorite, albite, and muscovite as major minerals. The Torrent de la Mina catchment has two slopes of contrasting aspect (north- and east-facing, respectively), both being very steep (mean 34°). The relief of the sideslopes of the catchment is very rugged, with rock outcrops and screes breaking the forest canopy. The relief of the highest part of the catchment is more moderate. The permanent plot is located at the base of a steep NW-facing slope, and has two zones of different slope: the lower part is relatively level (mean slope 7°) and the upper part is steeper (mean 23°). At Figaró the bedrock is phyllite, schists, and granodiorite. The sampling plots lie on steep slopes (mean between 20 and 36°) of widely different aspect.

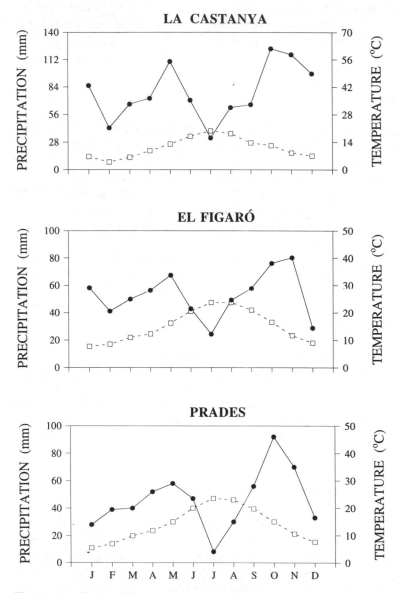

Fig. 2.3. Monthly distribution of precipitation (*solid symbols*) and temperature (*open symbols*) at Montseny (La Castanya and Figaró) and at Prades (Poblet)

2.2.4 Soils

At La Castanya, soils are rather shallow with maximum depths of 1.5–2 m, and a high level of spatial heterogeneity in the landscape because of the rugged topography. Ten soil profiles excavated in the north-facing slope of the Torrent de la Mina catchment yielded a mean depth to the bedrock of 60 cm. The organic layers (O horizon) are 3 cm deep on average. The L and F layers are usually well developed, but the H horizon is sometimes lacking. Most of the soils are colluvial with discontinuities in the distribution of the very abundant stones and little vertical distinction in morphological features. They are classified (Soil Survey Staff 1992) as Entisols (Lithic Xerorthents) or Inceptisols (Typic, Lithic or Dystric Xerochrepts). The main pedogenetic process is the formation of a cambic horizon, with moderate illuviation (Hereter 1990).

Major soil characteristics are shown in Table 2.2. They are stony soils with loamy textural classes, and the bulk density tends to increase with depth. Below the first centimeters of mineral soil, the concentration of organic carbon averages around 2% (Table 2.2). The soils are acidic, acidity being buffered mainly by silicate weathering and cation exchange. Calcium is the dominant exchangeable base cation, and it is especially abundant in the upper mineral soil. Base saturation is low – for a Mediterranean soil – under the first centimeters of mineral soil, and reaches the highest values in both A- and C-horizons. There is a significant positive relationship between the cation exchange capacity and the content of soil organic matter. Organic phosphorus is the dominant phosphorus form in all soil horizons.

2.2.5 Vegetation

At La Castanya, holm oak forest covers the lower 53% of the Torrent de la Mina catchment (TM0). On the north-facing slope, holm oak extends up to an altitude of 1000 m, while on the east-facing slope it reaches 1200 m. The headwaters of the catchment are occupied by heathlands and grasslands, with *Calluna vulgaris, Juniperus communis, Erica arborea, Erica scoparia,* and *Pteridium aquilinum* as major species. A beech forest on the north-facing slope covers 10% of the catchment. The TM9 catchment is wholly covered by a dense holm oak forest, undisturbed in the last decades. Throughout the holm oak forests within the Torrent de la Mina catchment, holm oak is virtually the only tree species. The tree canopy is usually 5–9 m high. Current holm oak stems are derived both from seed and from stump sprouts. On the north-facing slope of the Torrent de la Mina catchment 52% of stems having a diameter at breast height (dbh) of ≥5 cm originated from seed, but only 17% on the east-facing slope. Seed-derived stems dominate in the permanent plot, with 84% of tree stems. The understory is very sparse where the holm oak canopy has remained closed for some time, but a thick understory of the shrub *Erica arborea* is present where the tree canopy is open or was open

Table 2.2. Major soil characteristics of study sites in the Montseny and Prades experimental areas. (Means of n soil samples)

Soil property	Montseny			Prades			
	La Castanya[a]		Figaró[b]	Avic– upper slope[c]		Avic – lower slope[c]	
	0–4 cm (n=10)	4–44 cm (n=10)	0–20 cm (n=6)	0–10 cm (n=3)	10–40 cm (n=3)	0–20 cm (n=3)	20–45 cm (n=3)
Stoniness (%)	69.8	40.8	43.1	55.4	64.4	52.9	57.9
Texture	Sandy loam	Loam	Loam	Clay loam	Clay loam	Clay loam	Clay loam
Organic C (%)	9.6	2.3	1.7	6.0	1.4	5.6	1.4
Organic N (%)	0.49	0.22	0.12	0.38	0.10	0.39	0.14
pH water	5.3	4.6	5.8	6.2	5.7	7.3	6.6
pH KCl	4.9	4.1	4.7	5.4	4.3	6.6	5.2
Ca ($cmol_c$ kg^{-1})	9.5	1.5	–	50.7	16.8	38.7	13.2
Mg ($cmol_c$ kg^{-1})	5.8	1.4	–	4.09	2.12	3.65	2.71
Na ($cmol_c$ kg^{-1})	1.30	0.64	–	0.44	0.21	0.25	0.19
K ($cmol_c$ kg^{-1})	0.62	0.13	–	1.09	0.49	0.62	0.21
CEC ($cmol_c$ kg^{-1})	16.4	10.9	–	29.7	9.8	10.5	8.0
Base saturation (%)	100	37	–	100	100	100	100
Organic P (mg kg^{-1})	588[d]	441[d]	–	–	–	–	–
Extractable P (mg kg^{-1})	42.3	6.0	–	–	–	–	–

CEC, cation exchange capacity.

[a] Soil samples taken in autumn 1988 in a midslope position at 750 m a.s.l. on the north-facing slope of the Torrent de la Mina catchment. (From D. Bonilla.)

[b] Soil samples taken in each of 6 plots at 550 m a.s.l. on a NW-facing slope on phyllites and schists. Four individual samples per plot, composited per plot.

[c] From J. Piñol and J.M. Alcañiz (unpubl. data).

[d] From Hereter (1990); n=4.

until recently. Other frequent understory species are the shrubs (occasionally small trees) *Phillyrea latifolia* subsp. *media* and *Crataegus monogyna*. Major species in the sparse field layer are *Quercus ilex* seedlings, *Hedera helix*, *Rubus ulmifolius*, *Brachypodium sylvaticum*, *Melica uniflora*, *Pteridium aquilinum*, *Asplenium adiantum-nigrum*, and *Teucrium scorodonia*. Phytosociologically, these holm oak forests are classed as belonging to the association *Asplenio-Quercetum* (=*Quercetum mediterraneo-montanum*).

Research at the Figaró experimental area was done on a 900-ha forest made up of pure holm oak stands, pure Aleppo pine (*Pinus halepensis*) stands, and mixed holm oak–Aleppo pine stands. This forest is representative of the western sector of the Montseny massif at low and mid altitudes. The overstory is dominated by holm oak and/or Aleppo pine, with *Pinus pinea*, *Quercus cerrioides*, *Arbutus unedo* and *Erica arborea* occasionally joining in. Holm oak is the dominant tree species in most of the sampling plots whose results have been used in this book.

Holm oak forests at Montseny were heavily exploited for charcoal production until ca. 1955. Coppicing took place typically at intervals of 10–15 years (Gutiérrez 1996; Sect. 1.6). After charcoal production was discontinued, fellings were restarted in many holm oak forests after the mid-1970s for firewood production. Privately owned holm oak forests, such as those in our Figaró experimental area, are now managed for firewood through selection thinnings every 20–30 years (Retana et al. 1992). Selection thinning consists of a partial removal of standing trees, with the highest intensities of felling directed at the largest stems (dbh >15–20 cm) and at the abundant suppressed stems resprouted after the last thinning. Thinnings remove about 30–60% of the pre-thinning basal area and aboveground biomass (Retana et al. 1992).

The history of holm oak forests in the Torrent de la Mina catchment at La Castanya is somewhat more complex. Part of the permanent plot near the outlet of the catchment is relatively flat and it was cultivated in the past. It was abandoned, probably sometime between ca. 1880 and 1920, and colonized by seed-derived holm oaks that have never been cut, as evidenced by their single stems. The other, steeper, part of the plot was coppiced for firewood and charcoal more or less continuously, as all the rest of the holm oak forest within the catchment. The lower slopes of the catchment, mainly between altitudes of 700 and 800, m were subject to slash-and-burn agriculture, as reflected by place names and by the present structure of the forest, with an open canopy of low holm oak trees and a dense understory of the shrub *Erica arborea*. Since the 1950s the holm oak forests of the catchment have remain unmanaged, except for two small areas that were thinned, one around 1965 and the other in 1979.

Fire has not occurred in any of our forest sites at Montseny for decades. Grazing or browsing by domestic herbivores is very limited in our sites, except in the heathlands and grasslands in the upper part of the TM0 catchment.

2.3 Experimental Area at Prades

2.3.1 Experimental Layout

The Prades experimental area is located within the Poblet Forest (41° 13' N, 0° 55' E) in the Prades mountains (Fig. 2.1). It was established in 1980. Two major valleys harbour the experimental catchments and plots: the Titllar valley and the Torners valley (Fig. 2.4). Within the Titllar valley two north-facing, adjacent holm oak forested catchments were gaged in 1981, and have been continuously monitored for water and nutrient budgets since then. These catchments (Avic, 51.6 ha, and Teula, 38.5 ha) have been the subject of many studies on the hydrology and biogeochemistry of holm oak forest. In the Avic catchment intensive ecophysiological work on holm oak was conducted at two sites in the upper and lower parts of the catchment. The major characteristics of these study sites are summarized in Table 2.1.

A third catchment (Saucar) was gaged in the upper reaches of the Torners valley. On the slopes of the Torners valley three major plot-level experiments have been conducted in holm oak stands. In the east- to southeast-facing slope, one experiment addressed the effects of an experimental fire and/or clear-cutting on soil properties, nutrient losses, and stump resprouting. A second experiment on an adjacent site investigated the effects of irrigation and/or fertilization of tree growth, primary production, and nutrient cycling. On the opposite, north-facing slope of Torners a thinning experiment was conducted to assess tree and stand responses to decreased stem density. A fourth catchment (Castellfollit) was also monitored in an adjacent valley, but since it was covered mainly by Scots pine it is not dealt with in this book.

2.3.2 Climate

The climate of the Prades study sites is typically Mediterranean (more so than at Montseny), with a mean annual precipitation at the nearby Poblet Monastery of 537 mm (range 360–768 mm, period 1981–1995). The summer is markedly dry, and there are two precipitation maxima in autumn and spring (Fig. 2.3). An indication of the annual water deficit can be gained through the evaporation measured with a Piché evaporimeter at Poblet Monastery, which averages 1043 mm year^{-1}. The mean annual air temperature is 13.6 °C at Riudabella (485 m a.s.l., and 3.5 km to the north from the Avic catchment) and 10 °C at Prades (950 m a.s.l.).

2.3.3 Lithology and Relief

The bedrock at the study sites is mostly Palaeozoic phyllite with small amounts of metamorphic sandstones with microconglomerates. Phyllite and schists dominate at the Avic and Teula catchments, with the frequency of metamorphic sandstones increasing from Titllar towards Torners. Relief is very steep, with mean slopes ranging from 24 to 28° in the different experimental sites. Overall, the relief is less rugged than at La Castanya (Montseny), particularly in the Avic and Teula catchments where the forest cover is almost unbroken. More discontinuities exist at Torners, resulting mainly from scree areas.

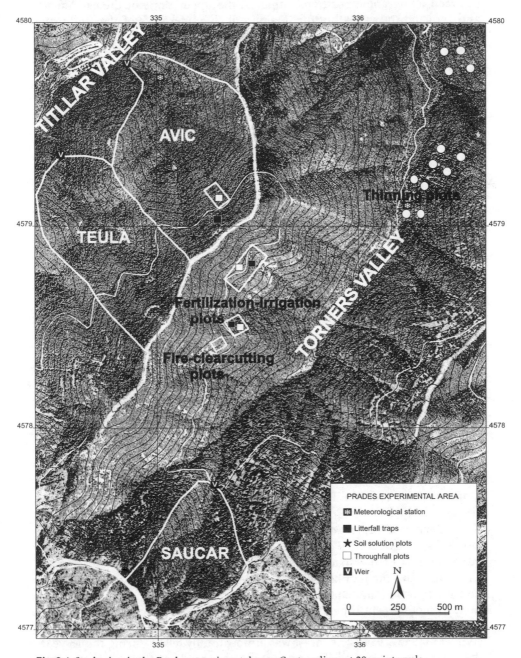

Fig. 2.4. Study sites in the Prades experimental area. Contour lines at 20-m intervals

2.3.4 Soils

The soils in the Prades experimental area are Inceptisols and Entisols. In the experimental catchments, the soils are Xerochrepts with subgroups differing between topographic positions: lithic on the upper slope of the catchments and typic on the valley bottom. Major soil characteristics of the study sites are shown in Table 2.2.

Soil depth usually ranges between 30 and 80 cm, but reaches 150 cm at valley bottom sites. There is a high content of coarse rock fragments, and the dominant soil texture is clay loam. Soil pH is much higher than at Montseny, particularly than at La Castanya. At Prades, soil pH on the lower slope is higher than on the upper slope. The cation exchange complex is saturated, Ca^{2+} being the predominant exchangeable cation. Small quantities of $CaCO_3$ (<0.3%) are detectable in the soil through gas chromatography after acidifying soil samples.

2.3.5 Vegetation

The vegetation at the Avic and Teula catchments is a closed-canopy holm oak forest. The upper parts of the catchments have a lower diversity of plant species and the vegetation can be included in the *Asplenio-Quercetum* association. In the lower parts, with a higher plant diversity and higher presence of termophilic species, it can be included in the *Viburno-Quercetum* association (Folch and Velasco 1978). The dominant species at all altitudes is holm oak, with *Arbutus unedo* and *Phillyrea latifolia* subsp. *media* as major companion species in the understory and, occasionally, also in the overstory (Lledó et al. 1992). At Torners, the same overall composition applies, with holm oak being even more dominant there. Overall, the overstory of holm oak forests in the Prades experimental area is less monospecific than at Montseny. At Prades, besides the evergreens *Arbutus unedo* and *Phillyrea latifolia*, *Pinus nigra* is sometimes present, and there is a minor but significant deciduous component made up of *Quercus cerrioides* (a marcescent to deciduous oak), *Acer monspessulanum*, and *Sorbus torminalis*.

The forest history at Prades is in broad terms similar to that described for Montseny. However, the Poblet Forest where our Prades experimental sites are located belonged for centuries to the Poblet Monastery. This provided some degree of protection, because though the forest was harvested, the monks saw it as a source of income and strove to use it wisely. There are frequent records of the struggle between villagers pressing for increasing the harvesting and the monks trying to limit it. The intensity of forest exploitation for firewood and charcoal generally increased until the eighteenth century. In more recent times, the Poblet Forest, and more specifically the Avic and Teula catchments, was heavily coppiced for charcoal after the Spanish civil war (1936–1939). The forest was almost totally felled during the 15 years that followed this period. The present forest is made of the stump resprouts that grew after such felling ceased around 1950. The holm oak forest in the Avic and Teula catchments has remained undisturbed since then (Lledó et al. 1992), while in Torners it was thinned in the 1960s.

As in Montseny, our Prades experimental areas have been subject neither to fire nor to grazing or browsing by domestic herbivores for decades.

References

Belillas C, Rodà F (1991) Nutrient budgets in a dry heathland watershed in northeast Spain. Biogeochemistry 13:137–157

Bolòs O de (1983) La vegetació del Montseny. Diputació de Barcelona, Barcelona

Folch R, Velasco E (1978) Dades cartogràfiques per a l'estudi de la vegetació de les Muntanyes de Prades. XVIII Assemblea Intercomarcal d'Estudiosos, L'Espluga de Francolí, 1974. Barcino, Barcelona

Gutiérrez C (1996) El carboneig. L'exemple del Montseny. Alta Fulla, Barcelona

Hereter A (1990) Els sòls forestals del massís del Montseny. PhD Thesis, University of Barcelona, Barcelona

Lledó MJ, Sánchez JR, Bellot J, Boronat J, Ibañez J, Escarré A (1992) Structure, biomass and production of a resprouted holm-oak (*Quercus ilex* L.) forest in NE Spain. Vegetatio 99/100:51–59

Mestres JM, Masalles RM (1988) El paisatge vegetal. In: Fuguet J (ed) La natura i l'home a les Muntanyes de Prades. Monografies III. Centre d'Estudis de la Conca de Barberà, Montblanc, pp 45–74

Piñol J, Terradas J, Àvila A, Rodà F (1995) Using catchments of contrasting hydrological conditions to explore climate change effects on water and nutrient flows in Mediterranean forests. In: Moreno JM, Oechel WC (eds) Global change and Mediterranean-type ecosystems. Springer, Berlin, pp 251–264

Poblet T, Pujadas J (1988) La geologia. In: Fuguet J (ed) La natura i l'home a les Muntanyes de Prades. Monografies III. Centre d'Estudis de la Conca de Barberà, Montblanc, pp 21–41

Retana J, Riba M, Castell C, Espelta JM (1992) Regeneration by sprouting of holm oak (*Quercus ilex* L.) stands exploited by selection thinning. Vegetatio 99/100:355–364

Soil Survey Staff (1992) Keys to soil taxonomy, 5th edn. SMSS Technical Monograph No 19. Pocahontas Press, Blacksburg, Virginia

Part 2   Stand Structure and Dynamics

3 Stand Structure, Aboveground Biomass and Production

Joan Josep Ibàñez, María José Lledó, Juan Rafael Sánchez and Ferran Rodà

3.1 Introduction

Holm oak forests are structurally diverse (Gracia and Retana 1996). Marked variations in the main structural variables, such as stem density, basal area and canopy height, are induced by gradients of environmental heterogeneity (David et al. 1979; Barbero 1988; Romane et al. 1988; Floret et al. 1989), caused by the rough topography of the region and by historical differences in human use (Di Castri 1973; Terradas 1991; Barbero et al. 1992). Moreover, holm oak can reproduce through seeds (acorns) or vegetatively (sprouting from stumps and, to a lesser extent, from roots). Stems of both origins are usually found in the same stand, but, due to the slow rate of holm oak seedling establishment under managed coppice forests (Espelta et al. 1995; Chap. 7), restocking of holm oak forests during the cutting cycle is left to occur by natural regeneration from stump resprouting. Since each stool can produce large numbers of sprouts after disturbances (Giovannini et al. 1992; Retana et al. 1992), stem density can be very high in stands that have been harvested or otherwise disturbed.

Management is thus an essential factor affecting structural development of holm oak forests. Even protected forests that are now largely unmanaged, as most of our experimental areas at Prades and Montseny, bear the imprint of past management in their present stand structure. These forests were coppiced for charcoal at short cutting intervals (typically 10–15 years) until ca. 1950 (Ferrés 1984; Lledó et al. 1992; Sect. 1.6). In most areas of Catalonia, including our experimental area at Figaró (Montseny), management of holm oak forests has now shifted to firewood production by selective cutting repeated at longer intervals (about 20–30 years). In contrast, in other areas of the Mediterranean Basin holm oak forests are managed mainly through clear-felling (Romane et al. 1988).

Holm oak forests often have a moderate aboveground biomass and a low net aboveground primary production. Reasons behind these facts will be explored in this chapter and elsewhere in this book (e.g. Chaps. 10, 12 and 13).

Ecological Studies, Vol. 137
Ferran Rodà et al. (eds) Ecology of Mediterranean
Evergreen Oak Forests
© Springer-Verlag, Berlin Heidelberg 1999

Let us only remark here that the consequences of having low production propagate throughout the ecosystem. Namely, carbon and nutrient fluxes between the vegetation and the soil will be correspondingly limited.

The objectives of this chapter are: (1) to describe the structure of the holm oak forests of the Montseny and Prades experimental areas, (2) to estimate their aboveground biomass and production, and (3) to place them within the wider context of Mediterranean holm oak forests. Biomass and production data will be used later in this book, e.g. for modelling ecosystem functioning (Chap. 12) and computing nutrient pools and fluxes (Chap. 18).

3.2 The Database

Data used in this chapter have been obtained from forest surveys carried out in our experimental areas described in Chapter 2. For Montseny, we will use data from the Torrent de la Mina catchment at La Castanya, where stem diameters greater than 5 cm were measured at 1.30 m from the ground (dbh). For Prades, we will use data from the Avic and Teula catchments and from the plot manipulation experiments in the Torners area (Chaps. 2, 13 and 23). In the latter case, only data from undisturbed plots are presented (i.e. from control plots or plots before treatment application). At Prades, given the generally shorter stature of the trees, stem diameters greater than 2 cm were measured at 50 cm from the ground (D_{50}). We define the tree layer as those stems of diameter ≥ 5 cm (either dbh or D_{50}, depending on site). Stems of diameters 2–5 cm are considered part of the shrub layer. Table 3.1 summarizes the characteristics of the sampling plots.

Aboveground biomass was estimated through dimensional analysis (Whittaker and Woodwell 1968). The allometric regressions for holm oak at Prades (Lledó 1990) and Montseny (Canadell et al. 1988) are as follows:

Prades: log dry wt. = $-1.01 + 2.34$ log D_{50}; $r^2 = 0.975, P < 0.0001, n = 20$;
Montseny: log dry wt. = $-1.05 + 2.46$ log D_{50}; $r^2 = 0.934, P < 0.0001, n = 70$;
                  log dry wt. = $-0.66 + 2.22$ log dbh; $r^2 = 0.908, P < 0.0001, n = 69$,

where dry wt. = total aboveground dry weight (kg); D_{50} = diameter measured at 50 cm above ground level (cm); dbh = diameter measured at 130 cm above ground level (cm). At Prades, aboveground biomass of the major companion tree species was estimated by the species-specific regressions of Lledó (1990).

Resurveys of the sampling plots furnished data on demographic changes and diameter growth rates of the trees, thus allowing the computation of biomass increments. Aboveground net primary production was taken as the sum of the annual woody biomass increment and the annual litterfall, which was collected in traps emptied monthly or every 2 weeks.

Table 3.1. Characteristics of plots used for density and biomass estimation of holm oak forests at Prades and Montseny. Stem diameters (≥ 5 cm) were measured at 50 cm from the ground at Prades and at 1.30 m at Montseny

Locality	Site	Sampling years	Altitude (m)	Number of plots	Plot area (m²)	Number of measured stems[a]	References
Prades	Avic	1981, 1986	750–950	69	25	729	Lledó (1990); Lledó et al. (1992)
	Teula	1986	825–1025	82	49	1804	Lledó (1990); Lledó et al. (1992)
	Torners S-facing	1988, 1991, 1992	900	24	36	1579[b]	Mayor and Rodà (1994); Mayor et al. (1994)
	Torners N-facing	1992	700	3	707	2679	Albeza et al. (1995); Martínez et al. (1996)
Montseny	Torrent de la Mina, whole catchment[c]	1985, 1988, 1991, 1995	700–1200	81	154	2400	F. Rodà and X. Mayor, (unpubl. data)
	Torrent de la Mina, closed-canopy plots	1985, 1988, 1991, 1995	800–1000	18	154	733	Mayor (1990); F. Rodà and X. Mayor, (unpubl. data)
	Permanent plot	1977, 1985, 1991, 1994	650–665	1	2300	462	Ferrés et al. (1984); Mayor (1990); F. Rodà, (unpubl. data)

[a] At first sampling date.
[b] Stems with $D_{50} \geq 2$ cm.
[c] Holm oak forest area only.

3.3 Stand Structure

Holm oak is the dominant species in both experimental areas. In Prades, it accounts for 77–89% of total arboreal basal area (Table 3.2), with *Arbutus unedo* and *Phillyrea latifolia* being the more abundant companion trees (Lledó et al. 1992). In Montseny, holm oak is practically the only species in the tree layer.

Our holm oak forests show a high stem density, specially in Prades, where the densities of the tree layer ($D_{50} \geq 5$ cm) at the Avic and Teula catchments average around 4400 stems ha^{-1}, compared to an average of about 2000 stems ha^{-1} (dbh ≥ 5 cm) in the Torrent de la Mina catchment (Table 3.2). Total density of stools (i.e. genetic individuals) of holm oak is much lower, being in the range of 700–1200 stools ha^{-1}. In Prades, almost all stems are stump sprouts, with an average of 4.0 sprouts stool^{-1} (SD = 2.4). In contrast, 84% of holm oak trees in the permanent plot at Montseny are derived from seed. The Torrent de la Mina catchment lies between these two extremes. In addition to tree-sized stems, there are a large number of shrub-sized stump resprouts, particularly at Prades where total densities (tree plus shrub layers) of living stems average about 9000 stems ha^{-1} in the Avic and Teula catchments and 18 500 stems ha^{-1} in the south-facing slope of Torners. Such extremely high densities are obviously the result of past coppicing, and make these forests difficult to penetrate by the wind and by humans.

Basal area averaged 16 m^2 ha^{-1} in the Montseny catchment in 1985 and 34–48 m^2 ha^{-1} in the Prades areas (Table 3.2). Nevertheless, it must be taken into account that basal area has been calculated at 50 cm from the ground in Prades and at 130 cm in Montseny. The relationship between dbh and D_{50} for

Table 3.2. Density (stems ha^{-1}), basal area (m^2 ha^{-1}) and aboveground biomass (Mg ha^{-1}) of the studied holm oak forests. Stem diameters were measured at 50 cm at Prades ($D_{50} \geq 5$ cm at Avic and Teula; $D_{50} \geq 2$ cm at Torners) and at 1.30 m (dbh ≥ 5cm) at Montseny

Locality	Site and year	Density		Basal area		Aboveground biomass
		Total	% Holm oak	Total	% Holm oak	
Prades	Avic 1981[a]	4226	71.6	34.3	80.2	} 104
	Teula 1986[a]	4490	79.5	34.4	77.3	
	Torners S-facing 1988	18495	81.2	48.4	89.5	128
Montseny	Torrent de la Mina, whole catchment 1985[b]	1911	99.1	16.4	99.4	79
	Torrent de la Mina, closed-canopy plots 1985	2645		21.5		104
	Permanent plot 1977	2008	99.8	25.7	99.9	160

[a] If the shrub layer ($D_{50} \geq 2$ cm) is included, averaged values for the Avic and Teula catchments are: density 9178 stems ha^{-1}, basal area 37.9 m^2 ha^{-1}, and aboveground biomass 113 Mg ha^{-1}.
[b] Holm oak forest area only.

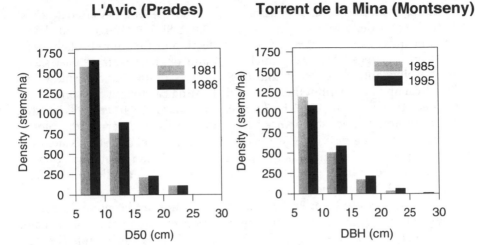

Fig. 3.1. Size distribution of stem diameters in Avic (Prades) and Torrent de la Mina (Montseny) in two different forest surveys

holm oak is: dbh = 0.833 D_{50} – 0.117 (r^2 = 0.945, n = 1202 stems on the south-facing slope of Torners). From this equation it follows than basal area at 50 cm from the ground is about 30% higher than at 130 cm.

The holm oak forests of Montseny and Prades may be structurally considered as coppices, though in Montseny holm oaks reach greater maximum diameters than those found in Prades (Fig. 3.1). The stem diameter distributions in both areas are reverse-J shaped and characterized by the abundance of small trees (Fig. 3.1): almost 50% of the stems in the tree layer have diameters below 10 cm, and less than 5% are above 20 cm. These distributions are consistent with the shade tolerance of holm oak, because uneven-aged stands of shade-tolerant species show steeply descending, monotonic, reverse-J shaped diameter distributions that can be approximated by the negative exponential and negative power functions (Lorimer and Krug 1983; Parker 1988; Abrams and Downs 1990).

3.4 Aboveground Biomass

Holm oak forests usually have moderate biomass. Aboveground biomass tends to be limited by the normally short stature and small diameters of holm oaks in managed or previously managed stands (Fig. 3.1). On the other hand, very high stem densities and the high density of the wood of holm oak cause the stand biomass to be higher than would be judged from tree sizes. The aboveground biomass of the tree layer in the Avic and Teula catchments at Prades averaged 104 Mg ha^{-1} in the mid-1980s. At Montseny and in 1985, the aboveground arboreal biomass averaged over the whole holm oak forest of the Torrent de la Mina catchment was 78.8 ± SE 4.8 Mg ha^{-1}. This lower mean biomass compared to Prades is a reflection of the more broken forest cover in the rugged Montseny catchment, and of the effects of past slash-and-burning and of a recent thinning in parts of this catchment. When closed-canopy plots within the Torrent de la Mina catchment are considered, the mean tree biomass (103.7 ± SE 8.3 Mg ha^{-1}; Mayor 1990) is similar to that at Prades. Our previously reported tree biomass for the permanent plot at Montseny (160 Mg ha^{-1} in 1978; Ferrés 1984; Escarré et al. 1987) is near the maximum for the Torrent de la Mina catchment. This can be attributed to the fact that the permanent plot is situated on a deep soil in a valley bottom and bears a high quality stand.

The holm oak forests of Prades and Montseny are similar not only in their aboveground biomass but also in the partition of this biomass into major tree components (Table 3.3). When the Avic and Teula catchments (Prades) are compared to the closed-canopy plots in the Torrent de la Mina catchment (Montseny), boles and large branches contribute about equally to aboveground biomass (52 versus 59%, respectively; Table 3.3). Leaf biomass is also similar at both sites: around 5–6 Mg ha^{-1}. Thus, the lower mean stature of holm oak trees at Prades is compensated by the higher stand density, and perhaps a higher wood density, giving similar biomass per unit area.

Table 3.3. Aboveground biomass (Mg ha^{-1}) in the experimental holm oak forests of Prades and Montseny

Forest layers	Prades	Montseny	
	(Avic–Teula)[a]	Closed-canopy plots[b]	Permanent plot[c]
Tree layer			
Boles and large branches (Ø ≥ 5 cm)	54.3	60.8	101.0
Small branches (Ø <5 cm)	44.0	37.7	53.4
Leaves	6.0	5.2	6.1
Aboveground biomass	104.2	103.7	160.4
Shrub layer			
Boles and large branches (Ø ≥ 5 cm)	0.8	n.a.	–[d]
Small branches (Ø <5 cm)	6.7	n.a.	4.2
Leaves	0.6	n.a.	0.4
Aboveground biomass	8.1	n.a.	4.6
Field layer	0.9	n.a.	0.1
Total aboveground biomass	113.2	> 103.7	165.1

n.a., Data not available.
Figures do not add up to totals because of rounding.
[a] Data for the year 1981. From Lledó (1990).
[b] Data for the year 1985 for 18 closed-canopy plots within the Torrent de la Mina catchment. From Mayor (1990) and F. Rodà and A. Escarré (unpubl. data).
[c] Data for the year 1977. From Ferrés et al. (1984).
[d] Included in small branches.

The shrub layer (stem diameters 2–5 cm) contributed little biomass in the two sites where it was estimated (Table 3.3), and the field layer contributed even less. Holm oak seedlings can be sometimes the dominant component in the field layer.

3.5 Stand Dynamics

3.5.1 Demographic Changes

In Montseny, mean tree density increased by 0.21% year^{-1} during the decade 1985–1995. This increase was the balance between an ingrowth (stems incorporated into the tree layer) of 0.45% year^{-1} and a mortality of 0.24% year^{-1}. This very limited recruitment may be explained because virtually all stems with dbh < 5 cm are suppressed stems having no or negligible diameter growth (Mayor and Rodà 1993). In Avic (Prades), there was a slight reduction in total stem density ($D_{50} \geq 2$ cm) of 0.25% year^{-1} from 1981 to 1986. On the south-facing slope of Torners, stem mortality between 1988 and 1991 averaged 1.5% year^{-1} (Mayor and Rodà 1994). These figures indicate that self-thinning is slowly underway in the dense Prades forests. At both Montseny and Prades mortality was concentrated in the smallest, suppressed stems. Mortality of canopy trees was negligible during the above periods.

3.5.2 Tree Growth

Diameter growth rates of holm oak are very small, averaging 0.6–1.1 mm year^{-1} in Montseny and 0.3–1.2 mm year^{-1} in Prades (Table 3.4). Holm oaks grow much faster when they are in the overstory than when they remain in the understory: in the Torrent de la Mina catchment, diameter growth of overstory trees triple that of understory trees, while in Prades, growth of dominant trees nearly doubles the average growth of all trees (Table 3.4). Stand basal area increased by 0.2 m^2 ha^{-1} year^{-1} between 1981 and 1986 in the Avic catchment at Prades (an increase of 0.8% year^{-1}), and by 0.32 m^2 ha^{-1} year^{-1} between 1985 and 1995 in the Torrent de la Mina catchment at Montseny (an increase of 2% year^{-1}). Our results confirm that holm oak is a slow-growing species (Cartan-Son et al. 1992; Lledó et al. 1992; Mayor and Rodà 1993, 1994). When it occurs at high densities as a result of past coppicing the outcome is a nearly stagnant forest where changes in stand structure are very slow. Chapters 12 and 23 deal with the forest responses to experimental thinnings.

Table 3.4. Mean growth rate in stem diameter (mm year^{-1}) of holm oak in the study sites. Mean values were obtained by dividing diameter increment by number of years in the period. Diameters were measured at 50 cm ($D_{50} \geq 2$ cm) at Prades and at 1.30 m (dbh ≥ 5 cm) at Montseny

Locality	Site	Period	No. of stems	Stem status	Diameter growth	References
Montseny	Torrent de la Mina	1985–1995	2197	All	0.92	F. Rodà and X. Mayor (unpubl. data)
		1985–1995	1749	Dominants	1.06	F. Rodà and X. Mayor (unpubl. data)
		1985–1995	448	Suppressed	0.35	F. Rodà and X. Mayor (unpubl. data)
	Permanent plot	1985–1994	431	All	0.60	F. Rodà (unpubl. data)
Prades	Avic	1981–1986	302	All	0.68	J.R. Sanchéz and M.J. Lledó (unpubl. data)
	Avic	1981–1986	141	$D_{50} \geq 5$ cm	1.2	J.R. Sanchéz and M.J.Lledó (unpubl. data)
	Torners S-facing	1988–1991	151	All	0.31[a]	Mayor et al. (1994)
	Torners S-facing	1988–1991	67	Dominants	0.56[a]	Mayor et al. (1994)
	Torners S-facing	1992	139	All	0.67[a]	Mayor et al. (1994)
	Torners N-facing	1992–1995	254	All	0.29	Gracia et al. (Chap. 23)

[a] Mean of three control plot means.

3.6 Aboveground Production

Net primary production in our experimental areas is rather low, as is usually the case in holm oak forests. In the Avic catchment at Prades, total aboveground production during the period 1981–1986 was only 5.0 Mg ha^{-1} year^{-1}. At Montseny, the mean aboveground primary production of the tree layer in 18 closed-canopy plots was 6.3 Mg ha^{-1} year^{-1}. These production rates are much lower than the averages for temperate deciduous (10.0 ± SD 2.8 Mg ha^{-1} year^{-1}) and temperate coniferous forests (8.4 ± SD 2.8 Mg ha^{-1} year^{-1}; Cole and Rapp 1981). Figure 3.2 shows how holm oak allocates the net aboveground production to the main tree components. About half of the annual aboveground production is devoted to leaf production, and 32–38% to production of woody tissues (wood and bark of boles and branches). Reproductive investment in male inflorescences and fruits consumes 8–11% of the aboveground production. Production of the shrub and field layer was very small at both sites. Net primary production has not been corrected for the minor consumption by phytophagous animals. There was no outbreak of defoliating insects in our experimental areas during the studied periods, and the consumed foliage was typically well below 10% of standing leaf biomass.

Mean fine litterfall in Prades varied between 2.2 and 5.8 Mg ha^{-1} year^{-1}, depending on period, site and topographic position. Our best current estimate of the mean litterfall in the Avic catchment (3.4 Mg ha^{-1} year^{-1}) is higher than the previously reported value (2.2 Mg ha^{-1} year^{-1}; Bellot et al. 1992) which was a 7-year average obtained in the upper part of the catchment where production is limited by thin soils. On the south-facing slope of Torners (Prades), the mean litterfall during 5 years was 4.2 Mg ha^{-1} year^{-1} (V. Diego, unpubl. data). In the Montseny catchment, litterfall production in

Fig. 3.2. Allocation of net aboveground primary production (Mg ha^{-1} year^{-1}) of the tree layer into major components for Prades (Avic; 1981–1986 for woody biomass increment; 1982–1992 for litterfall) and Montseny (closed-canopy plots in the Torrent de la Mina catchment; 1985–1988 for woody biomass increment; 1985–1987 for litterfall)

closed-canopy plots averaged 3.9 Mg ha^{-1} year^{-1} for a 2-year period (Mayor 1990; F. Rodà, unpublished data). In the permanent plot, litterfall during 3 years averaged 4.8 Mg ha^{-1} year^{-1} (Verdú 1984). In the different study sites and periods, leaves contributed 57–66% to total fine litterfall, as multi-year averages.

Mayor (1990) analyzed the contributions of litterfall and current woody biomass increment to primary production in closed-canopy plots at Montseny. Litterfall was the major component, accounting on average for 57% of the aboveground production, with woody biomass increment accounting for the remaining 43%. Despite such distribution, aboveground primary production was more strongly correlated across plots with biomass increment than with litterfall (r = 0.91 and r = 0.76, respectively; $P < 0.001$ in both cases; Fig. 3.3). There was no significant correlation between litterfall and biomass increment. These patterns can be interpreted assuming that production of fast-turnover canopy components (leaves, twigs, and fruits) has higher priority than secondary growth of woody tissues. As a consequence, spatial

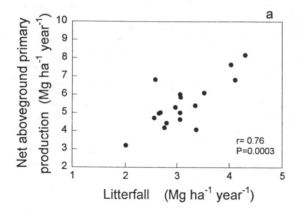

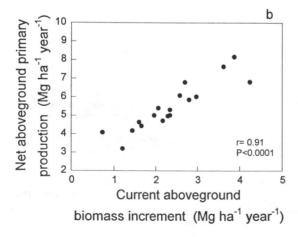

Fig. 3.3. Relationships between net aboveground primary production and a litterfall (1985–1986) and b current aboveground biomass increment (1985–1988), for 18 closed-canopy holm oak plots in the Torrent de la Mina catchment (Montseny)

variation among topographic positions is less for litterfall than for woody biomass increment, and spatial variations in aboveground primary production are better reflected by the latter.

Even with their low primary production, our holm oak forests accrete biomass at appreciable rates: between 1985 and 1995 the mean aboveground biomass of the tree layer in the Torrent de la Mina catchment increased from 78.8 to 96.5 Mg ha^{-1}. This amounts to a mean biomass accumulation of 1.8 Mg ha^{-1} year^{-1}, or to a carbon sequestration of 90 g C m^{-2} year^{-1} into woody tissues of living trees. Total sequestration of carbon by the ecosystem is probably higher since these relatively young forests are also accumulating, at least, dead wood. The mean annual increments in the woody biomass of living trees ranged from a low of 1.1 Mg ha^{-1} in the control plots of the south-facing slope of Torners (Prades) to a maximum of 4.7 Mg ha^{-1} obtained around the year 1980 in the permanent plot of Torrent de la Mina catchment (Montseny; Ferrés 1984). Intermediate values were found in Avic (Prades; Fig. 3.2), in closed-canopy plots across the Torrent de la Mina catchment (Fig. 3.2), and in later years in the permanent plot at Montseny (Mayor 1990).

The low rates of net aboveground primary production in our holm oak forests probably result from a combination of soil water limitation, intrinsic characteristics of the species (e.g. patterns of belowground allocation, see Chap. 4, and respiration costs, see Chap. 12), and the effects of long-term use and abuse by man on soil fertility and other forest properties. The role of water and nutrient availabilities in limiting primary production in holm oak forests is addressed in Chapter 13.

3.7 Structure, Biomass and Production in Mediterranean Holm Oak Forests

A review of published studies on holm oak forests across the Mediterranean Basin shows that their aboveground biomass varies greatly (Table 3.5). The most striking outlier is the forest of Supramonte di Orgosolo (Sardinia; Susmel et al. 1976). This magnificent forest shows the highest biomass (339 Mg ha^{-1}), and contains holm oaks with dbh > 60 cm. The holm oak forest of Le Rouquet (southern France; Lossaint and Rapp 1971) also has a biomass (269 Mg ha^{-1}) much higher than most holm oak stands (Table 3.5), a relevant fact since Le Rouquet is the only Mediterranean forest ecosystem listed in many worldwide compilations such as the IBP Woodland Data Set (Reichle 1981) or the review by Vogt et al. (1986). The high biomass and basal area of the Supramonte di Orgosolo and Le Rouquet holm oak forests are related to the long time these forests have remained without being felled (centuries in the former and ca. 150 years in the latter), much longer than the other holm oak forests studied (Table 3.5). This suggests a close relationship between

Table 3.5. Stand structure and aboveground biomass in several Mediterranean holm oak forests, compared to those at Montseny (Torrent de la Mina) and Prades (Avic and Teula)

Site	No. of plots	Dominant height (m)	Stems measured	Mean dbh (cm)	Basal area (m² ha⁻¹)	Density (stems ha⁻¹)	Biomass (Mg ha⁻¹) Aboveground	Biomass (Mg ha⁻¹) Leaves	Age (years)[a]	References
Supramonte di Orgosolo	–	>17	$Dbh \geq 17.5$	–	39.3	394	339	7.4	>200	Cartan-Sonand Romane (1992)
Monte Minardo	–	6–9	$Dbh \geq 5$	6–20[b]	–	9600	150	8.0	45	Cartan-Son and Romane (1992)
Miremma Natural Park–plot 1	1	8.7	?	–	37.8	13160	–	–	48	Cartan-Son and Romane (1992)
Miremma Natural Park–plot 2	1	6.2	?	–	36.2	6815	–	–	39	Cartan-Son and Romane (1992)
Le Rouquet	–	11	$Dbh \geq 5$	–	38.8	1427	269	7.0	ca. 150	Cartan-Son and Romane (1992)
Puéchabon	–	–	$Dbh \geq 4.5$	–	–	3858	64.7	–	–	Floret et al. (1989)
Gardiole de Rians-plot 2	1	6.6	$D_{10} \geq 2.5$[c]	–	19.8	11549	103.5	–	25–30	Cartan-Son and Romane (1992)
Gardiole de Rians-plot 5	1	6.4	$D_{10} \geq 2.5$	–	31.0	6900	104.6	–	60–65	Cartan-Son and Romane (1992)
La Bruguiere-plot 3	1	>2.6	$D_{50} \geq 1$[c]	2.4[c]	16.8	31000	–	–	15	Cartan-Son and Romane (1992)
La Bruguiere-plot 5	1	>3.7	$D_{50} \geq 1$	4.7[c]	27.2	12200	–	–	25	Cartan-Son and Romane (1992)
Torrent de la Mina, 1995[d]	79	4–13	$Dbh \geq 5$	10.4	19.7	1958	96.5	4.5	16–>50	F. Rodà et al. (unpubl. data)
Vallès Oriental[e]	108	8.6	$Dbh \geq 5$	11.2	18.9	1920	71.2	5.0	–	SIBosC[f]
Avic and Teula, 1981–1986	74	3–9	$D_{50} \geq 5$[c]	10.0[c]	34.3[g]	4590[g]	104.2	6.0	>35	Present work
Conca de Barberà[e]	14	6.6	$Dbh \geq 5$	9.91	16.5	2138	50.4	2.1	–	SIBosC[f]
Catalonia	1081	7.5	$Dbh \geq 5$	10.3	14.4	1745	51.7	3.1	–	SIBosC[f]

[a] Time since last harvest.
[b] Range.
[c] D_i = iameter at i cm from the ground.
[d] Data for the 53% of the catchment covered by holm oak forest.

[e] Vallès Oriental and Conca de Barberà are the administrative areas where Montseny and Prades are respectively placed.
[f] The Information System of Catalonian Forests (SIBosC) contains forest data on 10 500 sampling plots throughout Catalonia (one plot per square kilometer of forest area). All SIBosC data in this table are from plots where holm oak contributes at least 80% of total basal area, and the figures given are for holm oak only.
[g] Holm oak accounts for 27.0 m² ha⁻¹ and 3445 stems ha⁻¹.

stand age and structure (measured as either basal area or biomass) of holm oak forests.

Compared to these forests, the rest of the reported Italian and French holm oak forests show values in the range of those measured in Catalonia (NE Spain), where the arboreal aboveground biomass in 671 holm oak plots randomly located across the country varied between 15 and 284 Mg ha^{-1} (Information System of Catalonian Forests, SIBosC, unpubl. data). The Torrent de la Mina holm oak forest is quite representative of its geographical area (Vallès Oriental; Table 3.5), but the Prades stands show far higher biomass and basal area than the average for their geographic area (Conca de Barberà; Table 3.5). The catchments studied in Montseny and Prades have a substantially greater biomass than the vast majority of the holm oak forests of Catalonia. This is a consequence of the accumulation of biomass since the 1950s because little wood has been harvested from these protected catchments during the last decades, as opposed to most holm oak forests in Catalonia that are commercially exploited.

References

Abrams MD, Downs JA (1990) Successional replacement of old-growth white oak by mixed mesophytic hardwoods in southwestern Pennsylvania. Can J For Res 20:1864–1870

Albeza E, Sabaté S, Escarré A, Gracia CA (1995) A long-term thinning experiment on a *Quercus ilex* L. forest: changes in leaf characteristics and dynamics in response to different treatment intensities. In: Jenkins A, Ferrier RC, Kirby C (eds) Ecosystem manipulation experiments: scientific approaches, experimental design and relevant results. Ecosystems Research Report 20, European Commission, Luxembourg, pp 200–208

Barbero M (1988) Caractérisation de quelques structures et architectures forestières des arbres et arbustes à feuilles persistantes de l'étage méditerranéen. Rev For Fr 40:371–380

Barbero M, Loisel R, Quèzel P (1992) Biogeography, ecology and history of Mediterranean *Quercus ilex* ecosystems. Vegetatio 99/100:19–34

Bellot J, Sánchez JR, Lledó MJ, Martínez P, Escarré A (1992) Litterfall as a measure of primary production in Mediterranean holm-oak forest. Vegetatio 99/100: 69–76

Canadell J, Riba M, Andrés P (1988) Biomass equations for *Quercus ilex* L. in the Montseny massif, northeastern Spain. Forestry 61:137–147

Cartan-Son M, Romane F (1992) Standardized description of some experimental plots in the *Quercus ilex* L. ecosystems. Vegetatio 99/100:3–12

Cartan-Son M, Floret C, Galan MJ, Grandjanny M, Le Floc'h E, Maistre M, Perret P, Romane F (1992) Factors affecting radial growth of *Quercus ilex* in a coppice stand in southern France. Vegetatio 99/100:61–68

Cole DW, Rapp M (1981) Elemental cycling in forest ecosystems. In: Reichle DE (ed) Dynamic properties of forest ecosystems. Cambridge University Press, Cambridge, pp 341–409

David F, Poissonet P, Romane F (1979) Analyse de la structure horizontale d'un taillis de Bouleaux (*Betula verrucosa* Ehrh.) en Sologne. Acta Oecol Oecol Plant 14:237–247

Di Castri F (1973) Soil animals in latitudinal and topographical gradients of Mediterranean ecosystems. In: Di Castri F, Mooney HA (eds) Mediterranean type ecosystems. Origin and structure. Springer, Berlin, pp 171–190

Escarré A, Ferrés L, López R, Martín J, Rodà F, Terradas J (1987) Nutrient use strategy by ever-green-oak (*Quercus ilex* ssp. *ilex*) in NE Spain. In: Tenhunen JD, Catarino FM, Lange LO, Oechel WC (eds) Plant response to stress. Springer, Berlin, pp 429–435

Espelta JM, Riba M, Retana J (1995) Patterns of seedling recruitment in west Mediterranean coppiced holm-oak (*Quercus ilex* L.) forests influenced by canopy development. J Veg Sci 6: 465–472

Ferrés L (1984) Biomasa, producción y mineralomasas del encinar montano de La Castanya (Montseny). PhD Thesis, Autonomous University of Barcelona, Bellaterra

Ferrés L, Rodà F, Verdú AMC, Terradas J (1984) Circulación de nutrientes en algunos ecosiste-mas forestales del Montseny (Barcelona). Mediterr Ser Estud Biol 7:139–166

Floret C, Galan MJ, Le Floc'h E, Rapp M, Romane F (1989) Organisation de la structure, de la biomasse et de la minéralomasse d'un taillis ouvert de chêne vert (*Quercus ilex* L.). Acta Oecol Oecol Plant 10:245–262

Giovannini G, Perulli D, Piussi P, Salbitano F (1992) Ecology of vegetative regeneration after coppicing in macchia stands in central Italy. Vegetatio 99/100:331–343

Gracia M, Retana J (1996) Effect of site quality and thinning management on the structure of holm oak forests in northeast Spain. Ann Sci For 53:571–584

Lledó MJ (1990) Compartimentos y flujos biogeoquímicos en una cuenca de encinar del Monte Poblet. PhD Thesis, University of Alicante, Alicante

Lledó MJ, Sánchez JR, Bellot J, Boronat J, Ibàñez JJ, Escarré A (1992) Structure, biomass and production of a resprouted holm-oak (*Quercus ilex* L.) forest in N-E Spain. Vegetatio 99/100: 51–59

Lorimer CG, Krug AG (1983) Diameter distributions in even-aged stands of shade-tolerant and midtolerant tree species. Am Nat 109:331–345

Lossaint P, Rapp M (1971) Repartition de la matiére organique, productivité et cycles des éléments minéraux dans des écosystèmes de climat méditerranéen. In: Duvigneaud P (ed) Productivité des ecosystèmes forestiers. Actes coll Bruxelles, Unesco, Paris, pp 597–617

Martínez JM, Albeza E, Sabaté S, Bellot J, Escarré A, Gracia C (1996) Anàlisi de la resposta de *Quercus ilex* a tractaments d'aclarida selectiva: efectes en la producció i en la recuperació de la capçada foliar. Treballs Centre Hist Nat Conca Barberà 1:97–104

Mayor X (1990) El paper dels nutrients com a factors limitants de la producció primària de l'alzinar de la conca del Torrent de la Mina (Montseny). MSc Thesis, Autonomous University of Barcelona, Bellaterra

Mayor X, Rodá F (1993) Growth response of holm oak (*Quercus ilex*) to commercial thinning in the Montseny mountains (NE Spain). Ann Sci For 50: 247–256

Mayor X, Rodà F (1994) Effects of irrigation and fertilization on stem diameter growth in a Mediterranean holm oak forest. For Ecol Manage 68:119–126

Mayor X, Belmonte R, Rodrigo A, Rodà F, Piñol J (1994) Crecimiento diametral de la encina (*Quercus ilex* L.) en un año de abundante precipitación estival: efecto de la irrigación previa y de la fertilización. Orsis 9:13–23

Parker AJ (1988) Stand structure in subalpine forests of Yosemite National Park, California. For Sci 34:1047–1058

Reichle DE (ed) (1981) Dynamic properties of forest ecosystems. Cambridge University Press, Cambridge

Retana J, Riba M, Castell C, Espelta JM (1992) Regeneration by sprouting of holm-oak (*Quercus ilex*) stands exploited by selection thinning. Vegetatio 99/100:355–364

Romane F, Floret C, Galan M, Grandjanny M, Le Floc'h E, Maistre M, Perret P (1988) Quelques remarques sur les taillis de chênes verts. Répartition, histoire, biomasse. For Méditerr 10: 131–135

Susmel L, Viola F, Bassato G (1976) Ecologia della lecceta di Supramonte di Orgosolo (Sardegna Centro-Orientale). Ann Centro Econ Montana Venezie 10:1–216

Terradas J (1991) Altérations du paysage méditerranéen par l'homme. IMCOM 3:18–22

Verdú AMC (1984) Circulació de nutrients en tres ecosistemes forestals del Montseny: Caiguda de virosta i descomposició de la fullaraca. PhD Thesis, Autonomous University of Barcelona, Bellaterra

Vogt KA, Grier CC, Vogt DJ (1986) Production, turnover, and nutrient dynamics of above- and belowground detritus of world forests. Adv Ecol Res 15:303–377

Whittaker RH, Woodwell GM (1968) Dimension and production relations of trees and shrubs in the Brookhaven forest, New York. J Ecol 56:1–25

4 Structure and Dynamics of the Root System

Josep Canadell, Arezki Djema, Bernat López, Francisco Lloret,
Santiago Sabaté, Daniel Siscart and Carlos A. Gracia

4.1 Introduction

The belowground component of terrestrial ecosystems is much less understood than any of the aboveground components, yet important ecosystem
processes such as nutrient recycling, water storage, and long-term carbon accumulation occur largely in this compartment. For instance, belowground
structures accounted for up to 83% of the total biomass in 13 Mediterranean
woody communities (Hilbert and Canadell 1995), and belowground primary
production was 60–80% of the total net primary production in a variety of
woody systems (Coleman 1976; Ågren et al. 1980; Fogel 1985). Yet both root
biomass and production are infrequently studied and technical difficulties
make the measurements often inaccurate. Furthermore, plant root distribution
and maximum rooting depths play important roles in overall ecosystem function, but it was not until recently that ecosystem-level and global comprehensive studies have been undertaken (Canadell et al. 1996; Jackson et al. 1996).

In water-limited systems, as in the case of Mediterranean ecosystems, root
growth is among the deepest of all biome-types (Canadell et al. 1996), and
carbon allocation patterns strongly favour accumulation belowground (Oechel
and Lawrence 1981). This pattern occurs because of the development both of
deep root systems and of large subterranean structures such as lignotubers
(James 1984; Canadell and Zedler 1995; Canadell and López-Soria 1998).

Ultimately, the gaps in our understanding of root system attributes and
function are manifested in the poor implementation of root properties in
whole-ecosystem and biospheric/atmospheric models. These models often
represent the belowground compartment using a simple bucket model or
two or three arbitrary layers (Prentice et al. 1992; Melillo et al. 1993; Potter
et al. 1993). Furthermore, these models use maximum rooting depths between 1 and 2 m, yet we know that deep roots (> 3 m) are very common in
most biomes (Stone and Kalisz 1991; Canadell et al. 1996) and have important consequences for ecosystem processes (Silva et al. 1989; Nepstad et al.
1994; Richter and Markewitz 1995; Schulze et al. 1996).

Ecological Studies, Vol. 137
Ferran Rodà et al. (eds) Ecology of Mediterranean
Evergreen Oak Forests
© Springer-Verlag, Berlin Heidelberg 1999

In this chapter, we synthesize the information available on root system attributes of the Mediterranean evergreen holm oak (*Quercus ilex* L.), mainly, but not exclusively, studied in two experimental areas: Montseny (Barcelona) and Prades (Tarragona) in northeastern Spain (Chap. 2). We describe the root system structure and its distribution in the soil profile, the total root biomass, which we relate to the aboveground biomass, and the phenology of fine roots. We expect this information to contribute to a better understanding of holm oak ecology and to the efforts to model system responses under changing environmental conditions.

4.2 The Database

The data presented in this chapter are the result of various studies that have been carried out over the past 13 years: root excavation campaigns, root profiles on road cuts, soil coring, minirhizotron technology, and field observations.

At Montseny, the root systems of 32 single-stemmed holm oaks were excavated in the winter of 1985 in La Castanya (Chap. 2). Of these trees, 20 were excavated in a mesic site (north-facing slope) near the permanent plot, and 12 in a nearby xeric site (south-facing slope). In addition, three multi-stemmed holm oaks were excavated at the xeric site. At Prades, the root systems of multi-stemmed holm oaks were dug out during 1993 on the north-facing slope of Torners (Chap. 2), in the context of the thinning experiment described in Chapter 23. Here, we will use the data from the six trees excavated in the control (unthinned) plots. In all cases, roots were excavated down to 1-m depth or to the bedrock, whichever came first, and to an end diameter of 1 cm.

Biomass of roots of Ø < 1 cm was obtained by soil coring down to 0.6 m in the permanent plot at La Castanya. Profiles of roots of Ø > 1.5 mm were studied on various road cuts down to 1.5 m deep in several additional places in northeast Spain. At Prades, minirhizotron technology was used to study the dynamics of fine root for a 2.5-year period (1994–1996) in the same plots where root systems were excavated. Minirhizotron is an in situ non-destructive technique that minimizes point-to-point variation by allowing repeated measurements at the same point over time (Taylor 1987). In the stony Prades soils, rigid minirhizotron tubes would not allow a good soil-tube contact, thus preventing normal development of fine roots around the tubes. Instead, inflatable minirhizotrons were used which yielded good quality images (López et al. 1996).

4.3 Structure of the Root System

Taproots of holm oak usually grow quite rapidly into deep soil layers at early stages of seedling development. However, the dominance of a vertical single taproot in seedlings is often lost in mature trees. In the shallow soils of north-facing slopes in Montseny, most of the trees did not have a clear central taproot, while those growing on south-facing slopes, with deeper (> 1 m) and drier soils, frequently showed a large taproot. In Prades, where the climate is drier than in Montseny, 60% of the excavated trees had taproots. Thus, although the data are only correlational at this point, holm oak has taproots more often when growing in drier environments than in mesic sites. In any case, however, holm oak has the capacity to send roots quite deep in the soil profile when the substrate can be penetrated. Roots were observed at 3.0-m depth in a road cut on schists at Montseny and at a depth of 3.7 m in sandstone substrate elsewhere (Canadell et al. 1996). The presence of deep roots is thought to be one of the mechanisms for surviving summer drought. Maximum rooting depths of other evergreen tree species were between 2.7 and 40 m (Canadell et al. 1996) and 23 m for *Quercus wislizenii* in California (Lewis and Burgy 1964). Data for holm oak from deep root excavations (> 3 m) are not available. We have also observed plants of holm oak growing on very shallow soils with most of the root system growing in soft rocks (e.g. sandstone) or in-between the schistosity planes of slates with hardly any soil.

F. Lloret and D. Siscart (unpubl. data) described root density (Ø > 1.5 mm) distribution along 1.5-m-deep soil profiles in holm oak forests near Barcelona, taking advantage of recently made road cuts (Fig. 4.1). They compared two substrate types in which holm oak showed different drought tolerance to an extreme drought that took place in 1994. The root distribution profiles showed that more small roots were able to reach deeper layers when growing on schist – following exfoliation planes as preferred pathways – than when

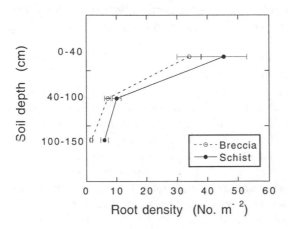

Fig. 4.1. Mean density (± SE) of holm oak roots of Ø > 1.5 mm at different soil depth intervals and substrates. Data for each substrate were obtained in a single locality from seven 5-m-wide by 1.5-m-high profiles located along the edge of a trail that were close to the transition between fault breccias and Palaeozoic schists

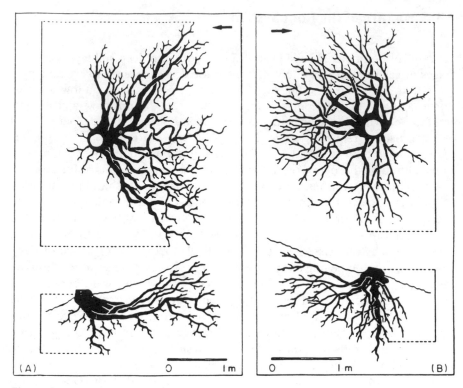

Fig. 4.2. Root systems (roots of Ø >1 cm) of single-stemmed holm oaks in A the mesic site and B the xeric site at La Castanya, Montseny. (Drawings by Mercè Cartañà)

growing on a compact matrix of breccias. Interestingly, holm oak trees growing on the schist soils were more tolerant to drought events. These observations suggest that trees having more roots in deep soil horizons (where water is stored for a longer time) are better prepared to withstand extreme drought events. Under these conditions, the interaction between climate and substrate type plays a critical role in reshaping communities and allowing deep-rooted individuals to survive.

The lateral root system of holm oak can be quite extensive. Lateral roots can grow far beyond the canopy projection and were observed at 5.9 m from the main stem in Prades (Djema 1995); large roots (Ø > 1 cm), however, do not usually extend further than 2.5 m. The smaller diameter roots account for the largest fraction of root length, which is important in terms of soil occupation and water and nutrient uptake. Most of the horizontal large roots grow between 20 and 60 cm from the soil surface, and only a few roots are found close to the surface, where dry–wet cycles are more frequent and therefore where harsher environmental conditions for root growth are found. It is worth noting that when trees grow on steep slopes, most of the lateral root system grows up-slope yielding a highly asymmetrical root distribution (Fig. 4.2). In a review of 96 Mediterranean woody species from Australia, California and Chile the average radius of root horizontal extension was 1.6 ± SD 1.5 m and the maximum extension was 7.0 m (Canadell and Zedler 1995).

In La Castanya, root grafts were occasionally observed between roots of the same tree, but never between roots of different trees. On the contrary, root grafts were somewhat common between holm oak trees in Prades, although no root grafting was observed between different species. Generally, root grafting is not very common in Mediterranean ecosystems (Keeley 1988); it was observed in only one of nine species excavated in the Chilean matorral (Hoffmann and Kummerow 1978).

4.4 Lignotubers

One of the most characteristic features of the root system of holm oak is the development of a large woody swollen structure at the stem base called the lignotuber (Canadell and Zedler 1995). This structure is genetically determined and appears early during seedling development. There are no studies on lignotuber development in this species, but it is known that Q. suber, another evergreen western Mediterranean oak, starts developing the lignotuber structure through an accumulation of bud clusters and starch that occurs close to the cotyledonary insertion (Molinas and Verdaguer 1993a,b).

The size of lignotubers depends on plant age, environmental and edaphic conditions, and, most of all, on the disturbance history that individual plants have experienced (e.g. fire, logging, extreme drought or cold events). In general, disturbances trigger the growth of the meristematic tissue of the lignotuber, which is further enlarged by the fusion of the stem bases of the new emerging shoots. The presence of large lignotubers was common in Montseny and Prades because both forests were intensively coppiced for charcoal production until the 1950s. At both sites, individual holm oaks often allocate as much as half of their total biomass to the lignotuber (Canadell and Rodà 1991; Djema 1995; Sabaté et al. 1998). In one case at Montseny, the lignotuber dry weight was 317 kg and accounted for 66% of the total tree biomass.

It is believed that lignotubers have a dual function. The first function, which is morphologically related, is to store concealed buds that will resprout after disturbances; the lignotuber generally stores a large number of buds that enable plants to regrow even in environments subjected to multiple disturbances. The second function is to store non-structural carbohydrates and nutrients that will support regrowth after disturbances. Our understanding regarding the latter function has been inferred largely from studies of plant growth (DeSouza et al. 1986; Castell et al. 1994), anatomical structure (Montenegro et al. 1983), and tissue analysis of lignotuber nutrient and carbohydrate contents (Mullete and Bamber 1978; Dell et al. 1985). To date, very few studies have provided direct experimental data (e.g. Canadell and López-Soria 1998; Sabaté et al. 1998). These studies indicate that nutrients and carbohydrates stored in the lignotuber play an important role in providing resources for rapid regrowth during the first years after a disturbance.

4.5 Allometric Relationships Between Above- and Belowground Biomass

Root excavation studies enable us to develop allometric equations that can be used to investigate the relationships between the aboveground and belowground parts of trees. They are also used to estimate large-root biomass of individual trees by measuring their aboveground dimensions. Table 4.1 shows these equations for Montseny and Prades. Several regressors were used depending on the tree structure. Diameter at breast height (dbh) and stem diameter at 50 cm (D_{50}) were good regressors to estimate the root biomass of single-stemmed trees, which is the dominant tree structure in the Montseny mesic site (Canadell et al. 1988). On the other hand, multi-stemmed trees are dominant in Prades and the root biomass of a stool is best estimated using the total cross-sectional area (CSA), at 50 cm from the ground, of all the stems it bears (Djema 1995).

It is worth noting that trees with dbh < 20 cm at Montseny had more root biomass in the xeric site than in the mesic site, as it would be predicted according to our current understanding of carbon allocation and soil water

Table 4.1. Allometric equations for total large-root biomass (kg dry weight, including lignotuber) for single-stemmed holm oaks at Montseny and multi-stemmed holm oaks at Prades

y^{a}	x	a	b	r^2	n
Montseny – mesic and xeric sites (single-stemmed trees)					
Total large roots	Dbh	– 1.047	2.191	0.73	32
Total large roots	D_{50}	– 1.687	2.623	0.79	32
Total large roots	TAB	– 0.212	0.894	0.83	30
Montseny – mesic site (single-stemmed trees)					
Total large roots	Dbh	– 1.393	2.451	0.81	20
Montseny – xeric site (single-stemmed trees)					
Total large roots	Dbh	– 0.448	1.734	0.71	12
Prades (multi-stemmed trees)					
Total large roots	CSA	– 55.210	0.605	0.91	6

dbh, Stem diameter at 1.3 m from the ground (cm); D_{50}, stem diameter at 50 cm from the ground (cm); TAB, total aboveground biomass (kg dry weight); CSA, cross-sectional area at D_{50} (cm^2) of all the stool stems.

[a] The model is log y = a + b log x for Montseny and y = a + b log x for Prades. Log indicates logarithm to base 10.

availability. Santantonio et al. (1977) compiled alometric regressions (log–log) of large-root biomass (kg) on dbh (cm) for a number of tree species. In the dbh range 10–20 cm, which contains most of our sample trees, values of large-root biomass (Ø > 5 cm) predicted by our equation for single-stemmed holm oak are very close to those derived from their mean regression for coniferous trees, and about 2.5 times higher than those from their mean regression for deciduous trees.

4.6 Root Biomass

The proportion of biomass allocated to belowground tissues by perennial plants usually increases as the environment becomes more severe (Rundel 1980). Thus, root systems may account for up to 90% of total plant biomass in many arctic plants and some desert shrubs (Rodin and Bazilevich 1967; Caldwell and Fernandez 1975).

In the Montseny mesic site, a holm oak stand dominated by single-stemmed trees had 28% of its total tree biomass belowground (Table 4.2). Of the 63 Mg dry wt. ha^{-1} of root biomass, 48% was in lignotubers and roots larger than 5 cm in diameter, 27% in roots of Ø 1–5 cm, and 24% in roots of Ø < 1 cm. The allocation of 28% of the total biomass into the belowground compartment is quite high, but it still falls within the range of 15–35% commonly found in temperate forest ecosystems (Rodin and Bazilevich 1967),

Table 4.2. Basal area (m^2 ha^{-1}), aboveground biomass (Mg dry wt. ha^{-1}) and various fractions of root biomass (Mg dry wt. ha^{-1}) for the Montseny mesic and xeric sites, and Prades

Site	Basal area	Above-ground biomass	Roots Ø > 5cm[a]	Roots Ø 1–5 cm	Roots Ø < 1 cm	Total roots	Root biomass (% of total)[b]
Montseny mesic[c]	26.6	160	30	17	16	63	28
Montseny xeric[d]	16.3	79	—— 91.0 ——		16[e]	108	58
Prades[f]	37.1	103	66	37	–	128[g]	55[g]

Data compiled from Canadell and Rodà (1991), Gracia et al. (1994), Djema (1995), and Sabaté et al. (1998).

[a] Biomass of roots of Ø >5 cm and lignotuber.
[b] Percentage of total root biomass respect to total aboveground and belowground biomass.
[c] Data for the permanent plot at La Castanya, using the tree sizes of 1978.
[d] Data for a 0.03-ha plot, tallied in 1990.
[e] Roots of Ø <1 cm were not measured at the xeric site, but their biomass was assumed equal to the mesic site.
[f] Data for three control plots on the north-facing slope of Torners, tallied in 1992.
[g] Does not include roots of Ø <1 cm.

and it falls at the high end of the range of 15–25% given by Harris et al. (1980). Thus, the Montseny holm oak forest is not strikingly different from temperate forest ecosystems in its pattern of root to shoot biomass allocation, at least in an undisturbed mesic site without major water stress during the dry season and dominated by single-stemmed trees. At the tree level, and not taking into account roots of $\emptyset < 1$ cm, the mean root:shoot ratio of single-stemmed holm oaks was 0.41 (SE 0.02, $n = 30$) for all excavated trees in Montseny.

A very different picture emerges for stands dominated by multi-stemmed holm oak (as in Prades and the Montseny xeric site). Through repeated re-sprouting after fire and coppicing, current stems are here much younger than the stools bearing them. Under these conditions, holm oak, a long-lived species, develops massive lignotubers in which belowground biomass keeps accumulating while aboveground biomass is burned or harvested at different frequencies. Belowground biomass in these multi-stemmed stands can exceed the aboveground biomass, as is the case at the Montseny xeric and Prades sites where 58 and 55% of total biomass is belowground, respectively (the latter figure representing an underestimation of the actual belowground biomass because roots of $\emptyset < 1$ cm were not accounted for; Canadell and Rodà 1991; Djema 1995). Belowground biomass accounted for 45 and 46% of the total biomass for two small- and medium-sized multi-stemmed trees in La Castanya, and 66% for the largest excavated tree.

4.7 Fine Roots

Fine roots of trees ($\emptyset < 2.5$ mm) are the most dynamic fraction of the root system, having turnover times of between a few weeks and more than 8 years (Hendrick and Pregitzer 1992). Fine roots are responsible for nutrient and water uptake. However, fine root distribution, phenology and turnover have only recently been studied for holm oak. The following account is based on repeated observations with inflatable minirhizotrons placed down to 60-cm depth at Prades (López et al. 1996, 1998).

Fine root density, averaged over 2 years of study (excluding the first 3 months after minirhizotron installation), was greatest in the upper soil layers, with 32% of the roots counted down to 60-cm depth being in the top 10 cm of soil, 60% in the top 20 cm and 80% in the top 30 cm. Root biomass, however, was more evenly distributed along the soil profile, with only 51% of the biomass in the top 30 cm. Only 7.4% of the fine root biomass was found between 50 and 60 cm deep. Deep fine roots, although representing small quantities of the total amount of biomass, may play a key role in extracting deep soil moisture during dry periods (Canadell et al. 1996; Hendrick and Pregitzer 1996), which are an important feature of the Mediterranean climate. Roots of $\emptyset < 0.5$ mm accounted at Prades for 95% of the total number of fine roots, which is in agreement with other species such as *Acer saccharum* (Hendrick and Pregitzer 1992; Burke and Raynal 1994) and *Picea sitchensis* (Ford and Deans 1977), and is similar to the root diameter distribution of a northern hardwood forest, where two thirds of the total root length was found in the 0.2- to 0.3-mm diameter class (Fahey and Hughes 1994).

To obtain biomass on a ground area basis from the minirhizotron observations, root growth was assumed to be isotropic, i.e. biomass density (g mm^{-2}; all biomass and production data are on dry weight basis), observed from the minirhizotron window, was supposed to be the same as the biomass density one would see from above the soil. The 2-year average fine-root biomass obtained was 94.8 ± 6.8 g m^{-2} (mean ± SE), which is similar to other reported fine root biomasses (white oak: 115 g m^{-2}, Aber et al. 1985; white pine: 97 g m^{-2}, Aber et al. 1985; European beech: 150 g m^{-2}, Ellenberg et al. 1986; American beech: < 100 g m^{-2}, Liu and Tyree 1997; lowlands of montane rainforest: 144 g m^{-2}, Cavelier 1992), but substantially lower than estimates from some northern hardwood forests (range: 510–990 g m^{-2}, Harris et al. 1977; McClaugherty et al. 1982; Joslin and Henderson 1987; Farrish 1991; Burke and Raynal 1994). A possible explanation for these differences is that a large portion of the soil volume at Prades is composed of coarse gravel. Root production between time t_i and time t_{i+1} was calculated by summing the biomass of new roots and the positive increments of biomass of existing roots. To obtain annual fine root production we summed the production of all the field campaigns for the given year and then averaged the 2 years of observations. Root production (Fig. 4.3, top) was found to be 500 ± 4.4 g m^{-2} year^{-1}, thus giving a turnover rate of 5.27 year^{-1} or a mean fine root lifespan of 68.3 days. However, longevity of individual roots ranged from 5 to more than 475 days.

The cost of fine root formation has been estimated assuming that 1 g of carbohydrates produces 0.68 g of new root tissues (Chap. 12), which means a cost of 7.35 kcal g^{-1} of new root. Similarly, the average maintenance cost has been estimated under the soil temperature at Prades as 50 cal g^{-1} day^{-1} (Chap. 12). Using these simplifications, the formation cost of fine roots was 3675 kcal m^{-2}, and the calculated cost of maintenance was 1730 kcal m^{-2} year^{-1} (Fig. 4.3, bottom). Fine root and leaf metabolic costs represent, in this forest, more than 60% of total carbon fixed in gross primary production. Fine root formation cost, on a ground area basis, is more than twice the leaf formation cost, while the maintenance of fine roots is only 15.8% of leaf maintenance cost. The high formation cost of fine roots is related to their high turnover. Death of fine roots represents a very active mechanism of carbon transport from the atmosphere to soil. This mechanism has been evaluated in Prades as 166 g C m^{-2} year^{-1}, 66% more than the carbon lost in leaf litterfall, making fine roots the most important channel of carbon loss from these holm oak trees.

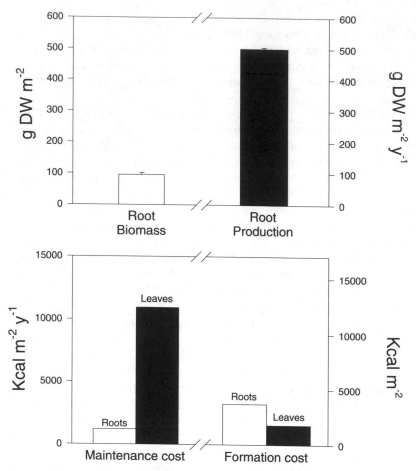

Fig. 4.3. *Above:* Fine root biomass and annual production; *below:* annual metabolic cost of fine root formation and maintenance compared with formation and maintenance costs of leaves

References

Aber JD, Melillo JM, Nadelhoffer KJ, McClaugherty CA, Pastor J (1985) Fine root turnover in forest ecosystems in relation to quantity and form of nitrogen availability: a comparision of two methods. Oecologia 66:317–321

Ågren GI, Axelsson B, Flower-Ellis JGK, Linder S, Persson H, Staaf H, Troeng E (1980) Annual carbon budget for a young Scots pine. Ecol Bull 32:307–313

Burke MK, Raynal DJ (1994) Fine root growth phenology, production, and turnover in a northern hardwood forest ecosystem. Plant Soil 162:135–146

Caldwell MM, Fernandez OA (1975) Dynamics of Great Basin shrub root systems. In: Hadley NF (ed) Environmental physiology of desert organisms. Halstead Press, New York, pp 38–51

Canadell J, López-Soria L (1998) Lignotuber reserves support regrowth following clipping of two Mediterranean shrubs. Funct Ecol 12:31–38

Canadell J, Rodà F (1991) Root biomass in a montane Mediterranean forest. Can J For Res 21: 1771–1778

Canadell J, Zedler P (1995) Underground structures of woody plants in Mediterranean ecosystems of Australia, California and Chile. In: Fox M, Kalin M, Zedler P (eds) Ecology and biogeography of Mediterranean ecosystems in Chile, California and Australia. Springer, Berlin, pp 177–210

Canadell J, Riba M, Andrés P (1988) Biomass equations for *Quercus ilex* L. in Montseny Massif, northeastern Spain. Forestry 61:137–147

Canadell J, Jackson RB, Ehleringer JR, Mooney HA, Sala OE, Schulze E-D (1996) Maximum rooting depth of vegetation types at the global scale. Oecologia 108:583–595

Castell C, Terradas J, Tenhunen JD (1994) Water relations, gas exchange, and growth of resprouts and mature plant shoots of *Arbutus unedo* L. and *Quercus ilex* L. Oecologia 98:201–211

Cavelier J (1992) Fine-root biomass and soil properties in a deciduous and a lower montane rain forest in Panama. Plant Soil 142:187–201

Coleman DC (1976) A review of root production processes and their influence on soil biota in terrestrial ecosystems. In: Macfadyen JMA (ed) The role of terrestrial and aquatic organisms in decomposition processes. Blackwell, Oxford

Dell B, Jones S, Wallace M (1985) Phosphorus accumulation by lignotubers of jarrah (*Eucalyptus marginata* Donn ex Sm.) seedlings grown in a range of soils. Plant Soil 86:225–232

DeSouza J, Silka PA, Davis SD (1986) Comparative physiology of burned and unburned *Rhus laurina* after chaparral wildfire. Oecologia 71:63–68

Djema A (1995) Cuantificación de la biomasa y mineralomasa subterránea de un bosque de *Quercus ilex* L. MSc Thesis, Instituto Agronómico Mediterráneo de Zaragoza, Zaragoza

Ellenberg H, Mayer R, Shauermann J (eds) (1986) Ökosystemforschung: Ergebnisse des Sollingprojekts 1966–1986. Eugen Ulmer, Stuttgart

Fahey TJ, Hughes JW (1994) Fine root dynamics in a northern hardwood forest ecosystem, Hubbard Brook Experimental Forest, NH. J Ecol 82:533–548

Farrish KW (1991) Spatial and temporal fine-root distribution in three Louisiana forest soils. Soil Sci Soc Am J 55:1752–1757

Fogel R (1985) Roots as primary producers in below-ground ecosystems. In: Fitter AH (ed) Ecological interactions in soil. Spec Publ No 4. British Ecological Society, London, pp 23–36

Ford ED, Deans JD (1977) Growth of a Sitka spruce plantation: spatial distribution and seasonal fluctuations of lengths, weights and carbohydrate concentrations of fine roots. Plant Soil 47: 463–485

Gracia CA, Sabaté S, Albeza E, Djema A, Tello E, Martínez JM, López B, León B, Bellot J (1994) Análisis de la respuesta de *Quercus ilex* L. a tratamientos de aclareo selectivo: producción, biomasa y tasa de renovación de hojas y raíces durante el primer año de tratamiento. Reunión de coordinación del Programa de restauración de la cubierta Vegetal de la Comunidad Valenciana, Valencia

Harris WF, Kinerson RS, Edwards NT (1977) Comparision of belowground biomass of natural deciduous forest and loblolly pine plantations. In: Marshall JK (ed) The belowground ecosystem: a synthesis of plant-associated processes. Range Science Department Science Series No 26. Colorado State University, Fort Collins, pp 29–38

Harris WF, Santantonio D, McGinty D (1980) The dynamic belowground ecosystem. In: Waring RH (ed) Forests: fresh perspectives from ecosystem analysis. Oregon State University Press, Oregon, pp 118–129

Hendrick RL, Pregitzer KS (1992) Spatial variation in tree root distribution and growth associated with minirhizotrons. Plant Soil 143:283–288

Hendrick RL, Pregitzer KS (1996) Temporal and depth-related patterns of fine root dynamics in northern hardwood forests. J Ecol 84:167–176

Hilbert DW, Canadell J (1995) Biomass partitioning and resource allocation of plants from Mediterranean-type ecosystems: possible responses to elevated atmospheric CO_2. In: Oechel WC, Moreno J (eds) Global change and Mediterranean-type ecosystems. Springer, Berlin, pp 76–101

Hoffmann A, Kummerow J (1978) Root studies in the Chilean matorral. Oecologia 32:57–69

Jackson RB, Canadell J, Ehleringer JR, Mooney HA, Sala OE, Schulze E-D (1996) A global analysis of root distributions for terrestrial biomes. Oecologia 108:398–411

James S (1984) Lignotubers and burls -their structure, function and ecological significance in Mediterranean ecosystems. Bot Rev 50:225–266

Joslin JD, Henderson GS (1987) Organic matter and nutrients associated with fine root turnover in a white oak stand. For Sci 33:330–346

Keeley JE (1988) Population variation in root grafting and a hypothesis. Oikos 52:364–366

Lewis DC, Burgy RH (1964) The relationship between oak tree roots and groundwater in fractured rock as determined by tritium tracing. J Geophys Res 69:2579–2588

Liu X, Tyree MT (1997) Root carbohydrate reserves, mineral nutrient concentrations and biomass in a healthy and a declining sugar maple (*Acer saccharum*) stand. Tree Physiol 17:179–185

López B, Sabaté S, Gracia CA (1996) An inflatable minirhizotron system for stony soils. Plant Soil 179:255–260

López B, Sabaté S, Gracia CA (1998) Fine roots dynamics in a Mediterranean forest: effects of drought and stem density. Tree Physiol 18:601–606

McClaugherty CA, Aber JD, Melillo JM (1982) The role of fine roots in the organic matter and nitrogen budgets of two forested ecosystems. Ecology 63:1481–1490

Melillo JM, McGuire AD, Kicklighter DW, Moore B III, Vorosmarty CJ, Schloss AL (1993) Global climate change and terrestrial net primary production. Nature 363:234–240

Molinas ML, Verdaguer D (1993a) Lignotuber ontogeny in the cork-oak (*Quercus suber*; Fagaceae). I. Late embryo. Am J Bot 80:172–181

Molinas ML, Verdaguer D (1993b) Lignotuber ontogeny in the cork-oak (*Quercus suber*; Fagaceae). II. Germination and young seedling. Am J Bot 80:182–191

Montenegro G, Avila G, Schatte P (1983) Presence and development of lignotubers in shrubs of the Chilean matorral. Can J Bot 61:1804–1808

Mullete KJ, Bamber RK (1978) Studies of the lignotubers of *Eucalyptus gummifera* (Gaertn. & Hoch.). III. Inheritance and chemical composition. Aust J Bot 26:23–28

Nepstad DC, de Carvalho CR, Davidson EA, Jipp PH, Lefebvre PA, Negreiros GH, da Silva ED, Stone TA, Trumbore SE, Vieira S (1994) The role of deep roots in the hydrological and carbon cycles of Amazonian forests and pastures. Nature 372:666–669

Oechel WC, Lawrence L (1981) Carbon allocation and utilization. In: Miller PC (ed) Resource use by chaparral and matorral. Springer, Berlin, pp 185–235

Potter CS, Randerson JT, Field CB, Matson PA, Vitousek PM, Mooney HA, Klooster SA (1993) Terrestrial ecosystem production: a process model based on global satellite and surface data. Global Biogeochem Cycles 7:811–841

Prentice IC, Cramer W, Harrison SP, Leemans R, Monserud RA, Solomon AM (1992) A global biome model based on plant physiology and dominance, soil properties and climate. J Biogeogr 19:117–134

Richter DD, Makewitz D (1995) How deep is soil? BioScience 45:600–609

Rodin LE, Bazilevich NI (1967) Production and mineral cycling in terrestrial vegetation. Oliver and Boyd, London

Rundel PW (1980) Adaptations of Mediterranean-climate oaks to environmental stress. In: Ecology, management, and utilization of California oaks. USDA For Serv Gen Tech Rep PSW-44:43–54

Sabaté S, Djema A, Gracia C, López B (1998) Effects of thinning on belowground biomass of a growth stagnated old coppice Mediterranean *Quercus ilex* L. forest. Tree Physiol (submitted)

Santantonio D, Hermann RK, Overton WS (1977) Root biomass studies in forest ecosystems. Pedobiologia 17:1–31

Schulze E-D, Bauer G, Buchmann N, Canadell J, Ehleringer JR, Jackson RB, Jobbagy E, Loreti J, Mooney HA, Oesterheld M, Sala OE (1996) Water availability, rooting depth, and vegetation zones along an aridity gradient in Patagonia. Oecologia 108:503–511

Silva S, Whitford WG, Jarrell WM, Virginia RA (1989) The microarthropod fauna associated with a deep rooted legume, *Prosopis glandulosa*, in the Chihuahuan desert. Biol Fertil Soils 7:330–335

Stone EL, Kalisz PJ (1991) On the maximum extent of tree roots. For Ecol Manage 46:59–102

Taylor HM (1987) Minirhizotron observation tubes. In: Taylor HM (ed) Methods and applications for measuring rhizosphere dynamics. ASA Spec Publ 50. ASA, CSSA, and SSSA, Madison

5 Resprouting Dynamics

Josep Maria Espelta, Santiago Sabaté and Javier Retana

5.1 Introduction

Resprouting, i.e. the production of sprouts from buds on pre-existing plant organs, is one of the most important mechanisms of plant regeneration under natural and anthropogenic disturbances. This response has been widely observed in many plant communities around the world, but it is probably in Mediterranean regions that it has received most attention (Keeley and Zedler 1978). In fact, the resprouting ability of many Mediterranean shrubs and trees has been one of the most important keystones to building up the paradigm of the resilience (sensu Westman 1986) and autosuccessional nature (Hanes 1971) of these communities after disturbances. However, the evolutionary meaning and the selection forces that lead to the resprouting habit have been the focus of a long controversy. Since fire is one of the most typical disturbances in Mediterranean areas, and resprouting is a common response to fire, early studies pointed out that it was an adaptation, a trait selected by fire (Naveh 1974). Recent reviews have criticized this concept arguing that: (1) resprouting is an old and widespread trait among many angiosperm species, present in taxa not regularly exposed to fire (Axelrod 1989); and (2) other selective factors, such as grazing or drought, may also lead to a resprouting habit (James 1984; Keeley 1986). Thus, resprouting should be considered as an attribute evolved in response to multiple possible selective forces, a sort of reiteration by which a plant reverts to a juvenile state after damage. In the case of Mediterranean species, the resprouting habit is usually accompanied by other characteristic life history traits, such as longevity, slow growth and some degree of shade tolerance during juvenile stages (Keeley 1986).

Most present-day holm oak stands in the western Mediterranean Basin have originated through resprouting after disturbances such as fire, overgrazing, firewood extraction or charcoal production. This phenomenon has produced a certain homogeneity on the structure of these forests, usually characterized by high-density stands of relatively small stump sprouts

Ecological Studies, Vol. 137
Ferran Rodà et al. (eds) Ecology of Mediterranean
Evergreen Oak Forests
© Springer-Verlag, Berlin Heidelberg 1999

(Chap. 3). To understand their current structure as well as to predict their dynamics, it is necessary to have a complete view of the main features involved in the resprouting process of holm oak. Forest regeneration through resprouting is considered a complex phenomenon involving different organization levels (Ford and Newbould 1970; Pysek 1991; Riba 1991), the development and dynamics of sprout cohorts (ramets) being a basic process whereby the later fate of individuals (genets) and, hence, of the whole population are determined. In this chapter we describe the resprouting process in holm oak stands after the most common disturbances, scaling up from the morphological and physiological traits of sprouts (the sprout level), through the dynamics of sprouts on the stool (the stool level) and finally to the consequences for the forest structure and its maintenance (the stand level).

5.2 Resprouting Response of Holm Oak after Disturbances

Holm oak (*Quercus ilex* L.) resprouts vigorously after disturbances, with sprouts occurring by activation of dormant buds located in the stump, the root–crown or to a lesser extent the roots (Lobreaux 1987; Giovannini et al. 1992; Retana et al. 1992). The resprouting vigour of holm oak after disturbance is much higher than that observed in other Mediterranean resprouters (López Soria and Castell 1992). Moreover, mean number of sprouts per stool after cutting holm oak trees (Fig. 5.1A) is also higher than in other *Quercus* species, such as *Q. coccinea*, *Q. prinus* and *Q. velutina* (Ross et al. 1986), and *Q. alba*, *Q. falcata* and *Q. stellata* (Lowell et al. 1987). The high resistance of holm oak to disturbances through resprouting has been argued as one of the main causes explaining the spread of this species all over the western Mediterranean Basin 8000 years B.P. (Pons and Quézel 1985). According to different authors (see Romane et al. 1988), this expansion was probably driven by intense management combining fire, coppicing and domestic livestock, which involved in large areas the substitution of other *Quercus* species by holm oak.

A good example of the aforementioned potential of holm oak to resprout is provided by the words of V. de Larminat, a French forester who defined this species in 1893 as "resistant for centuries to all types of man-caused disturbance". Notwithstanding this optimistic view, we know that not all holm oak trees resprout after disturbances, with the causes of mortality being basically related to three types of major influences: (1) the source, intensity and season of disturbance, (2) edaphic and climatic factors, and (3) the size (and probably age) of individuals.

Although we lack comparative studies addressing the influence of disturbance type on the likelihood of holm oak individuals to resprout, Table 5.1 provides the ranges of holm oak mortality after four kinds of disturbances in separate studies. In all cases, survivorship through resprouting is higher than

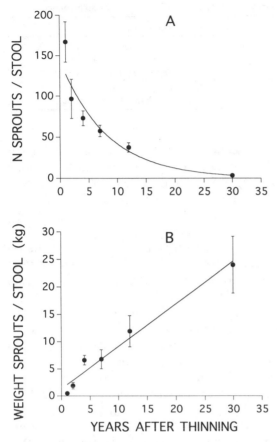

Fig. 5.1. Change in A mean number of holm oak sprouts per stool, and B total weight of holm oak sprouts per stool at different years after thinning. *Bars* = ± 1 SE

80%. Fire is the type of disturbance which causes the greatest genet mortality, probably because it can destroy the bud bank present in root–crowns, directly charring it or indirectly damaging superficial roots and stool metabolism. Mortality of holm oak individuals in the Montseny massif (Chap. 2) was indeed directly correlated with fire intensity (Obón 1997). On the other hand, grazing, clear-cutting and drought appear to have a much lesser effect on the survivorship of the bud bank and, consequently, on the resprouting success. The season of the disturbance event may also affect resprouting. Ducrey and Turrel (1992) conclude that when disturbance occurs during the dormant season, new sprouts have greater diameter and height growth, compared to sprouts originated after disturbances during the growth season, probably because higher amounts of reserves are present in the stool during the dormant season (see Kay and Canham 1991).

Table 5.1. Mortality ranges found in holm oak forests after different types of disturbance

Disturbance type	Mortality range (%)	References
Wildfire	12–15	López-Soria and Castell (1992)
Coppicing	3–5	Retana et al. (1992)
		Ducrey and Boisserie (1992)
Drought	4–6	Lloret and Siscart (1995)
Herbivory by cows	2–3	J.M. Espelta (unpubl. data)

Apart from the source of disturbance, the chance of resprouting also seems related to soil properties, stand exposure and climatic conditions following the disturbance event. After fire, López-Soria and Castell (1992) observed a higher percentage of holm oaks resprouting in deep vs. shallow soils and north vs. south slopes, which was supposed to be related to higher soil moisture in the former conditions. The precipitation regime after a given disturbance has also been considered to have a paramount effect on the resprouting success of many Mediterranean resprouters, and several studies have described a close relationship between the number and amount of rain events and the production of new sprouts (Riba 1991).

The success of resprouting is also closely linked with characteristics at the individual (genet) level. Thus, mortality (lack of resprouting) tends to be higher in small-sized individuals, probably because they suffer greater damage, and they have a smaller bud bank and a reduced amount of reserves in the stool (Trabaud 1987; Obón 1997). The number and growth of new sprouts are also positively linked with the previous size of individuals (Rundel et al. 1987; Retana et al. 1992; Lloret and López Soria 1993). In holm oak, mean number of living sprouts per stool increases with stool area, probably because larger stools have more buds in the individual bud-bank and higher belowground reserves (Djema 1995). Concerning the age of stools, conclusive results have not been produced yet, but some authors point out a negative effect of ageing on two basic features of the resprouting process: total number of sprouts and mean sprout height (Ducrey and Boisserie 1992).

5.3 Main Sprout Features

Resprouting after disturbances occurs by the activation of dormant buds in the stool, probably driven by a combined effect of changes in light availability, temperature and hormonal levels due to the destruction of aerial parts (Champagnat 1989). After bud onset, height growth of holm oak sprouts increases quickly and total lengths between 0.8 and 1 m are commonly achieved during the first growing season (Sabaté 1993), reflecting vigorous height growth rates compared to the poor performance of adult trees (Castell et al. 1994) and seedlings (Espelta et al. 1995; Chap. 7). This preferential growth in height exhibited by sprouts is also reflected by the fact that leaves in the main axis tend to be larger than those produced on lateral branches (Sabaté 1993). Sprouts also show a higher specific leaf area and nutrient content (e.g. N, P, K) in leaves and stems compared to the values exhibited by adult trees (Castell et al. 1994). Differences between sprouts and adult trees in leaf morphometry and nutrient content have been assumed to occur because of the balance between above- and belowground biomass occurring in the resprouting individuals, favouring a greater amount of resources available to sprouts and thus higher photosynthetic and growth rates (Canadell 1995). Nevertheless, these differences tend to narrow quickly during the regeneration process, due both to the tendency to re-equilibrate biomass partitioning between roots and aerial parts and to the ontogenic ageing occurring during sprout development (Sabaté 1993; Castell et al. 1994)

Despite these general features of holm oak sprouts, variations may arise according to the type and intensity of the disturbance leading to resprouting. Among other things, different disturbances may vary in their effects on soil resource availability (Christensen 1987). Similarities and differences between the effects of disturbance-type on the resprouting process of holm oak have been observed by Sabaté (1993) in three experimental treatments in Prades (Chap. 2; see Chap. 22 for a description of the experiment): clear-cutting followed by burning, clear-cutting with the slash remaining on site, and clear-cutting with the slash removed (firebreak area). Sprouts in the firebreak showed lower weight and length, while sprouts from the burned area had higher final mean weight (Table 5.2). These differences could be related to nutritional aspects: a short-term increase in nutrient availability in the burnt area (Chap. 22; Christensen 1987), and a decreased availability in the firebreak, where growth was presumably more dependent on previously stored reserves. Moreover, leaves of sprouts from the burnt area tended to increase mean leaf area (mainly during their second year of growth), probably in response also to the short-term pulse in resource (nutrients and water) availability per unit of leaf area after the fire (Serrasolsas 1994). The main struc-

Table 5.2. Structural traits (mean ± SE) of sprouts growing after experimental disturbances at Prades. Sprouts were measured in the second summer after disturbance

Structural trait	Disturbance applied		
	Burning[a]	Clear-cutting[b]	Firebreak[c]
Leaf area / basal area (cm^2/cm^2)	6105 ± 2222	4529 ± 529	3298 ± 25
Leaf weight / stem weight (g/g)	0.62 ± 0.02	0.73 ± 0.03	0.58 ± 0.07
Mean stem weight of main axis (g)	72.0 ± 1.9	43.9 ± 7.5	39.6 ± 12.7
Mean stem length of main axis (cm)	73.3 ± 14.9	73.4 ± 2.79	61.5 ± 13.1

[a] Clearcut, slash left on site and burned when dry.
[b] Clearcut, slash left on site, unburned.
[c] Clearcut, slash removed, unburned.

tural relations of the sprouts growing in the three areas (Table 5.2) show the negative effect of removing the slash on sprout growth: the leaf area supported per unit of stem basal area is smaller than that of sprouts growing in the burnt area and in the area with slash on site.

Differences in the nutrient content of leaves between sprouts growing after these disturbances have also been observed. Sprouts from the burnt area generally have a lower N concentration, probably because of the high N losses during the fire (Chap. 22). Some differences have also been observed with K: sprouts from the firebreak area show lower K concentration (Sabaté 1993). Thus, nutrients other than N could play a key role during the first growth stages, such as K, involved in osmotic processes and in the expansion of leaves and stems (Marshner 1995). This is supported by our results, since despite the lower N concentration in the burned area, higher stem biomass was achieved after burning than after clear-cutting.

5.4 Sprout Dynamics in the Stool

Once sprouts have started to develop, interaction between them in the same stool occurs. This process is particularly evident during the second year after disturbance, when second axis growth appears and mortality of sprouts in the stools is observed (Ducrey 1992; Retana et al. 1992). At Montseny, the mean number of living sprouts per stool decreases exponentially along the regeneration cycle, from a mean of 167.0 sprouts per stool during the first year of regeneration to 3.5 sprouts per stool at 30 years from the last thinning (Fig. 5.1A; Retana et al. 1992).

At the stool level, death of sprouts throughout the cutting cycle is compensated by growth of the surviving ones. Thus, the total living sprout weight per stool increases linearly with time from last thinning ($r^2 = 0.99$, $P = 0.0001$; Fig. 5.1B), at an average rate of 0.9 kg year^{-1} stool^{-1}. Similar patterns of biomass accumulation have been described in other Mediterranean woody plant communities (Merino and Martín Vicente 1981; Riba 1991), where the maximum biomass accumulation is found 20–30 years after disturbance. In holm oak, accumulation of biomass also depends on coppiced basal area in the stump (Retana et al. 1992). This relationship between diameter of the stumps and growth of sprouts has also been shown by Lowell et al. (1987) for different species of Quercus, and could be a consequence of the reduction of the degree of inhibition or interference of the standing biomass, both for aerial and soil resources (see Riba 1991).

Height and biomass growth of sprouts follow different patterns along the regeneration process (Fig. 5.2). Mean sprout height increases during the first years after disturbance, but progressively slows down (Fig. 5.2A). This fact is associated with a decrease in the relative growth rate in height (Fig. 5.2B)

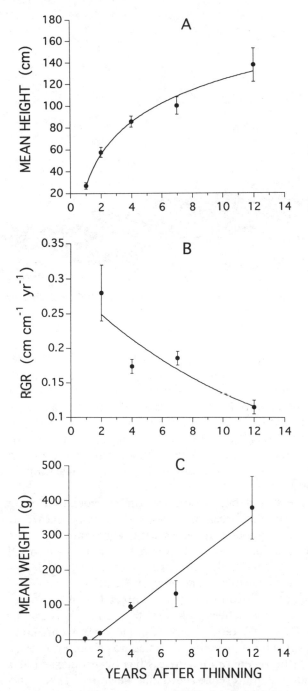

Fig. 5.2. Changes in A mean sprout height, B sprout relative growth rate (RGR) in height, and C mean sprout weight, at different years after thinning. *Bars* = ±1 SE

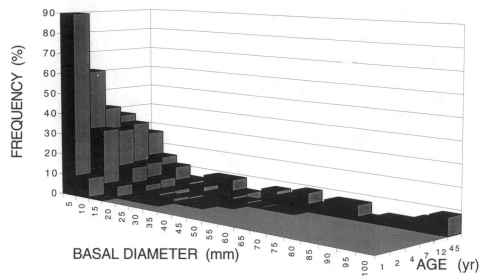

Fig. 5.3. Size (basal diameter, in mm) distributions (%) of living holm oak sprouts at different years after last thinning

observed by Retana et al. (1992) in holm oak, and also in other Mediterranean shrub species (Riba 1991). According to Hara (1988), the general pattern of change of the mean height should be a sigmoidal function. In the case of holm oak, the absence of an initial period of reduced growth may be related to the large belowground biomass accumulated in the stool and roots (Chap. 4; Canadell and Rodà 1991; Djema 1995), which allows a quick recovery of the aerial structures. Instead, mean weight of holm oak sprouts shows a linear relationship with time from last thinning (Fig. 5.2C), at least during the first 12 years after disturbance.

The size structure of the sprout population and the changes that occur through time are reflected in the basal diameter distributions shown in Fig. 5.3. The range of sprout diameters increases with regenerative age (i.e. time since disturbance): at 1 and 2 years, most sprouts have a diameter of <1 cm; from 4 to 12 years a great part of the sprouts are in the 1–3 cm classes, while at 30 years many sprouts have diameters between 3 and 14 cm.

During the first year of regeneration there is little mortality of resprouts; in the second year, mortality approaches 42–56%, with dead sprouts mainly found in the smallest size classes (Retana et al. 1992). These results are comparable to those found in French coppiced holm oak stands (Lobreaux 1987). Mortality of sprouts along the regeneration process has been regarded as a consequence of competition or interference (Ford and Newbould 1970; MacDonald and Powell 1983; Ferm and Kauppi 1990). This would also apply to holm oak. Despite no study having focused on the nature of competition (aerial or belowground) which causes the death of sprouts on holm oak stools, Castell (1992) demonstrated for another Mediterranean resprouter (*Arbutus unedo*) that there are greater differences in light availability between dominant and suppressed sprouts (aerial competition) than those present in their water relations (belowground competition).

According to Retana et al. (1992), the existence of interference among holm oak sprouts growing on the same stool throughout the regenerative period has been demonstrated by analyzing the between-sprout variability of relative growth rate in height: sprout populations for all regenerative ages show a similar pattern, growth rate being higher in the smallest classes, which also have the greatest variability. These plastic responses to crowding are caused by shading (Weiner et al. 1990): small plants of a population have to allocate their scarce production to height growth at the expense of diameter growth (Hara 1984). This rapid development of a high degree of asymmetry in the first regeneration years may be responsible for the self-thinning process at the stool level, which usually starts in the second year after disturbance.

Sprouts developing within the first 15 years of regeneration belong to a single cohort which appeared during the first year after cutting. This is quite the same for other tree and shrub species regenerating by resprouting, such as *Castanea sativa* (Ford and Newbould 1970) and *Betula pubescens* (Kauppi et al. 1988). After this initial wave, new sprouts appear from dormant buds: Retana et al. (1992) found 5–6 young sprouts per stool in a 30-year-old plot. These new sprouts ranged from 1 to 18 years old, and had basal diameters of 0.4–11.8 mm. The presence of new sprouts indicates a change in the internal dynamics of stools, and should be regarded as a consequence of the decreasing inhibitory effect on bud activation by the sprouts already present (Riba 1991). This is probably due to the mortality of medium-sized sprouts and the creation of better light conditions in the lowest part of the canopy. The development of a single resprouting cohort during a so long post-disturbance period (12 years) has also been observed in other *Quercus* (Ross et al. 1986), as well as in other shrub species (Noble 1984; Tappeiner et al. 1984; Riba 1991; Vilà 1993), but it cannot be generalizable to all Mediterranean sprouting species. Several authors (Lacey 1983; Malanson and Westman 1985; Keeley 1986; Mesléard and Lepart 1989) have observed a continuous sprout production in other woody species. The importance of this new wave of regeneration at the whole-stool level is slight in terms of the biomass involved, but it may have a greater importance to the future forest development if it can function as advanced regeneration (see Ross et al. 1986).

5.5 Consequences of Resprouting Dynamics at the Stand Level

As shown in the previous section, competition within the stool during the regeneration cycle causes a reduction in the number of resprouts, while the
mean biomass of survivors increases. As growth of the remaining sprouts
progresses, it has been presumed that they become more independent, and a
division of the stool into more or less independent functional stems could
occur (Aymard and Fredon 1986). On the other hand, competition between
stools is also expected to appear progressively as the stand becomes crowded,
both above- and belowground. Although we lack direct results about the extent of this inter-stool competition in holm oak stands, some clues have been
provided by our studies. In this context, the growth of new holm oak sprouts
not only appears to be related to the basal area coppiced within the stool
(stems removed) but also increases with the amount of basal area removed in
the stand (Retana et al. 1992; see also Chap. 23). These results are similar to
the higher growth rates of sprouts in heavy vs. light coppiced stands found by
Ducrey and Boisserie (1992). On the other hand, Mayor and Rodà (1993)
found at Montseny that growth of the remaining stems in holm oak individuals not completely cut slows down 12 years after thinning, a reduction
linked to the fact that canopy closure is almost complete at that time. In most
holm oak stands at Montseny, canopy closure takes place between 10 and 15
years after thinning, involving dramatic changes in stand dynamics (Espelta
et al. 1995). In this sense, canopy closure is accompanied not only by a reduction in the growth of adult trees (as found by Mayor and Rodà 1993), but
also by a progressive decrease in sprout growth (Ducrey and Turrel 1992;
Espelta 1993) and the inhibition of seedling recruitment (Espelta et al. 1995).
This process continues throughout the whole regeneration cycle and, so,
heavy mortality of large sprouts in old plots has been observed (Retana et al.
1992). This late mortality of some stems may hint at the eventual death of
holm oak individuals (stools) which is rarely found even in the oldest stands
(30 years; Retana et al. 1992). On the other hand, the present management
cycles (ca. 25–30 years) may also prevent the mortality of holm oak individuals, by diminishing the strength of inter-stool competition just before
death of large stems could occur.
    Mortality of sprouts along the whole regeneration cycle coupled with
stand ageing would eventually lead to holm oak stands formed by individuals with a single sprout of great diameter. The time necessary to achieve this
state as well as the final features of the stand will be closely linked with site
characteristics (i.e. exposure, slope, soil depth) and the nature of the disturbance which promoted the resprouting process. The structure of holm oak
stands at Montseny is indeed influenced both by their management history
and by site characteristics (Gracia and Retana 1996). Thus, due to the sprout
dynamics discussed in the preceding section and to the growth of remaining

stems, stands exploited by selection thinning show in 30 years a structure relatively similar to mature or undisturbed forests, i.e. with few stems per stool. In contrast, stands exploited by clear-cutting show after the same time a very different pattern, with a high number of smaller sprouts in each stool. Nevertheless, both patterns can be very influenced by site characteristics, which will finally shape the forest structure. For instance, old forests on poor sites still exhibit a high density of small stems, probably due to less interference between sprouts for light as the canopy remains partially open.

Despite, under the present management and disturbance regime, the regeneration of holm oak forests seems to be assured through resprouting, and thus recruitment of new individuals from seeds is probably not very important, this may in the long term compromise the genetic diversity and the very continuity of these forests. Thus, some authors have claimed that the persistence of holm oak individuals only via resprouting might eventually lead to genet senescence and decreasing productivity (Floret et al. 1987; Ducrey 1992). Moreover, from the genetic point of view and although holm oak shows substantial genetic variability within populations (Michaud et al. 1992), a very long life span coupled with preferential regeneration through resprouting can finally constrain genetic differentiation, and limit the response to environmental change.

References

Axelrod DI (1989) Age and origin of chaparral. In: Keeley SC (ed) The California chaparral. Paradigms reexamined. Natural History Museum of Los Angeles County, Los Angeles, pp 7–19

Aymard M, Fredon JJ (1986) Etude des relations entre une racine et les rejets de la souche chez *Castanea sativa* Mill. Ann Sci For 43:351–364

Canadell J (1995) Vegetative regeneration *of Arbutus unedo* L. and *Erica arborea* L. after disturbances: the role of root structures in the acquisition and use of resources. PhD Thesis, Autonomous University of Barcelona, Bellaterra

Canadell J, Rodà F (1991) Root biomass of *Quercus ilex* in a montane Mediterranean forest. Can J For Res 21:1771–1778

Castell C (1992) Ecofisiologia d'individus adults i rebrots de dues espècies esclerofil·les mediterrànies: *Arbutus unedo* i *Quercus ilex*. PhD Thesis, Autonomous University of Barcelona, Bellaterra

Castell C, Terradas J, Tenhunen JD (1994) Water relations, gas exchange, and growth of resprouts and mature plant shoots of *Arbutus unedo* L. and *Quercus ilex* L. Oecologia 98:201–211

Champagnat P (1989) Rest and activity in vegetative buds of trees. Ann Sci For 46 Suppl: 9–26

Christensen NL (1987) The biogeochemical consequences of fire and their effects on the vegetation of coastal plain of southeastern United States. In: The role of fire in ecological systems. SPB Academic Publishing, The Hague, pp 1–21

Djema A (1995) Cuantificación de la biomasa y mineralomasa subterránea de un bosque de *Quercus ilex* L. MSc Thesis, Instituto Agronómico del Mediterráneo, Zaragoza

Ducrey M (1992) Quelle sylviculture et quel avenir pour les taillis de chêne vert (*Quercus ilex* L.) de la region méditerranéenne française. Rev For Fr 44:12–34

Ducrey M, Boisserie M (1992) Recrû naturel dans des taillis de chêne vert (*Quercus ilex* L.) à la suite d'explotations partielles. Ann Sci For 49:91–109

Ducrey M, Turrel M (1992) Influence of cutting methods and dates on stump sprouting in holm oak (*Quercus ilex* L.) coppice. Ann Sci For 49:449–464

Espelta JM (1993) Establiment i supervivència de plantules en els alzinars del Montseny. MSc Thesis, Autonomous University of Barcelona, Bellaterra

Espelta JM, Riba M, Retana J (1995) Patterns of seedling recruitment in west-Mediterranean *Quercus ilex* forests influenced by canopy development. J Veg Sci 6:465–472

Ferm A, Kauppi A (1990) Coppicing as a means for increasing hardwood biomass production. Biomass 22:107–121

Floret C, Galan NJ, Le Flo'ch E, Orshan G, Romane F (1987) Local characterization of vegetation through growth forms: Mediterranean *Quercus ilex* coppice as an example. Vegetatio 71:3–11

Ford ED, Newbould PJ (1970) Stand structure and dry weight production through the sweet chestnut (*Castanea sativa*) coppice cycle. J Ecol 58:275–296

Giovannini G, Perulli D, Piussi P, Salbitano F (1992) Ecology of vegetative regeneration after coppicing in *macchia* stands in central Italy. Vegetatio 99/100:331–343

Gracia M, Retana J (1996) Effect of site quality and thinning management on the structure of holm oak forests in northeast Spain. Ann Sci For 53:571–584

Hanes T (1971) Succession after fire in the chaparral of southern California. Ecol Monogr 41:27–42

Hara T (1984) A stochastic model and the moment dynamics of the growth and size distribution in plant populations. J Theor Biol 109:173–190

Hara T (1988) Dynamics of size structure in plant populations. Trends Ecol Evol 3:129–133

James S (1984) Lignotubers and burls – their structure, function and ecological significance in Mediterranean ecosystems. Bot Rev 50:225–266

Kauppi A, Rinne P, Ferm A (1988) Sprouting ability and significance for coppicing of dormant buds on *Betula pubescens* stumps. Scand J For Res 3:343–354

Kay JS, Canham CD (1991) Effects of time and frequency of cutting on hardwood root reserves and sprout growth. For Sci 37:524–539

Keeley JE (1986) Resilience of Mediterranean shrub communities to fires. In: Keeley SC (ed) The California chaparral. Paradigms reexamined. Natural History Museum of Los Angeles County, Los Angeles, pp 107–113

Keeley JE, Zedler PH (1978) Reproduction of chaparral shrubs after fire: a comparison of sprouting and seeding strategies. Am Midl Nat 99:142–161

Lacey CJ (1983) Development of large plate-like lignotubers in *Eucalyptus botryoides* in relation to environmental factors. Aust J Bot 31:105–118

Lloret F, López-Soria L (1993) Resprouting of *Erica multiflora* after experimental fire treatments. J Veg Sci 4:367–374

Lloret F, Siscart D (1995) Los efectos demográficos de la sequía en poblaciones de encina. Cuadernos SECF 2:77–81

Lobreaux M (1987) Quelques aspects de la régénération par semis, par rejets de souche et après depressage du taillis de chêne-vert (*Quercus ilex* L.). ENITEF third year report, CNRS Montpellier, INRA, Avignon

López-Soria L, Castell C (1992) Comparative genet survival after fire in woody Mediterranean species. Oecologia 91:493–499

Lowell KE, Mitchell RJ, Johnson PS, Garrett HE, Cox GS (1987) Predicting growth and "success" of coppice-regenerated oak stems. For Sci 33:740–749

MacDonald JE, Powell GR (1983) Relationships between stump sprouting and parent-tree diameter in sugar maple in the 1st year following clear-cutting. Can J For Res 13:390–394

Malanson GP, Westman WE (1985) Postfire succession in Californian coastal sage scrub: the rule of continual basal sprouting. Am Midl Nat 113:309–318

Marschner H (1995) Nutrition of higher plants. Academic Press, London

Mayor X, Rodà F (1993) Growth response of holm oak (*Quercus ilex* L.) to commercial thinning in the Montseny mountains (NE Spain). Ann Sci For 50:247–256

Merino J, Martín Vicente A (1981) Biomass, productivity and succession in the scrub of Doñana biological reserve in south west Spain. In: Margaris NS, Mooney HA (eds) Components of productivity of Mediterranean-climate regions. Junk Publishers, The Hague, pp 197–203

Mesléard F, Lepart J (1989) Continuous basal resprouting from lignotuber: *Arbutus unedo* and *Erica arborea*, as woody Mediterranean examples. Oecologia 80:127–131

Michaud H, Lumaret R, Romane F (1992) Variation in genetic structure and reproductive biology of holm oak populations. Vegetatio 99/100:107–113

Naveh E (1974) Effects of fire in the Mediterranean region. In: Kozlowski TT, Ahlgren CE (eds) Fire and ecosystems. Academic Press, New York, pp 401–434

Noble IR (1984) Mortality of lignotuberous seedlings of *Eucalyptus* species after an intense fire in montane forest. Aust J Ecol 9:47–50

Obón B (1997) Recuperació de la vegetació després del gran incendi del estiu de 1994 a Gualba (Vallès Oriental). MSc Thesis, University of Lleida, Lleida

Pons A, Quézel P (1985) The history of the flora and vegetation and past and present human disturbance in the Mediterranean area. In: Gómez-Campo C (ed) Plant conservation in the Mediterranean area. Junk Publishers, Dordrecht, pp 25–43

Pysek P (1991) Sprout demography and intraclonal competition in *Lycium barbatum*, a clonal shrub, during an early phase of revegetation. Fol Geobot Phytotaxon 26:141–169

Retana J, Riba M, Castell C, Espelta JM (1992) Regeneration by sprouting of holm oak (*Quercus ilex*) stands exploited by selection thinning. Vegetatio 99/100:355–364

Riba M (1991) Estudi de la regeneració per rebrotada en poblacions d'*Erica arborea* sotmeses a tallades. PhD Thesis, Autonomous University of Barcelona, Bellaterra

Romane F, Floret C, Galan M, Grandjanny M, Le Flo'ch E, Maistre M, Perret P (1988) Quelques remarques sur les taillis de chênes verts. Repartition, histoire, biomasse. For Méditerr 10:131–135

Ross MS, Sharik TL, Smith DW (1986) Oak regeneration after clear felling in southeast Virginia. For Sci 32:157–159

Rundel PW, Baker GA, Parsons DJ, Stohlgren TJ (1987) Postfire demography of resprouting and seedling establishment by *Adenostoma fasciculatum* in the California chaparral. In: Tenhunen JD, Catarino FM, Lange OL, Oechel WC (eds) Plant response to stress. Functional analysis in Mediterranean ecosystems. Springer, Berlin, pp 575–596

Sabaté S (1993) Canopy structure and nutrient content in a *Quercus ilex* L. forest of Prades mountains: effects of natural and experimental manipulation of growth conditions. PhD Thesis, University of Barcelona, Barcelona

Serrasolsas I (1994) Fertilitat de sòls forestals afectats pel foc. Dinàmica del nitrogen i el fòsfor. PhD Thesis, University of Barcelona, Barcelona

Tappeiner JC, Harrington TB, Walstad JD (1984) Predicting recovery of tanoak (*Lithocarpus densiflorus*) and Pacific madrone (*Arbutus menziesii*) after cutting or burning. Weed Sci 32:413–417

Trabaud L (1987) Natural and prescribed fire: survival strategies of plants and equilibrium in Mediterranean ecosystems. In: Tenhunen JD, Catarino FM, Lange OL, Oechel WC (eds) Plant response to stress. Functional analysis in Mediterranean ecosystems. Springer, Berlin, pp 607–621

Vilà M (1993) Efecte de la competència en la rebrotada, en el creixement i en la floració d'*Erica multiflora* sotmesa a diferents pertorbacions. PhD Thesis, Autonomous University of Barcelona, Bellaterra

Weiner J, Berntson GM, Thomas SC (1990) Competition and growth form in a woodland annual. J Ecol 78:459–469

Westman WE (1986) Resilience: concepts and measures. In: Dell B, Hopkins AJM, Lamont BB (eds) Resilience in Mediterranean-type ecosystems. Junk Publishers, Dordrecht, pp 5–19

6 Acorn Ecology

Daniel Siscart, Victoria Diego and Francisco Lloret

6.1 Introduction

The significance of acorn production in holm oak forests can be addressed at different levels. At the individual tree level it is one of the final results of the classical trade-off between reproductive and vegetative effort (Bazzaz and Ackerly 1992; Stearns 1992). Although an iteroparous, long-living species may integrate over all its life the consequences of interannual environmental variability on individual fitness, the pattern of seed production is expected to reflect a balance between the losses of growth potential and the gains of sexual reproduction and dispersal success (Stearns 1992). Another related trade-off refers to the relationship between acorn size and seedling fate. A generally positive relationship between both parameters is expected, although interactions with other environmental factors, such as water availability, may be important (Rice et al. 1993). This individual scope can be expanded to the population level, in which fruit production is an essential component of population dynamics because both long-term persistence of populations and colonization of new locations will depend on seed availability (Fenner 1992; Crawley and Long 1995).

Acorns are also important components of food webs in oak forests because of their large size and easy consumption by many animals. Acorn predation by mammals (Felhamer et al. 1989; Jedrzejewska et al. 1994), birds (Hannon et al. 1987; DeGange et al. 1989; Servello and Kirkpatrick 1989; Johnson et al. 1993), and insects (Hails and Crawley 1991) has been widely studied in relation to the population dynamics of these animals, many of which have been shown to be dependent on the acorn crop. Some secondary metabolites, such as tannins and phenolic substances, have been shown to act as chemical defences in the acorns of oaks (Basden and Dalvi 1987; Steele et al. 1993). Although studies on the food web associated with holm oak (*Quercus ilex* L.) acorns are few and partial (Rupérez 1957; Tellería et al. 1991), acorn availability may be specially important in montane holm oak forests, such as those of our experimental areas at Montseny and Prades (NE Spain; Chap. 2), because of the low diversity of food sources for consumers.

Ecological Studies, Vol. 137
Ferran Rodà et al. (eds) Ecology of Mediterranean
Evergreen Oak Forests
© Springer-Verlag, Berlin Heidelberg 1999

On the other hand, some studies have addressed the reverse relation, that is, the effect of predation on the establishment of new cohorts of trees (Matsuda 1985; Jensen and Nielsen 1986; Crawley and Long 1995). Some of these studies suggest that tritrophic interactions, such as mammals consuming acorns infested by weevils, may also be relevant to the dynamics of oak forests (Crawley and Long 1995). These interactions will occur over a relatively short period of time, because the life span of acorns is short before they decay.

At the ecosystem level, fruit production may achieve a sizeable proportion of the annual production in temperate forests (Pregitzer and Burton 1991). Therefore, studies on the functional performance of these ecosystems must consider acorn production as an important component of both energy and nutrient balances.

In this chapter, we analyze the main abiotic and biotic factors contributing to the observed pattern of acorn production and acorn fate before germination in Catalonian holm oak ecosystems, and particularly in the Montseny and Prades forests. Abiotic factors mainly include climatic conditions and soil resource availability. Biotic factors include both seed infestation by insects, which can affect the development and abscission of acorns (Crawley and Long 1995), and predation-dispersal of acorns by vertebrates, which potentially may control the recruitment of new genets both inside forests and in adjacent open areas (Crawley 1992).

6.2 Acorn Production

6.2.1 Reproductive Biology of Holm Oak

Quercus ilex is a monoecious species with unisexual inflorescences. Staminate flowers are borne in catkins. Female inflorescences have 1–8 flowers and are borne at the top of a stalk growing from leaf axils. Pistillate flowers have an ovary, from which only one ovule develops after fertilization (Rupérez 1957). The development of both male and female inflorescences is asynchronous in each tree. From April to June, depending on altitude, catkins develop before pistillate inflorescences do. After fertilization by wind, acorns begin to grow. Acorn development is very slow during summer and fruits are very small (less than 0.5 cm width) in early September, when a fast growth period begins. From late November to early December, acorns begin to dry and turn brown. Mean weight of mature acorns ranges from 2 to 4 g. They drop normally from December to January, although this varies depending on altitude, continentality of the locality, and weather conditions.

6.2.2 Pattern of Acorn Production

High interannual variability of fruit production is a well-known pattern in many forest species and particularly in oaks (Sork and Bramble 1993). A pattern of synchronous high production in periodic years, named mast years, has been described in temperate forests (Kelly 1994). Masting is not a simple response to weather events but arises as an evolved reproductive strategy (Silvertown 1980; Sork et al. 1993). Resource availability, attraction of seed dispersers, increased efficiency of wind pollination, and predator satiation are some of the main hypotheses proposed to explain masting, but at present there is no general consensus about its evolutionary significance (Kelly 1994). Mast years have been observed in several *Quercus* species (Sork et al. 1993; Koenig et al. 1994), and some observations are consistent with the wind pollination and predator satiation hypotheses (Koenig et al. 1994).

At the plant level, we studied acorn production by individual trees in the lowlands near Montseny. We found some degree of alternance between years (Fig. 6.1). This pattern has been also observed in other species of *Quercus* (Kelly 1994; Crawley and Long 1995). Sork et al. (1993) found negative correlations between the acorn production of the current year and the acorn crop of the previous years. This regular pattern of production may be modified by extreme climatic conditions. For instance, the low acorn production we found in 1994 (Fig. 6.1) may be the result of crop alternance, but also a consequence of the extreme drought of that year, which specially affected the crop of the trees living in the Airesol locality, where the soil was very shallow.

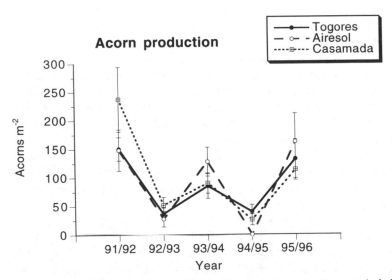

Fig. 6.1. Mean acorn production per tree (± SE) along 5 consecutive years in holm oaks from three localities (Togores, Airesol and Casamada) in lowlands near the Montseny massif. Fallen acorns were collected from two 1.5-m² plots below the crown of each of eight trees per locality

Table 6.1. Interannual variation of acorn production (g m^{-2} year^{-1}, including cups) in different holm oak forests of the NW Mediterranean Basin[a]

Year	Le Rouquet	Mallorca – Menut	Prades – Avic	Prades – Torners	Montseny 1	Montseny 2
	1965–1972	1988–1990	1982–1988	1989–1994	1978-1981	1985–1987
1	15.0	76	32.1	0.1	14.5	29.9
2	48.6	19	26.6	21.8	88.2	17.2
3	10.7	–	8.6	64.5	72.5	–
4	37.8	–	3.6	108.5	–	–
5	116.1	–	2.0	13.7	–	–
6	36.9	–	24.6	–	–	–
7	93.6	–	12.4	–	–	–
Mean	51.2	47.5	15.7	41.7	58.4	23.6
SE	14.9	28.5	4.5	19.9	22.4	6.4
CV (%)	77.0	84.9	76.3	106.4	66.5	38.1

CV, Coefficient of variation.

[a] Acorns collected in litter traps. Data from: Le Rouquet (southern France; Loissant and Rapp 1978), Mallorca–Menut (Xamena 1994), Prades–Avic (Bellot et al. 1992), Prades–Torners (three control plots; V. Diego, pers. observ.), Montseny 1 (permanent plot at La Castanya; Verdú 1984), and Montseny 2 (Torrent de la Mina, 18 plots; X. Mayor, pers. comm.).

At the stand level, data from Montseny and Prades do not show a clear pattern of regular alternance in the biomass of acorn production (Table 6.1). The only other multi-year study in Table 6.1, that at Le Rouquet (southern France), did not found a clear alternance either. The interannual variability was very high in all studies (Table 6.1). The number of acorns produced per square meter also shows a high interannual variability. For instance, in untreated plots in Prades, the mean acorn crop ranged from ca. 10 to 400 acorns m^{-2} over 3 years (Table 6.2).

Interannual variation in acorn production may be partially explained by weather conditions. Climatic effects, however, are not immediate, and significant correlations between the acorn crop and the environmental conditions of previous years are difficult to find (Sork et al. 1993; Crawley and Long 1995). In addition, climatic parameters are not only associated with the resources that the tree allocates to acorn growth. Climate may also influence flower abundance and flower development to fruit, as well as wind pollination (Sork et al. 1993; Koenig and Knops 1995). In several plots located in the Torners valley at Prades, acorn crop biomass was positively correlated to rainfall of the previous period from October to June along the 1989–1993 years ($r = 0.98$ $P = 0.017$, $n = 4$ years). Biomass of male inflorescences and acorn production showed a positive correlation across plots, in the 2 years in which they were measured simultaneously in litterfall (1990–1991: $r = 0.71$, $P = 0.0001$; 1991–1992: $r = 0.73$, $P = 0.0001$; $n = 24$ plots, including different fertilization and irrigation treatments), as observed in other oak species (Sork and Bramble 1993). This suggests that acorn production may be limited by low flower availability in spring.

Table 6.2. Mean (± SE) number of acorns, percentage of aborted acorns, and mean (± SE) acorn size, including aborted fruits, of holm oak fruit fall in plots receiving nitrogen vs. no nitrogen fertilizer, irrigation vs. no irrigation, and control plots during 3 years[a] of study at Prades. For each row $n = 12$ plots, except in the control plots where $n = 3$[b]

Plots receiving	1989–1990			1990–1991			1991–1992		
	Number of acorns ($m^2 \, year^{-1}$)	Abortion (%)	Acorn size (g)	Number of acorns ($m^2 \, year^{-1}$)	Abortion (%)	Acorn size (g)	Number acorns ($m^2 \, year^{-1}$)	Abortion (%)	Acorn size (g)
No nitrogen	26 ± 12	98	0.08 ± 0.03	323 ± 53	71	0.20 ± 0.03	305 ± 52	64	0.18 ± 0.01
Nitrogen	49 ± 12**	99	0.04 ± 0.01	571 ± 82*	71	0.20 ± 0.04	499 ± 116*	61*	0.18 ± 0.01
No irrigation	12 ± 3	100	0.08 ± 0.02	462 ± 95	87	0.09 ± 0.02	582 ± 105	64	0.18 ± 0.01
Irrigation	62 ± 14***	97	0.04 ± 0.01	432 ± 57	55**	0.31 ± 0.02***	222 ± 29***	61	0.19 ± 0.01
Control plots	11 ± 9	100	0.04 ± 0.03	209 ± 86	80	0.13 ± 0.04	406 ± 149	64	0.19 ± 0.04

[a] Years defined from 1 April to 31 March.
[b] ANOVA signification: * = $P<0.05$; ** = $P<0.01$; *** = $P<0.001$.

The same abiotic factors that determine primary production will also influence to some extent acorn production. An experiment to evaluate the effect of nitrogen, phosphorus and water on primary production was carried out in Prades. The experiment was based on a factorial combination of irrigation, nitrogen fertilization, and phosphorus fertilization (see Chap. 13). Acorn production was estimated by litterfall traps sampled monthly from April 1989 to March 1994.

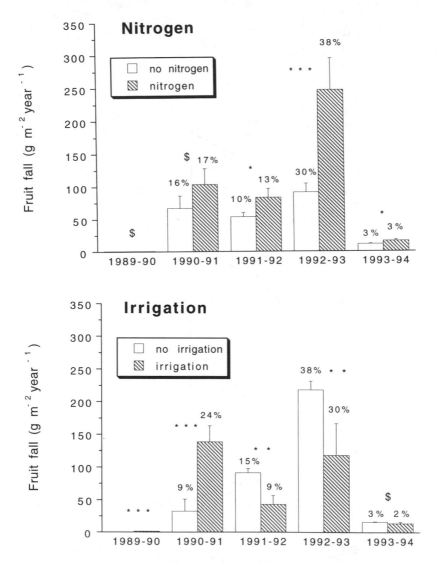

Fig. 6.2. Fruit fall (mean + SE) during 5 years for holm oak in plots receiving nitrogen vs. no nitrogen fertilizer (*top*) and irrigation vs. no irrigation (*bottom*) at Prades. Percentages of fruit fall over total litterfall are indicated. For each bar n=12 plots. ANOVA signification: \$ $P< 0.1$; * $P < 0.05$; ** $P < 0.01$; *** $P < 0.001$

Nitrogen addition increased the production of both number and biomass of fruits (Fig. 6.2) in most years, while phosphorus addition only produced a non-significant increase of the crop biomass in two of the years. A positive effect of irrigation treatment was only observed in years with high summer drought (1989–1990 and 1990–1991; Fig. 6.2), in which it increased acorn production both by decreasing summer abortion and by allowing larger fruit sizes (Table 6.2). The effects of nitrogen and water availability, however, can hardly be considered separately. Nitrogen fertilization increased acorn production especially in 1992–1993 (Fig. 6.2), which was a very rainy year. In the very dry year 1989–1990, fruit production was extremely low, and although the number of acorns was significantly higher in nitrogen-fertilized plots (Table 6.2), there was no effect of nitrogen addition on acorn biomass production (Fig. 6.2). This result may have been also influenced by the short time existing between fertilization (March) and flowering (May).

Therefore, the importance of the resources controlling the crop size changes between years depending on the climatic conditions. Probably, the initial flower production is mainly determined by nutrient availability and also by climatic cues occurring in the winter–spring period. Acorn development seems to be very dependent on water availability. The hydric stress occurring during summer produces very high levels of acorn abortion (Stephenson 1981), while surviving acorns do not increase their size during this period.

6.3 Acorn Predation

6.3.1 Pre-dispersal Predation

A large number of animals can consume acorns before they drop from the tree (pre-dispersal predation). The most important are insects that can infect a large proportion of the crop and may produce a premature fall of undeveloped acorns or may consume the embryos of the seed. This consumption may sometimes be partial, allowing germination but decreasing the probabilities of seedling success. These different effects on seed fate often depend on the infecting species. In the holm oak forests of northeastern Spain, including the Montseny and Prades massifs, the main insects feeding on acorns are the larvae of the weevil *Curculio glandium* (O. Coleoptera, Fam. Curcilionidae), those of the wasp *Callirhytis glandium* (O. Hymenoptera, Fam. Cynipidae) and of the moth *Carpocapsa amplana* (O. Lepidoptera, Fam. Yponomeutidae).

Curculio glandium is the main pre-dispersal consumer in the studied holm oak forests, with infection percentages ranging from 11 to 100% for individual trees and different years, with an overall mean of 49%. This weevil has a life history strongly dependent on the reproductive cycle of several *Quercus* species. Females inoculate one to three eggs inside the young acorns from late August to November. The larvae consume the resources stored in the acorn.

Weevil infection stops the development of the acorns, which dry and drop prematurely from the trees. This response of the plant to infection may have an important effect on the larvae populations and may allow trees to decrease the resource losses from predation. The infected acorns are smaller than healthy ones, and the amount of available food in the acorn strongly influences the weight of the larvae. Acorns smaller than approximately 0.5 g fresh wt. normally show a complete destruction of the endosperm because of direct consumption or digging by the larvae. The weight of the larvae living in these small acorns is lower than 50 mg live wt., while in larger acorns larvae can weigh up to 120 mg. Larger infected acorns usually show some unaltered endosperm and germination may occur in acorns over 1.5 g if the embryo has not been damaged. The size of the acorns at the time of the inoculation would determine the size structure of weevil populations.

After dropping, infected acorns are a potential food for other consumers, such as wood mice (*Apodemus sylvaticus*) (Crawley and Long 1995), which in this way may also influence weevil populations. This effect may be particularly important because, in the absence of other food sources, prematurely dropped acorns are expected to be more intensely predated than later ones.

Two to four weeks after the acorn drops from the tree, larvae complete their development, leave the acorn and bury into the soil at 10- to 30-cm depth. The length of the complete life cycle seems variable because adults may emerge from the soil at least 1 or 2 years after burying (Crawley and

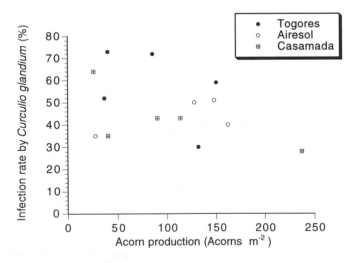

Fig. 6.3. Acorn infection rate by *Curculio glandium* in holm oak from three lowland localities (Togores, Airesol and Casamada) near Montseny. Data obtained from 1991–1992 to 1995–1996 as indicated in Fig. 6.1

Long 1995). The infection rate does not appear clearly associated with acorn production (Fig. 6.3), suggesting the existence of mechanisms, such as diapause and migration, that make the infection rate somewhat independent from the high interannual variability in the acorn crop.

The infection rates by other insect species are lower, and data from holm oak populations in the lowlands near Montseny indicate that they range from 0 to 16% (mean 1.7%, SE 0.33) in *Callirhytis glandium*, and from 0 to 25% (mean 4.9%, SE 0.42) in *Carpocapsa* sp. In both cases, infection typically damages the embryos and germination is not possible.

6.3.2 Post-dispersal Predation

In the studied holm oak forests, many animals may consume the acorns after they drop from trees. They include birds, such as jay (*Garrulus glandarius*), magpie (*Pica pica*) and wood pigeon *(Columba palumbus)*, and mammals, such as wood mice (*Apodemus sylvaticus*), Mediterranean mice (*Mus spretus*), squirrel (*Sciurus vulgaris*) and wild boar (*Sus scrofa*) (Rossell 1988). Observations over several years in holm oak populations in the lowlands near Montseny have shown that wood mice and Mediterranean mice are the only effective post-dispersal consumers of acorns.

Both mice species consume a great proportion of the dropped crop. Predation may also occur on the remaining part of the acorn after seedling emergence, and in these cases it does not necessarily imply the death of the new plant. Acorns may be consumed in the same place where they are found, or they may be carried to a safer place for the animal, where they are consumed immediately. Occasionally, they may be carried to the rodent belowground-refuge, where they are consumed or stored. When the size of the acorn stock makes difficult their immediate consumption or their storage in the belowground refuges, they can also be buried. Burying avoids desiccation and acorns belowground remain appropriate for consumption longer than those remaining at the soil surface. Some of these buried acorns remain in the soil enough time to allow seedling emergence in March-April. In this way, rodent activity increases the dispersal distance of acorns (Iida 1996) and, when buried, favours their germination.

Acorn predation is especially intense in shrublands close to holm oak forests because of the small acorn production of these areas and the relatively high density of rodent populations. We have observed that the removal rate of acorns by rodents is very high in secondary woodlands developing from previous pine forests that burned 12–20 years ago. Most acorns are removed on the first night after being labelled and experimentally placed on the ground (Fig. 6.4), and almost all of them disappear in a few days. Some of these acorns are consumed in the same spot where they were placed. We have not found seedlings emerging from removed, labelled acorns, suggesting high rates of consumption. Acorns placed in clearings more than 3 m in diameter take longer to be removed and some of them remain untouched. In

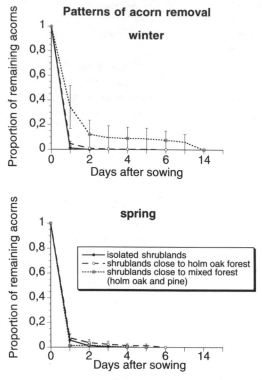

Fig. 6.4. Acorn removal (± SE) after sowing on the soil in three different habitats: isolated shrublands, shrublands close to holm oak forest, and shrublands close to mixed pine–holm oak forest. Values were obtained from 144 acorns placed in four stands per habitat in two consecutive seasons: winter and spring 1993

these open areas the number of captured mice is also much lower, supporting the hypothesis of lower predation pressure in these areas. The distance to areas forested with holm oak is less relevant to predation rates than the distribution of vegetation cover inside the stands.

In dense holm oak forests, the effect of rodents on acorns may be less than in open habitats, such as shrublands. Mice were only rarely found in winter samplings at La Castanya (Montseny), when acorns had recently dropped from the trees. Despite the large production of acorns in these forests, they are only available for consumption for a few months before they rot. If other food sources are rare, as occurs in La Castanya due to the scarce understory, rodent populations are not able to maintain high densities throughout the year, thus limiting predation on the acorn crop.

6.4 Abiotic and Biotic Interactions

The results obtained in Prades support the hypothesis that acorn production is determined by both water and nutrient availability. Interannual variability in rainfall may partially explain the high interannual variability in acorn production. Acorn fate, however, is the result of more complex interactions, which include the amount of acorn production, consumption by animals, and dispersal. Consumer densities are in turn influenced at some spatial and temporal scales by acorn production, as well as by other food sources.

In low crop years, even low density populations of rodents may consume nearly all the available acorns. In high crop years, the survival probability of acorns increases strongly (Crawley and Long 1995) and some acorns are expected to escape predation even when rodent population densities are high (satiation hypothesis; Silvertown 1980). Rodents can be considered as generalist consumers (Obrtel and Holisova 1983; Hansson 1985); their populations do not seem to be regulated by the acorn crop but by the availability of other food resources during the months in which acorns are not available. In forests with higher diversity of food sources for rodents throughout the year (Hoffmeyer and Hansson 1974), or in areas with more heterogeneous landscapes providing a higher diversity of food supply (Tellería et al. 1991; Montgomery and Dowie 1993), rodent populations are expected to increase enormously in years of high production of acorns. Then, these populations can potentially deplete other resources which eventually will regulate their populations. Therefore, through rodent predation, acorn survival is less likely to occur in heterogeneous landscapes with higher rodent densities, especially in low crop years.

On the other hand, the weevil *Curculio glandium* is the main specialist consumer of acorns, potentially affecting a very high proportion of seeds. Weevil populations are expected to be regulated by the acorn production of previous years (Dempster and Pollard 1981), and in this case the number of produced acorns appears more relevant than their total biomass. This species shows winter diapause, and a wide interannual variability in acorn crop is expected to favour strategies of longer diapause (Crawley and Long 1995) or wider dispersion. On the other hand, in the long term, alternate low acorn crops may favour the maintenance of relatively low density populations of weevils, increasing the escape from infection in high crop years (Crawley and Long 1995).

Viability of acorns by escaping predation in pure or mixed holm oak forests depends to some degree on both number and size of the seed production in previous years through its interaction with the features of the life history and the population regulation of potential consumers. Nevertheless, the spatial distribution of acorn crops and other food sources and their effects on animal movements strongly contribute to avoiding a rigid correspondence between holm oak seed production and consumer activity. In addition to abiotic conditions determining germination, the success of seeds in producing new plants is the result of complex temporal and spatial interactions between highly variable acorn production and moderate regulation of consumer populations.

References

Basden KW, Dalvi RR (1987) Determination of total phenolics in acorns from different species of oak. Vet Hum Toxicol 29:305-306

Bazzaz FA, Ackerly DD (1992) Reproductive allocation and reproductive effort in plants. In: Fenner M (ed) Seeds. The ecology of regeneration in plant communities. CAB, Wallingford, pp 1-26

Bellot J, Sánchez JR, Lledó MJ, Martínez P, Escarré A (1992) Litterfall as a measure of primary production in Mediterranean holm-oak forest. Vegetatio 99/100:69-76

Crawley, MJ (1992) Seed predators and plant population dynamics. In: Fenner M (ed) The ecology of regeneration in plant communities. CAB International, Wallingford, pp 157-192

Crawley MJ Long CR (1995) Alternate bearing, predator satiation and seedling recruitment in Quercus robur L. J Ecol 83:683-696

DeGange AR, Fitzpatrick JW, Lay JN, Woolfenden GE (1989) Acorn harvesting by Florida scrub jays. Ecology 70:348-356

Dempster JP, Pollard E (1981) Fluctuations in resource availability and insect populations. Oecologia 50:412-416

Felhamer GA, Kilbane TP, Sharp DW (1989) Cumulative effect of winter on acorn yield and deer body weight. J Wild Manage 53:292-295

Fenner M (ed) (1992) Seeds. The ecology of regeneration in plant communities. CAB International, Wallingford

Hails RS, Crawley MJ (1991) The population dynamics of an alien insect: Andricus quercuscalicis (Hymenoptera: Cynipidae). J Anim Ecol 60:545-562

Hannon SJ, Mume RL, Koening WD, Spon S, Pitelka FA (1987) Poor acorn crop, dominance and decline in numbers of acorn woodpeckers. J Anim Ecol 56:197-207

Hansson L (1985) The food of bank voles, wood mice and yellow-necked mice. In: Flowerdew SR, Gurnell J, Gipps JMW (eds) The ecology of woodland rodents: bank voles and wood mice. Symp Zool Soc Lond 55:1-148

Hoffmeyer I, Hansson L (1974) Variabiliy in number and distribution of Apodemus flavicollis and A. sylvaticus in south Sweden. Z Saeugetierkd 39:15-23

Iida S (1996) Quantitative analysis of acorn transportation by rodents using magnetic locator. Vegetatio 124:39-43

Jedrzejewska B, Okarma H, Jedrzejewska W, Milkowski L (1994) Effects of exploitation and protection on forest structure, ungulate density and wolf predation in Bielowieza Primeval Forest, Poland. J Appl Ecol 31:664-676

Jensen TS, Nielsen OF (1986) Rodent acorn dispersal can explain the observed succession rate of oaks into the heathland. Oecologia 70:214-221

Johnson WC, Thomas L, Adkisson CS (1993) Dietary circumvention of acorn tannins by blue jays. Oecologia 94:159-164

Kelly D (1994) The evolutionary ecology of mast seeding. Trends Ecol Evol 9:465-470

Koenig WD, Knops J (1995) Why do oaks produce boom-and-bust seed crop? Calif Agric 49:7-12

Koenig WD, Mumme RL, Carmen WJ, Stanback MT (1994) Acorn production by oaks in central coastal California: variation within and among years. Ecology 75:99-109

Lossaint, P, Rapp M (1978) La fôret méditerranéenne de chênes verts (Quercus ilex L.). In: Lamotte M, Bourlière F (eds) Problèmes d´ecologie: structure et fonctionnement des ecosystèmes terrestres. Masson, Paris, pp 129-185

Matsuda K (1985) Studies on the early phase of regeneration of a konara oak (*Quercus serrata* Thumb.) secondary forest. 2. The establishment of current-year seedlings on the forest floor. Jpn J Ecol 35:145–152

Montgomery WI, Dowie M (1993) The distribution and population regulation of the wood mouse *Apodemus sylvaticus* on field boundaries of pastoral farmland. J Appl Ecol 30:783–791

Obrtel R, Holisova V (1983) Winter and spring diets of three coexisting *Apodemus* spp. Folia Zool, 32:291–302

Pregitzer KS, Burton AJ (1991) Sugar maple seed production and nitrogen in litterfall. Can J For Res 21:1148–1153

Rice KJ, Gordon DR, Hardison JL, Welker JM (1993) Phenotypic variation in seedlings of a "keystones" tree species (*Quercus douglasii*): the interactive effect of acorn source and competitive environment. Oecologia 96:537–547

Rossell C (1988) La població de senglar al Montseny. Introducció a la biologia de l'espècie. Diputació de Barcelona, Barcelona

Rupérez A (1957) La encina y sus tratamientos. Manero Graf, Madrid

Servello FA, Kirkpatrick RL (1989) Nutritional value of acorns for ruffed grouse. J Wildl Manage 53:26–29

Silvertown JW (1980) The evolutionary ecology of mast seeding in trees. Biol J Linn Soc 14:235–250

Sork VL, Bramble J (1993) Prediction of acorn crop in 3 species of North American oaks: *Quercus alba, Q. rubra* and *Q. velutina*. Ann Sci For 50 Suppl:128s–136s

Sork VL, Bramble J, Sexton O (1993) Ecology of mast-fruiting in three species of North American deciduous oaks. Ecology 74:528–541

Stearns SC (1992) The evolution of life histories. Oxford University Press, Oxford

Steele MA, Knowles T, Bridle K, Simms EL (1993) Tannins and partial consumption of acorns: implications for dispersal of oaks by seed predators. Am Midl Nat 130:229–238

Stephenson AG (1981) Flower and fruit abortion: proximate causes and ultimate functions. Annu Rev Ecol Syst 12:253–279

Tellería JL, Santos T, Alcantara M (1991) Abundance and food-searching intensity of wood mice (*Apodemus sylvaticus*) in fragmented forests. J Mammal 72:183–187

Verdú AMC (1984) Circulació de nutrients en tres ecosistemes forestals del Montseny: Caiguda de virosta i descomposició de la fullaraca. PhD Thesis, Autonomous University of Barcelona, Bellaterra

Xamena J (1994) Análisis de algunos compartimentos y flujos biogeoquímicos en el encinar mediterráneo de Menut (Serra de Tramuntana). PhD Thesis, University of the Balearic Islands, Palma de Mallorca

7 Seedling Recruitment

Javier Retana, Josep Maria Espelta, Marc Gracia and Miquel Riba

7.1 Introduction

Processes operating during the seed, seedling and juvenile phases are crucial for understanding patterns, dynamics and succession in plant communities (Schupp 1990; Grime and Hillier 1992). Many recent studies in temperate forests have emphasized the difficulties seedlings encounter in establishing themselves and the problems hardwood tree species face for survival, in particular the species of the genus *Quercus* (Lorimer 1984; Ross et al. 1986; Crow 1988, 1992; Johnson 1992; Keeley 1992; Ward 1992). This apparent paradox presented by late-successional species which, while dominant in the landscape, do not regenerate easily has long been noticed (Crow 1988), but the reasons for this are still poorly understood. The reasons for failure to recruit may include many factors, such as adequacy of seed dispersal (McClanahan 1986; Willson 1992), seed predation by animals (Crawley 1992), abiotic stresses such as drought or light limitation (Pons 1992; Espelta et al. 1995), and availability of suitable microsites for seed germination and seedling establishment (Crawley 1990). Furthermore, the regeneration failure of recent oak forests has also been attributed to the fact that the distribution of oaks overlaps to a large extent zones of intense human activity and thus oak forests have suffered from human impact (Matsuda et al. 1989).

Anthropogenic factors have played a major role in shaping the vegetation structure in the Mediterranean Basin (Naveh and Lieberman 1984; Fox and Fox 1986). There is no other region in the world where the development of ecosystems has been so closely associated with human social systems and for so long (Blondel and Aronson 1995). Thus, it is difficult to analyze dynamics of Mediterranean forests, in particular holm oak (*Quercus ilex* L.) forests, without taking into account disturbances produced or favoured by man. Holm oak forests have long been severely and continuously modified by wood cutting (coppicing for firewood or charcoal production), repeated fires, and grazing by sheep and goats (Barbero et al. 1992). A large proportion of these forests have been maintained as coppices, where cuttings at regular in-

Ecological Studies, Vol. 137
Ferran Rodà et al. (eds) Ecology of Mediterranean
Evergreen Oak Forests
© Springer-Verlag, Berlin Heidelberg 1999

tervals (about every 20–25 years) are followed by sprout regeneration of stumps rather than sexual reproduction (Chap. 5; Ducrey and Boisserie 1992; Ducrey and Turrell 1992; Retana et al. 1992). Under this management, the seedling bank plays only a minor role in stand regeneration because seedlings are less competitive than stump sprouts. Nevertheless, as has been suggested for other oak forests (Matsuda et al. 1989; Nowacki et al. 1990), the importance of seedlings may be viewed in relation to ecological phenomena such as the persistence of plant populations, genetic variability or potential forest decline. This chapter focuses on the effects that the heterogeneity of the physical environment and the main natural and man-made disturbances of Mediterranean ecosystems have upon seedling recruitment in holm oak forests.

7.2 Variability of Holm Oak Recruitment According to Environmental Conditions

The inherent heterogeneity of Mediterranean landscapes (Naveh and Lieberman 1984) is evident across different temporal and spatial scales (Barbero 1988; Romane et al. 1988; Floret et al. 1989). In the Mediterranean region, the regional climate is an important factor in environmental heterogeneity: there is a general north-to-south gradient of hydric stress for organisms, which is associated with a rising gradient of radiation and temperature. There is therefore a transition from more mesic conditions in the north to more xeric ones in the south. This determines the distribution of tree species. Recent studies (Tretiach 1993) suggest that holm oak is, to a certain extent, more sensitive to drought than expected. Within the Mediterranean region, the dominance of holm oak decreases as the water deficit increases, but the role of climatic features is less evident in relation to seedling density: no correlation has been found between water availability (measured as total annual rainfall) and density of holm oak seedlings, for a wide range of old holm oak stands distributed throughout Catalonia (NE Spain; data from the Forest Inventory of Catalonia).

A smaller scale of heterogeneity occurs in the species composition associated with gradients of aspect, elevation or soil depth (Aschmann 1984; Quézel and Barbero 1989; Pigott and Pigott 1993). The distribution of holm oak seedlings follows topographical patterns more closely than regional gradients. In holm oak stands of the Figaró area at Montseny (see Chap. 2 for site description), the evaluation of holm oak regeneration throughout the elevation gradient (between 300 and 750 m) indicates that the density of holm oak seedlings (number ha^{-1}) increases with elevation (m) (seedling density = 39.6 × elevation − 15261.0, r = 0.79, P = 0.001). Aspect also influences holm oak regeneration because of the difficulties that holm oak recruits experience withstanding water stress. The density of holm oak seed-

Table 7.1. Density of holm oak seedlings[a] (mean number of seedlings ha^{-1} ± SE) in plots of different aspect in two study areas at Montseny (El Figaró and La Castanya)

Study area	North	East–west	South
El Figaró	5938 ± 1915 ($n = 10$)	3678 ± 1461 ($n = 15$)	1736 ± 433 ($n = 18$)
La Castanya	18024 ± 7777 ($n = 17$)	8279±2724 ($n = 11$)	3409 ± 1116 ($n = 36$)

[a] Seed-derived individuals ≤0.5 m tall.

lings is thus greatest on north-facing slopes (i.e., more mesic areas), average on east/west-facing slopes, and lowest on south-facing slopes (i.e., more xeric areas; Table 7.1).

The availability of water rather than that of light may also control the patterns of survival and growth of holm oak seedlings in various microsites during early phases following establishment. Espelta (1996) found greater seedling survival in bare or litter microsites than in stony areas, and lower seedling growth on mounds than in depressions or flat areas. This trend, also suggested by Crow (1992) for *Quercus rubra* seedlings, may be related to differences in water availability: stony sites are drier than bare or litter sites, while mounds are also drier than depressions or flat areas.

7.3 Responses of Holm Oak Seedlings to Different Light and Water Levels

It is supposed that the recruitment of tree individuals is controlled by changes in light levels in relation to the shade tolerance of species (Denslow 1987; Platt and Strong 1989; Whitmore 1989; Bazzaz and Wayne 1994). However, Mediterranean ecosystems are characterized by the scarcity of available soil water during the warm season. This implies that water availability is also of prime importance for the distribution of forest species (Pigott and Pigott 1993). Espelta (1996) tested the response of potted holm oak seedlings to different light and water conditions. Under controlled conditions, he analyzed the effect of light intensity and water availability on survival, growth, biomass allocation, and the water and gas exchange relations of holm oak seedlings. A factorial experiment was carried out with five distinct light and water levels. The five light levels were chosen within the range of light variation found in holm oak forests since the last disturbance (Espelta et al. 1995), while the five water levels simulated the broad gradient found across the geographic range of holm oak (Barbero et al. 1992; Chap. 1).

The survival rate of potted holm oak seedlings fell under increasing incident radiation and decreasing water availability (Fig. 7.1). Both factors inter-

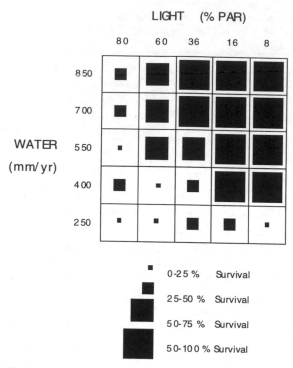

Fig. 7.1. Survival of potted holm oak seedlings under different light and water treatments

acted: the negative effect of the water deficit was enhanced at increased light levels. Along the light gradient, the threshold between high and low survival rates was found at a photosynthetically active radiation (PAR) level of 36% of the incident PAR. The higher survival rates of holm oak seedlings in medium or low light environments agree with the higher survival recorded for this species in closed-canopy stands (Espelta et al. 1995), and with the pattern described for other Mediterranean resprouters (Keeley 1992). The relative growth rate (RGR) of holm oak seedling biomass was higher with increased water availability, showing significant differences for water inputs up to 700 L m^{-2} year^{-1} (Fig. 7.2B). The poor growth rates exhibited by seedlings growing at low light levels were improved with increased light availability, although maximum growth rates were not achieved until the 36% PAR level was reached (Fig. 7.2A).

The question needs to be considered as to which anatomical and physiological characteristics are responsible for the optimum performance (survival + growth) of holm oak seedlings under intermediate light levels and medium to high watering levels. It seems clear that both morphological and physiological components of growth are involved (see Causton and Venus 1981). In the same experiment, Espelta (1996) analyzed the changes along the light and water gradients of the morphological (LAR, leaf area ratio)

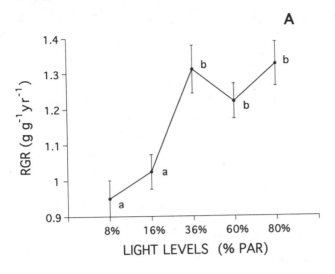

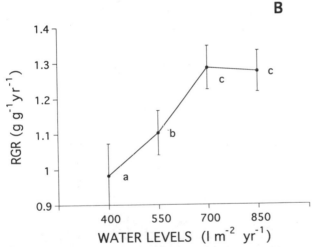

Fig. 7.2. Relative growth rate (RGR) in weight of potted holm oak seedlings under different **A** light levels and **B** water levels. The lowest water treatment has not been included because of its low survival rate (see Fig. 7.1). *Bars* are 1 SE. *Letters* indicate significantly different values ($P<0.05$) according to the Fisher LSD F-test

and physiological components (NAR, net assimilation rate). NAR of holm oak seedlings changed significantly along both light and water gradients, while LAR was only affected by light (Fig. 7.3). The fact that water availability did not modify the leaf area of seedlings but greatly increased their NAR (up to the 700 L m^{-2} year^{-1} level) indicates that the growth increment of well-watered plants is not due to an increment in their assimilation surface but to an increase in their photosynthetic capacity. In fact, well-watered plants exhibited the highest photosynthetic rates during most of the year (Espelta

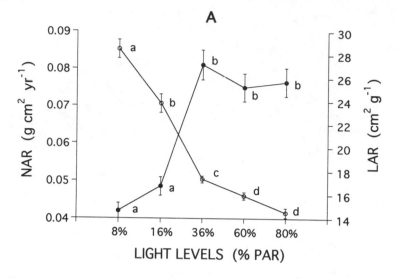

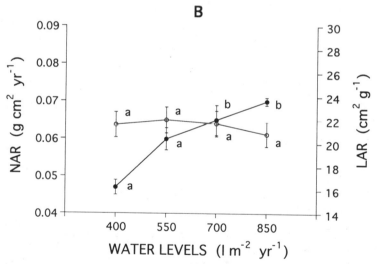

Fig. 7.3. Changes in leaf area ratio (LAR, *white dots*) and net assimilation rate (NAR, *black dots*) of potted holm oak seedlings throughout **A** light and **B** water gradients. *Bars* are 1 SE. *Letters* indicate significantly different values ($P<0.05$) according to the Fisher LSD F-test

1996). Concerning light, holm oak seedlings exhibited the typical inverse relationship between NAR and LAR: NAR increased with light, while LAR decreased (Fig. 7.3). The fact that no differences in seedling RGR with light were found above 36% of incident light might be a consequence of the lack of variation in NAR and LAR beyond this light level. This inability of holm oak seedlings to increase their growth rates in high light levels because of the lack of improvement in their assimilation rates has also been previously reported for shade-tolerant species (Walters et al. 1993; Kitajima 1994; Osunk-joya et al. 1994).

Seedling characteristics which affect carbon assimilation, namely leaf area ratio and photosynthetic rates, may be highly responsible for the poor performance of holm oak seedlings at high radiation levels. Nevertheless, we must take into account other plant features that might also affect plasticity. In this sense, the pattern of biomass partitioning in the roots, stem and leaves of the holm oak appears soon after acorn germination, and the plasticity of such traits may influence the response of seedlings to light and water availability (Leishman and Westoby 1994). Holm oak, as other *Quercus* species, produces heavy acorns without photosynthetic cotyledons. Both these features are commonly interpreted as being adaptations to poor light environments (Kitajima 1994; Leishmann and Westoby 1994; Osunkoya et al. 1994). The main characteristic of biomass partitioning in this species is the large amount of resources devoted to roots in all environments. Thus, it has been reported that root biomass can account for nearly 55% of the whole seedling weight in shady environments, and up to 70% in high light levels (Broncano 1995). Due to the increasing respiration rates under high light conditions (Givnish 1988), a great investment in non-photosynthetic structures (mainly roots) may lead to important construction and maintenance costs (Lambers and Poorter 1992), resulting in a small net carbon gain. Thus, as previously reported for other shade-tolerant species (Kitajima 1994), the optimum performance of holm oak seedlings under moderate light regimes seems to be the result of multiple features, both anatomical and physiological, which result in a high survival rate but which constrain growth rates.

7.4 Effects of Disturbance on Seedling Density and Dynamics

Man-induced disturbances play an important role in holm oak seedling population dynamics. The two major disturbances affecting holm oak forests in NE Spain are thinning and fire. Grazing has declined in importance in recent decades as domestic herbivores, which have a much greater impact on holm oak than wild ones (Cuartas and García-González 1992), are no longer common in these forests. Both thinning and fire have direct effects on seedlings. Thinning causes high seedling mortality through dragging of tree trunks and, more importantly, through the presence of slash piles, which cover and finally kill many seedlings. Yet, the direct effects of fire are even more highly destructive, as most of the aerial parts of seedlings are burned or die in the extreme heat. These disturbances also have indirect effects on the seedlings via changes in the post-disturbance environment. When disturbances such as fire or thinning cause the opening of the canopy, most of the factors that affect seedling recruitment change drastically. The availability of light is increased, associated with exposure to environmental extremes resulting from shelter destruction. Since summer midday temperatures in Mediterranean areas can be very high, a large proportion of holm oak seedlings are exposed to harsh post-disturbance conditions (intense heat and water stress).

The mean densities of holm oak seedlings in recently disturbed (burned or thinned) and undisturbed holm oak stands in the Montseny massif were determined in order to establish the direct effects of both disturbance types. In burned areas of Montseny, nearly all seedlings died after the fire (a reduction of 97.5% in the seedling population, from a mean 73 750 to 1875 seedlings ha^{-1}). The only seedlings capable of withstanding fire and surviving

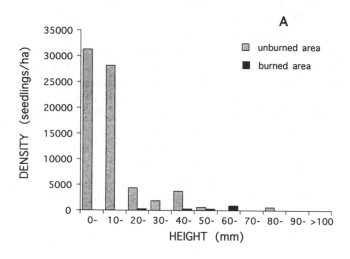

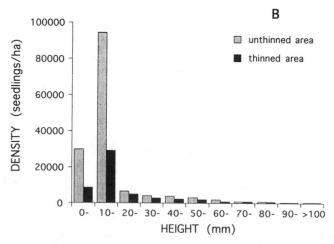

Fig. 7.4. Height distribution of holm oak seedlings in disturbed and undisturbed stands, in **A** burned and unburned stands and **B** thinned and unthinned stands of the Montseny massif

1 year after the fire were very large seedlings (Fig. 7.4A). This would seem to indicate that the survival of holm oak individuals following a fire, as that of other Mediterranean resprouting species (Moreno and Oechel 1994), is strongly related to plant size before the fire. Mortality resulting from thinning in the eight sampled plots of the Montseny massif was considerably lower. Seedling density fell from 83 819 to 51 805 seedlings ha^{-1} (a 38.2% fall in density). Seedlings of all sizes were affected by thinning (Fig. 7.4B), although the mortality of small seedlings was greater than that of larger seedlings.

The specific features of each disturbance type also affect the establishment of new seedlings following the event. Thus, the long-term consequences of fire are more damaging than those of thinning. Acorn survival and germination after fire are both unlikely because uncovered acorns are sensitive to dehydration and destruction by heat. Thus, the arrival of acorns and the establishment of new seedlings in burned stands (other than the episodic dispersal of acorns by animals in areas adjacent to the unburned area) are delayed until new resprouts reach the reproductive age, which can be anywhere between 4 and 8 years (pers. obs.). In contrast, harvesting preserves, at least when it is carried out by selection thinning, nearly one third of uncut trees (ca. 1000 stems ha^{-1}; Retana et al. 1992). Thus, an immediately abundant seed source can be expected in such thinned stands.

Some authors (Crow 1988; Ward 1992) have suggested that disturbance promotes oak seedling regeneration and have described a significant positive correlation between the extent of oak regeneration and the amount of light reaching the forest floor. In holm oak forests, however, the observed pattern is quite different (Fig. 7.5), as the density of holm oak seedlings decreases

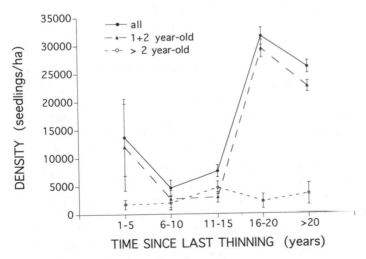

Fig. 7.5. Density of all, 1- and 2-year-old, and >2-year-old holm oak seedlings at different times since last thinning (years). Data represent mean ± SE values of 6, 11, 11, 7 and 12 plots for 1–5, 6–10, 11–15, 16–20 and >20 year categories, respectively

during the initial years following thinning: seedling recruitment is highest in old-cut stands (16–20 and >20 years-since-last-thinning stands) and lowest in medium-cut stands (6–11 years-since-last-thinning stands). As shown in Fig. 7.5, this pattern is mainly due to differences in density of 1- and 2-year-old seedlings. Exposure to full sunlight in recently cut stands not only has a negative effect on survival of already established holm oak seedlings, but also can prevent seed germination. The fact that holm oak acorns are large seeds has long been thought to provide an adaptive advantage for seedling establishing under shade (Westoby et al. 1992; Bazzaz and Miao 1993; Leishman and Westoby 1994). Acorns have a short life span when exposed to full sunlight because they dry out very quickly (Vuillemin 1982), and they germinate better under tree cover, with reduced light levels and the associated increase in soil moisture content (Bran et al. 1990). Broncano (1995) has confirmed experimentally these results: the percentage of holm oak acorn germination at low light levels (8% PAR) is considerably greater than that at full sunlight (23 vs. 1% acorn germination, respectively). The higher overall seedling density in older stands results from a remarkable increase in the recruitment of 1- and 2-year-old seedlings rather than the variation of older ones (Fig. 7.5). Holm oak has always been considered a shade-tolerant species (Ceballos and Ruíz de la Torre 1979; Bran et al. 1990), which can survive under low light by means of efficient light interception and reduced respiration. However, probably the most important reason for such abundant regeneration under covered conditions is the avoidance of desiccation due to high moisture levels rather than higher light-use efficiency.

7.5 The Sapling Stage: Is There a Lack of Recruitment in Holm Oak Forests Under Present Management?

Various authors (Lorimer 1984; Crow 1988; Ward 1992) suggest that oaks can be established under a sylvicultural system that greatly reduces overstory densities and creates gaps in the canopy, favouring successful release of seedlings. Nevertheless, a high number of seedlings before thinning in recently cut holm oak forests is no guarantee of good regeneration. Young seedlings are not able to outgrow the new stump sprouts, which grow much more rapidly than seedlings during the initial years following the disturbance (Fig. 7.6), chiefly by virtue of the root systems of the parent trees. Only vigorous holm oak saplings present prior to thinning can successfully compete with sprouts and reach the forest canopy. Nevertheless, in well-developed holm oak forests, present conditions of genet density and thinning management do not allow seedlings older than 15 years to survive (Espelta et al. 1995).

Holm oak lies at one end of the spectrum of those species identified by Canham (1988) that can survive under canopy cover: those for which shade

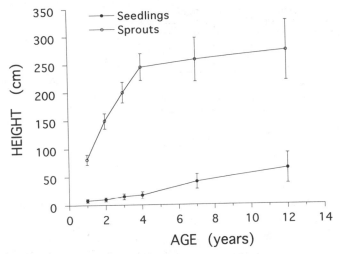

Fig. 7.6. Mean height (± SD) of even-aged seedlings (*n* = 437, 22, 22, 24, 10 and 10 for 1-, 2-, 3-, 4-, 7- and 12-year-old stands, respectively) and sprouts (n = 10 for all stands) in plots from 1 to 12 years after thinning at Figaró. Differences between seedlings and sprouts are significant for all ages (Student's t-test, P<0.0001)

tolerance functions mainly as a mechanism for persistence in the understory, without significant net growth underneath a closed canopy. For holm oak, the most favourable germination sites do not coincide with the most favourable sites for seedling survival and growth (there is a seed–seedling conflict sensu Schupp 1995). While the beneficial effects of shade are likely to include an improved moisture and temperature regime for seed germination and early seedling survival, there are also negative effects, which include insufficient light for growth under dense canopies. When there is not enough light in the understory, seedlings are only able to survive for several years. Results shown in Fig. 7.5 support this hypothesis: while young holm oak seedlings are abundant in closed stands, old seedlings are rare. Old seedlings have greater survival rates under intermediate levels of canopy closure: maximum percentages of 3- to 6-year-old, 7- to 10-year-old, and >10-year-old seedlings are found 6–15 years after thinning the stand (Table 7.2), when light levels are within the range of 15–35% of incident light (Espelta et al. 1995). Once the canopy closes and the light reaching the forest floor is reduced to less than 10%, light becomes a limiting factor for growth and further sapling survival is prevented.

From these results it can be concluded that the sapling stage will only be attained in stands where sufficient light reaches the forest canopy. In open sites, such as Aleppo pine (*Pinus halepensis* Mill.) forests and also in old fields, the establishment of young holm oak individuals is hindered by the high hydric stress, but, after achieving a certain size, holm oak saplings show enhanced survival and growth as enough light reaches the understory. Espelta

Table 7.2. Percentage of seedlings in each age class found in plots of different years since last thinning at Figaró and La Castanya

Years since last thinning	Seedling age (years)				
	n	1–2	3–6	7–10	>10
1–5	6	56.7	31.7	11.0	0.6
6–10	11	36.9	41.7	19.9	1.5
11–15	11	36.0	41.2	20.2	2.6
16–20	7	85.7	6.6	7.5	0.2
>20	12	79.4	10.7	9.3	0.6

n, number of plots sampled.

(1996) reports that holm oak saplings are more abundant in Aleppo pine stands, with a mean density of 359 saplings ha^{-1}, representing 37.5% of holm oak recruits (= seedlings + saplings) sampled in this forest type, and in mixed holm oak–Aleppo pine stands (575 holm oak saplings ha^{-1}, representing 35% of recruits) than in holm oak stands (259 saplings ha^{-1}, representing 6.7% of recruits). This progressive recruitment of holm oaks under an overstory of Aleppo pine supports the traditional idea of succession from pines to oaks, and the expected long-term holm oak dominance in most sites: when both species are found together in undisturbed forests, Aleppo pine trees are usually replaced by holm oaks as young pines are less able to establish themselves in a shaded understory.

Problems arise when the holm oak forest is already established, resulting in a recruitment bottleneck at the sapling stage. The question needs to be asked as to whether the renewal of individuals in holm oak forests under the present conditions is really necessary. From the point of view of wood production, and under the present management practices, the answer is that the recruitment of new individuals of seed-origin is probably not very important. Individuals of seed-origin usually represent a small percentage of stem density (less than 10% at Figaró; 17% on the east-facing slope of the Torrent de la Mina catchment at Montseny, but 52% on its north-facing slope; J. Retana and F. Rodà, unpubl. data). Conversely, percentages of resprouting individuals after thinning (Retana et al. 1992) and fire (López Soria and Castell 1992) are very high, and are sufficient to completely regenerate the forest after a few years (Chap. 5). However, qualitatively speaking, the answer is not so clear cut. Stump resprouting is only a "half-regeneration", since no new genotypes are incorporated into the forest (Ducrey and Turrel 1992). Regeneration exclusively by sprouts has been considered an undesirable management practice because of stool ageing and the risk of forest decline (Lowell et al. 1987; Floret et al. 1989). Some authors (Ducrey 1992; Floret et al. 1992) have claimed that the perpetuation of individuals via the sprouting mechanism might eventually lead to genet senescence, although little information is available about the physiological mechanisms occurring within holm oak stools. The progressive senescence of coppice stumps and roots might have brought about the low production observed in some holm oak stands (Floret et al. 1989) and led to the eventual degradation of these forests. Nevertheless, little is known about the long-term history of holm oak forests, but Barbero (1988) considers that some of the present holm oak stands date from the Middle Ages, and show no evident signs of decline. In agreement with this opinion, holm oak in our experimental area at Prades shows a marked growth response when site conditions are improved either through additions of soil resources (Chap. 13) or through thinning (Chap. 23).

References

Aschmann H (1984) A restrictive definition of Mediterranean climates. Bull Soc Bot Fr 131:21–30

Barbero M (1988) Caractérisation de quelques structures et architectures forestières des arbres et arbustes à feuilles persistantes de l'étage méditerranéen. Rev For Fr 40:371–380

Barbero M, Loisel R, Quézel P (1992) Biogeography, ecology and history of Mediterranean *Quercus ilex* ecosystems. Vegetatio 99/100:19–34

Bazzaz FA, Miao SL (1993) Successional status, seed size, and responses of tree seedlings to CO_2, light, and nutrients. Ecology 74:104–112

Bazzaz FA, Wayne PM (1994) Coping with environmental heterogeneity: the physiological ecology of tree seedling regeneration across the gap–understory continuum. In: Caldwell MM, Pearcy RW (eds) Exploitation of environmental heterogeneity by plants. Academic Press, San Diego, pp 349–390

Blondel J, Aronson J (1995) Biodiversity and ecosystem function in the Mediterranean Basin: human and non-human determinants. In: Davis GW, Richardson DM (eds) Mediterranean-type ecosystems. The function of biodiversity. Springer, Berlin, pp 43–120

Bran D, Lobreaux O, Maistre M, Perret P, Romane F (1990) Germination of *Quercus ilex* and *Q. pubescens* in a *Q. ilex* coppice. Long-term consequences. Vegetatio 87:45–50

Broncano MJ (1995) Estudio experimental de los factores que afectan a la germinación y el establecimiento del pino carrasco y la encina. MSc Thesis, Autonomous University of Barcelona, Barcelona

Canham CD (1988) Growth and canopy architecture of shade-tolerant trees: response to canopy gaps. Ecology 69:786–795

Causton DR, Venus JC (1981) The biometry of plant growth. Edward Arnold, London

Ceballos L, Ruíz de la Torre J (1979) Arboles y arbustos. Publicaciones ETSIM, Madrid

Crawley MJ (1990) Life history and environment. In: Crawley MJ (ed) Plant ecology. Blackwell Scientific Publications, Oxford, pp 253–291

Crawley MJ (1992) Seed predators and plant population dynamics. In: Fenner M (ed) Seeds. The ecology of regeneration in plant communities. CAB International, Wallingford, pp 157–192

Crow TR (1988) Reproductive mode and mechanisms for self-replacement of northern red oak (*Quercus rubra*) – a review. For Sci 34:19–40

Crow TR (1992) Population dynamics and growth patterns for a cohort of northern red oak (*Quercus rubra*) seedlings. Oecologia 91:192–200

Cuartas P, García-González R (1992) *Quercus ilex* browse utilization by Caprini in Sierra of Cazorla and Segura (Spain). Vegetatio 99/100:317–330

Denslow JS (1987) Tropical rainforest gaps and tree species diversity. Annu Rev Ecol Syst 18: 431–451

Ducrey M (1992) Quelle sylviculture et quel avenir pour les taillis de chêne vert (*Quercus ilex* L.) de la région méditerranéenne. Rev For Fr 44:12–34

Ducrey M, Boisserie M (1992) Recrû naturel dans des taillis de chêne vert (*Quercus ilex*) à la suite d'exploitations partielles. Ann Sci For 49:91–109

Ducrey M, Turrel M (1992) Influence of cutting methods and dates on stump sprouting in holm oak (*Quercus ilex* L.) coppice. Ann Sci For 49:449–464

Espelta JM (1996) La regeneració de boscos d'alzina (*Quercus ilex* L.) i pi blanc (*Pinus halepensis* Mill.): estudi experimental de la resposta de les plàntules a la intensitat de llum i a la disponibilitat d'aigua. PhD Thesis, Autonomous University of Barcelona, Barcelona

Espelta JM, Riba M, Retana J (1995) Patterns of seedling recruitment in west Mediterranean coppiced holm-oak (*Quercus ilex* L.) forests as influenced by canopy development. J Veg Sci 6:465–472

Floret C, Galan MJ, Le Floc'h E, Rapp M, Romane F (1989) Organisation de la structure, de la biomasse et de la minéralomasse d'un taillis ouvert de chêne vert (*Quercus ilex* L.). Acta Oecol Oecol Plant 10:245–262

Floret C, Galan MJ, Le Floc'h E, Romane F (1992) Dynamics of holm-oak (*Quercus ilex*) coppices after clearcutting in southern France. Flora and life cycles changes. Vegetatio 99/100:97–105

Fox BJ, Fox MD (1986) Resilience of animal and plant communities to human disturbance. In: Dell B, Hopkins AJM, Lamont BB (eds) Resilience in Mediterranean-type ecosystems. Junk Publishers, Dordrecht, pp 39–64

Givnish TJ (1988) Adaptation to sun and shade: a whole-plant perspective. Aust J Plant Physiol 15:63–92

Grime JP, Hillier SH (1992) The contribution of seedling regeneration to the structure and dynamics of plant communities and larger units of landscape. In: Fenner M (ed) Seeds. The ecology of regeneration in plant communities. CAB International, Wallingford, pp 349–364

Johnson PS (1992) Oak overstory/reproduction relations in two xeric ecosystems in Michigan. For Ecol Manage 48:233–248

Keeley JE (1992) Recruitment of seedlings and vegetative sprouts in unburned chaparral. Ecology 73:1194–1208

Kitajima K (1994) Relative importance of photosynthetic traits and allocation patterns as correlates of seedling shade tolerance of 13 tropical trees. Oecologia 98:419–428

Lambers H, Poorter H (1992) Inherent variation in growth rate between higher plants: a search for physiological causes and ecological consequences. Adv Ecol Res 23:187–261

Leishman MR, Westoby M (1994) The role of large seed size in shaded conditions: experimental evidence. Funct Ecol 8:205–214

López Soria L, Castell C (1992) Comparative genet survival after fire in woody Mediterranean species. Oecologia 91:493–499

Lorimer CG (1984) Development of the red maple understory in northeastern oak forest. For Sci 30:3–22

Lowell KE, Mitchell RJ, Johnson PS, Garrett HE, Cox GS (1987) Predicting growth and "success" of coppice-regenerated oak stems. For Sci 33:740–749

Matsuda K, McBride R, Kimura J (1989) Seedling growth form of oaks. Ann Bot 64:439–446

McClanahan TR (1986) The effect of a seed source on primary succession in a forest ecosystem. Vegetatio 65:175–178

Moreno JM, Oechel WC (1994) Fire intensity as a determinant factor of postfire plant recovery in southern California chaparral. In: Moreno JM, Oechel WC (eds) The role of fire in Mediterranean-type ecosystems. Springer, Berlin, pp 26–45

Naveh Z, Lieberman AS (1984) Landscape ecology. Theory and application. Springer, Berlin

Nowacki GJ, Abrams MD, Lorimer CG (1990) Composition, structure, and historical development of northern red oak stands along an edaphic gradient in north-central Wisconsin. For Sci 36:276–292

Osunkjoya OO, Ash JE, Hopkins MS, Graham AW (1994) Influence of seed size and seedling ecological attributes on shade-tolerance of rain-forest tree species in northern Queensland. J Ecol 82:149–163

Pigott CD, Pigott S (1993) Water as a determinant of the distribution of trees at the boundary of the Mediterranean zone. J Ecol 81:557–566

Platt WJ, Strong DR (1989) Gaps in forest ecology. Ecology 70:535–576

Pons TL (1992) Seed responses to light. In: Fenner M (ed) Seeds. The ecology of regeneration in plant communities. CAB International, Wallingford, pp 259-284

Quézel P, Barbero M (1989) Zonation altitudinale des structures forestières de végétation en Californie méditerranéenne. Leur interprétation en fonction des méthodes utilisées sur le pourtour méditerranéen. Ann Sci For 46:233-250

Retana J, Riba M, Castell C, Espelta JM (1992) Regeneration by sprouting of holm-oak (*Quercus ilex*) stands exploited by selection thinning. Vegetatio 99/100:355-364

Romane F, Floret C, Galan M, Grandjanny M, Le Floc'h E, Maistre M, Perret P (1988) Quelques remarques sur les taillis de chênes verts. Répartition, histoire, biomasse. For Méditerr 10: 131-135

Ross MS, Sharik TL, Smith DWM (1986) Oak regeneration after clear felling in southwest Virginia. For Sci 32:157-169

Schupp EW (1990) Annual variation in seedfall, postdispersal predation, and recruitment of a neotropical tree. Ecology 71:504-515

Schupp EW (1995) Seed-seedling conflicts, habitat choice, and patterns of plant recruitment. Am J Bot 82:399-409

Tretiach M (1993) Photosynthesis and transpiration of evergreen Mediterranean and deciduous trees in an ecotone during a growing season. Acta Oecol 14:341-360

Vuillemin J (1982) Ecophysiologie comparée du développement initial de *Quercus pubescens* et de *Quercus ilex*. I. Développement des semis in situ. Ecol Mediterr 8:139-146

Walters MB, Kruger EL, Reich PB (1993) Growth, biomass distribution and CO_2 exchange of northern hardwood seedlings in high and low light: relationships with successional status and shade tolerance. Oecologia 94:7-16

Ward JS (1992) Response of woody regeneration to thinning mature upland oak stands in Connecticut, USA. For Ecol Manage 49:219-231

Westoby M, Jurado E, Leishman M (1992) Comparative evolutionary ecology of seed size. Trends Ecol Evol 7:368-372

Whitmore TC (1989) Canopy gaps and the two major groups of forest trees. Ecology 70:536-538

Willson MF (1992) The ecology of seed dispersal. In: Fenner M (ed) Seeds. The ecology of regeneration in plant communities. CAB International, Wallingford, pp 61-86

8 A Model of Stand Dynamics for Holm Oak–Aleppo Pine Forests

Miguel A. Zavala

8.1 Introduction

Understanding the mechanisms controlling forest species dynamics and composition is a fundamental issue in silviculture and plant ecology (Oliver and Larson 1990; Crawley 1997). Over the last few decades a growing literature on ecophysiology, population biology and ecosystem ecology has contributed to the empirical expertise developed by practical forestry during the last century. As a whole these disciplines have documented some of the most important processes that take place in forest ecosystems at different levels of biological organization, from leaf physiology to ecosystem function. Nevertheless, there is a rather limited understanding of how species differences scale from physiology to whole plant performance and how these whole plant processes interact with competition to determine stand dynamics and composition.

Mediterranean-type ecosystems have been frequently used for the study of plant physiological adaptations to water stress (Rambal 1984; Tenhunen et al. 1987), ecosystem-level processes (Kruger et al. 1983; Rodà et al. 1990) and more recently to address issues relevant to population ecology (Keeley 1991; Espelta et al. 1995). For this reason, these systems are an attractive scenario for integrating ecological processes across levels of biological organization.

In this chapter I discuss the role of modelling in relating regional forest dynamics and composition to stand and individual level processes. I also introduce two different modelling approaches that predict forest composition, each calibrated from data gathered at two different scales of observation (the stand and the individual). The first approach consists of a correlational probabilistic model that has been calibrated with stand data from forest inventories and that provides a statistical description of forest structure across the landscape. The second approach uses an analytically tractable simulator of stand dynamics, calibrated from experimental data, to explain regional forest composition in terms of the population ecology of individuals.

Ecological Studies, Vol. 137
Ferran Rodà et al. (eds) Ecology of Mediterranean
Evergreen Oak Forests
© Springer-Verlag, Berlin Heidelberg 1999

8.2 Background

8.2.1 Models of Forest Dynamics

As in other sciences, mathematical modelling has played a major role in the development of ecology (e.g. MacArthur 1972; Tilman 1982). In plant ecology, models of coexistence have been used primarily to explain or to predict community pattern and dynamics in relation to observations carried out at the population level (for a review see Czarán and Bartha 1992; Pacala 1997). These models have typically been classified in two broad categories depending on whether competition among individuals is evaluated by a measure of local population density (Horn 1975; Pacala and Silander 1985) or through the effect of resources on plant performance (Tilman 1982; Shugart 1984). For the purpose of this contribution I will refer to the latter type of forest dynamics models – models that consider explicitly species' responses to changes in resource availability (Shugart 1984; Pacala et al. 1996).

Mechanistic or resource-based models of forest dynamics can be broadly defined as stochastic formulations that compute the performance of each tree as a function of resource availability and keep track of the variations in resource levels induced by supply rates and tree consumption (Shugart 1984; Botkin 1993). The original JABOWA (Botkin et al. 1972) and FORET (Shugart and West 1977) models have been modified and applied to a wide range of forest types, such as subtropical forests (Shugart 1984), temperate forests (Prentice and Leemans 1990) and chaparral (Malanson and O'Leary 1995). Thise type of models considers individual trees that compete for resources within a cell scaled to the size of a dominant tree's canopy. Within this cell, resource availability is spatially constant and varies in time according to resource supply and tree interception (light) or tree uptake (water and nitrogen). Different species respond differentially to the resource levels based on their growth functions (Shugart 1984) or both growth and mortality functions (Prentice and Leemans 1990). A recently developed model of forest dynamics has introduced important structural differences in regard to the widely used JABOWA–FORET simulators. SORTIE (Pacala et al. 1996) is a spatially explicit, calibrated model in which light availability is computed separately for each individual and considers the shade cast by each neighboring tree and daily solar trajectory. Recruitment is also truly spatial and depends on the spatial distribution of the parental trees and their reproductive potential. Analyses of this model have shown that spatial detail is critical to adequately describe some aspects of forest dynamics such as standing crop or species diversity (Pacala and Deutschman 1995). Likewise, SORTIE has exemplified the need for developing models intimately calibrated to data and simple enough to be predictive and biologically interpretable.

Individual-based simulators have the advantage of tracing patterns at the stand level to individual level processes, hence allowing us to interpret

changes in species abundance in terms of environmental variation and competitive interactions (see Kobe 1996; Pacala et al. 1996). The mathematical
detail required to characterize individual variability, however, increases
model complexity and hinders on the mathematical and biological interpret-
ability of the model. Therefore, the challenge for developing explanatory and
predictive models is to achieve an adequate balance between biological real-
ism and model complexity.

8.2.2 The Case of Mediterranean Forests

Models of forest dynamics developed to explain successional patterns in
temperate forests have provided important insights into the mechanisms
structuring plant communities (Horn 1975; Shugart 1984; Pacala et al. 1996).
These models have been implemented specifically for temperate systems and
some of their underlying assumptions are clearly inadequate in the context
of other forest types.

In the case of Mediterranean forests, topographic factors, disturbances
and agro-silvicultural systems provide recruitment opportunities at very
different spatial and temporal scales from the forest gaps that characterize
temperate forest dynamics. This contrasts with the inherent gap phase
structure common to most JABOWA–FORET-type models and raises doubts
about the possibility of using these models for Mediterranean systems. The
main factors affecting species recruitment in Mediterranean forests are also
very different from those typically considered in existing models of forest
dynamics. In Mediterranean forests, the ability to tolerate drought stress
during the seedling stage and the differential ability of species to re-establish
after disturbance are considered major determinants of species distributions
(Retana et al. 1992; Pigott and Pigott 1993; Espelta et al. 1995). Although the
effects of water stress on the physiology of these species have received considerable attention (e.g. Terradas and Savé 1992; Castell et al. 1994; Sala and
Tenhunen 1994; Chaps. 10 and 11) the mechanisms linking water limitation
and whole tree performance are not well understood and have not been explicitly incorporated in any model of stand dynamics.

With the purpose of developing models of stand dynamics specifically
suited to the specific conditions of Mediterranean plant communities I initi-
ated a series of studies on mixed holm oak (*Quercus ilex* L.)–Aleppo pine
(*Pinus halepensis* Mill.) forests. These two species dominate the overstory of
extensive areas of the western Mediterranean Basin and their population
ecology has been the subject of extensive research during recent years (e.g.
Retana et al. 1992; Espelta 1996; Chaps. 5–7). These studies have shown that
the structure of mixed holm oak–Aleppo pine forests throughout the landscape consists of a compositional gradient of mixed and monospecific stands
related to physiographic factors and stage of forest development after disturbance (Retana et al. 1996). Through time, the relative abundances of the two
species within a stand are expected to change according to the local thinning

regimes and water availability. Nevertheless, the specific environmental factors to which holm oak and Aleppo pine respond and the population-level mechanisms that control changes in species dominance over space and time have not been adequately described.

8.3 Top-Down Models of Stand Composition in Mixed Holm Oak–Aleppo Pine Forests

The investigation of statistical methods to interpret changes in community composition in terms of species' responses to environmental gradients has been central to the development of vegetation science (Whittaker 1975; Ter Braak and Prentice 1988). Gradient analyses have provided important insights into the factors controlling plant community composition in different biomes (e.g. Whittaker 1956; Peet and Loucks 1977). In the Mediterranean region, however, these analyses have often produced unsatisfactory results, apparently due to anthropogenic influence on vegetation structure (Romane 1987; Floret et al. 1989; Retana et al. 1996).

In order to detect the main sources of correlation between forest structure and environmental factors in this region and to account for the specific patterns of variability of these stands, a series of statistical methods was developed. The mathematical derivation of these methods is beyond the scope of this chapter and only some relevant features are reported here. The approach used follows the general definition of statistical inference and likelihood (see Edwards 1972) and it does not make any a priori assumption regarding the structure of the data. Therefore, it allows us to explore the effects of different model formulations not only on the mean but also on any other aspects of variability in the statistical distribution that underlies our observations. In the example shown here, I describe stand composition in holm oak–Aleppo pine forests in relation to drought length (DL) (Walter and Lieth 1960). This index indicates the number of months in which Turc's potential evapotranspiration (Turc 1958) exceeds rainfall for a given plot and integrates the effect of local climate and exposure on vegetation. Information about forest structure was obtained from more than 1000 plots of the Forest and Ecological Inventory of Catalonia (Gracia 1992) and climatic data were computed through multiple regression models that estimate local measurements as a function of altitudinal, latitudinal and longitudinal deviations from the closest meteorological station.

According to conventional wisdom, the mean basal area of holm oak and Aleppo pine covaried inversely along the DL gradient according to a sigmoid function. However, the residual variation in species basal area along the aridity gradient followed a characteristic U-shaped pattern that suggests strong patterns of interspecific segregation between these two species (Fig. 8.1). This figure considers the frequency distribution of holm oak pro-

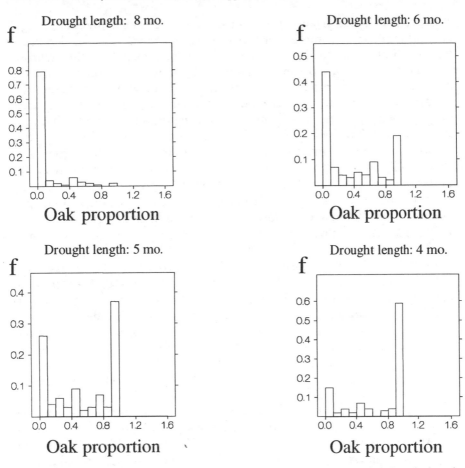

Fig. 8.1. Distribution of holm oak proportion along a gradient of decreasing drought length (*DL*, in months) in Catalonia (Spain). *Horizontal axes* represent holm oak proportion in terms of basal area for each inventory plot. *Vertical axes* (*f*) measure relative frequency of the variable "holm oak proportion" for a given *DL* in the whole region studied. The sequence of graphs show that distribution of holm oak proportion in mixed pine–oak stands changes along the *DL* gradient according to a U-shaped function

portion for different values of *DL*. It can be observed that in dry sites the function peaks in the left extreme because of the increasing number of monospecific pine stands, while in hydric sites (*DL* smaller than 5) the largest values are found in the right extreme, as most of the stands are dominated by holm oak. These patterns are not as intuitive as the trends observed in the mean and can be interpreted according to different biological and anthropogenic processes, many of which can bias the outcome of competition. For example, silvicultural practices may hasten succession by favoring the performance of the best local competitor. Alternatively, ecological processes alone such as species spatial segregation or founder effect could generate these distributions.

The statistical significance of these patterns was evaluated by calibrating a top-down probabilistic model for this system. This model was developed for two main purposes. On the one hand, it permits detection of the environmental factors that have the most significant effect on forest composition. On the other hand, if the model is successfully tested in a variety of locations, it could be used to predict species' relative abundance at a given site as a function of local climate and management. Following this idea, a model that considers the total basal area in the plot and the proportion of each species within the stand was calibrated. The variation in mean holm oak proportion along a soil moisture gradient can be characterized as a sigmoid response similar to the one described in a model of logistic regression. However, to my knowledge no mathematical transformation exists that transforms a U-shaped function to the binomial distribution assumed in a logistic regression model. For this reason, it was necessary to derive a probabilistic process that generates a pattern characterized by U-shaped residuals. Briefly, this process considers an index (E) that measures the holm oak proportion ("oakiness") of a given site and summarizes all the underlying processes that affect stand composition at equilibrium, like soil moisture, management history and altitude. The proportion of holm oak for a given value of this index can be calculated according to a simple response function. In this particular example, E can be assumed to be normally distributed, with a constant standard deviation and a mean that increases linearly with soil moisture. The resulting model is then described by four parameters: the slope of the logistic function that describes the variation in mean holm oak proportion, the standard deviation of the index and the intercept and slope that define the dependence of the index mean on soil moisture. Given this probabilistic process and the underlying hypotheses (model parameters) it is then possible to estimate the values of the parameters that maximize the likelihood of occurrence in relation to our data (see Edwards 1972). The significance of the effect of soil moisture on holm oak relative abundance can be shown with a statistical test that compares the likelihood of two alternative models, one of them without an explicit dependence on soil moisture (ratio likelihood test; Edwards 1972; 1 df, $P < 0.001$).

The flexibility of this model permits to analyze properly heteroscedastic data patterns, and to relate stand variability to factors that cannot be described by continuous random variables such as historical management. Top-down models can be used primarily to identify the environmental factors that determine large-scale species distributions (Landsberg 1986; Chaparro 1996) and for explanatory analyses of landscape structure. Ultimately these models can be used together with geographic information to generate dynamic spatial maps that render error-bounded predictions on the effect of different planning strategies on regional forest structure (see Grossman 1991).

8.4 Mechanistic Models of Stand Dynamics in Mixed Holm Oak–Aleppo Pine Forests

Correlational top-down models provide a quantitative description of large-scale forest structure in relation to environmental gradients. However, ecologists and foresters are mainly concerned with the identification of the species-specific traits that control stand composition. In Mediterranean forests, the interaction of soil moisture and radiation appears as the main determinant of stand development (Olazábal 1916; González Vázquez 1947; Espelta 1996), so patterns in stand dynamics and composition might be explained in terms of species-specific responses to light intensity and soil moisture.

In the case of mixed holm oak–Aleppo pine forests these two species show throughout their life cycle remarkable differences in the way they respond to light and water availability, particularly during the seedling and sapling stages (Espelta et al. 1995). The effect of these traits on stand composition for different levels of management and rainfall can be investigated with an analytical model of stand dynamics calibrated from experimental data. The model determines stand dynamics for patches situated along a rainfall gradient; within a stand, however, soil moisture is spatially uniform. Plants are tracked individually starting from a given number of adult pines and oaks that are randomly dispersed in the population. In each rotation, management removes a fraction of the standing crop and changes the competitive environment where seedlings and saplings develop (Fig. 8.2). At the next management rotation, stand composition will change according to the effect that soil moisture and light have on seedling and juvenile mortality. Next, the individuals that survive the seedling and the sapling stage are considered

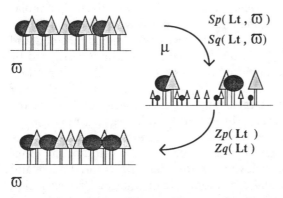

Fig. 8.2. Schematic flow chart of the analytical model described in the text. *Top left* shows initial stand composition prior to thinning. Management (*middle*) removes a given percentage of individuals (μ). *Bottom left* indicates the most likely stand composition based on seedling and sapling probability of survival in the post-thinning environment. Functions describing seedling (Z_p, Z_q) and sapling (S_p, S_q) survivorship depend on light availability and local water balance (ϖ)

adults that reproduce at a constant rate and disperse their seeds to random locations within the stand.

Seedling mortality functions were calibrated from experimental studies by Espelta (1996). These studies indicate a positive dependence of holm oak and Aleppo pine survival on water availability and a strong negative effect of light on holm oak and only a moderately negative effect on Aleppo pine. Sapling mortality probability is modelled as a logistic function of light availability, with the inflection point reflecting the maximum degree of shade, measured as the percentage of full sunlight that a sapling can tolerate: Lp for the pine and Lo for the oak ($Lo > Lp$). In each generation, light is measured as a percentage of total photosynthetic available radiation (PAR) and follows an exponential decay that depends on each species' basal area and light extinction coefficients (Beer's law). For simplicity, the effect of holm oak resprouting and sapling mortality dependence on water are not considered in these simulations, but their inclusion does not change the broad qualitative results presented here. The deterministic form of this model can be written as:

$$P_{t+1} = - \mu_p \cdot P_t + P_t \cdot f_p \cdot Z_p (L_t, \varpi) . S_p \cdot (L_t)$$
$$Q_{t+1} = - \mu_q \cdot Q_t + Q_t \cdot f_q \cdot Z_q (L_t, \varpi) \cdot S_q (L_t)$$
$$L_t = k \cdot \exp(-\alpha_p \cdot P_t - \alpha_q \cdot Q_t),$$

where, P_t and Q_t are pine and oak genet density respectively at time t, f_p and f_q are pine and oak fecundity, L_t is light (%) at time t, ϖ indicates annual rainfall (mm), Z_p and Z_q are the probability of pine and oak sapling survival, S_p and S_q indicate the probability of pine and oak seedling survival, k, α_p and α_q are the parameters that control light decay, and finally μ_p and μ_q represent pine and oak adult mortality rate (thinning intensity). The time step represents the time that it takes for a sapling to establish as an adult, and coincides with the rotation cycle of the stand.

A complete solution of this model includes the trajectory over time for each species. Most commonly, however, these models are solved at singular points such as the equilibrium state that takes place when the population sizes of both species remain stationary. In this way, the dependence of stand composition at equilibrium on ($\mu_p = \mu_q = \mu$) and ϖ will be investigated by analyzing the invasibility conditions for the system. Concisely, this method evaluates the possibility that one species can invade a monospecific stand of the other species and vice versa. Depending on which species can invade the other, several outcomes are possible. For example, both species can coexist at equilibrium when either of them can invade a monospecific stand of the other. If neither of the species can invade a monospecific stand of the other species, then there is mutual exclusion or founder effect and final stand composition depends on the initial conditions. Finally, competitive exclusion occurs when one species can invade a monoculture of the other species, but the opposite is not true.

The conditions for invasibility can be formally quantified by the population growth rate of one species in a stand dominated by the other species, technically named the eigenvalues of the Jacobian matrix of the mono-specific system. A value greater than one indicates population expansion, one indicates equilibrium and less than one indicates population decrease. Given that the eigenvalues are functions of μ and ϖ, the qualitative dynamics of the model in relation to these two variables can be studied by a combination of graphical and analytical methods. Figure 8.3 indicates for a specific example how to predict graphically the outcome of competition for different levels of thinning and water availability. The horizontal axis indicates light dynamics in the model and the vertical axis (β) represents the population growth rate as determined by the invasion matrix. The shape of the functions that describe β for a given species depends on μ and ϖ; thus, these two parameters can change the relative positions of the population growth functions. In this example annual rainfall averages 600 mm and thinning removes 40% of the standing basal area at each rotation, making two stable equilibria possible.

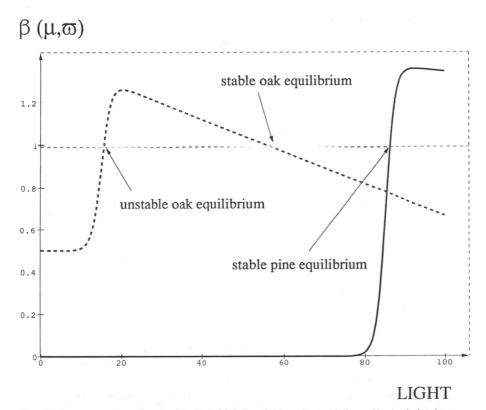

Fig. 8.3. Representation of a graphical method that depicts the model's qualitative behaviour as a function of mortality rate or thinning intensity (μ) and rainfall (ϖ). *Horizontal axis* indicates light dynamics in the model and *vertical axis* (β) represents holm oak (*dotted line*) and Aleppo pine (*solid line*) population growth. ϖ equals 600 mm and μ removes 40% of the standing basal area at each rotation cycle. The system exhibits founder effect: either of the two species has a growth rate less than one when the other reaches a stable equilibrium

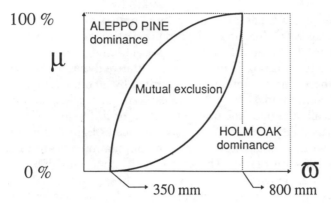

Fig. 8.4. Representation of the dynamic behaviour of the model as a function of thinning intensity (μ) and annual rainfall (ϖ) based on a limited number of simulations

The oak equilibrium shows that for this rainfall and disturbance rate, pines cannot invade a pure oak stand because when oaks are at equilibrium (growth rate equal to one) pine growth rate is smaller than one. Similarly, the pine equilibrium shows that oaks are also unable to invade a monospecific pine stand. Thus, the model exhibits mutual exclusion such that stand composition at equilibrium is ultimately driven by initial conditions.

The results of graphical and analytical analyses can be summarized as follows (Fig. 8.4). Pine dominance takes place in a narrow region that ranges from the most xeric areas of the gradient with no disturbance to moderately dry areas with high levels of disturbance. A broad region where either species can dominate, depending on the initial conditions (mutual exclusion), is found at intermediate values of precipitation and management. Finally, a region of holm oak dominance occurs in sites ranging from mesic with low disturbance rates to areas of elevated rainfall and moderate disturbance rates.

8.5 Conclusions: Theoretical and Practical Implications

Regional patterns in forest composition and mechanistic models suggest that the realized niches of holm oak and Aleppo pine along an aridity gradient can be described as two overlapping curves, with pine preponderance in the lower extreme and increasing oak dominance at higher soil moisture levels. A similar pattern has been documented for other mixed pine–oak forests (Whittaker 1975; Oliver and Larson 1990) and agrees with paleoecological and physiological studies that consider Mediterranean pines a stable component of Iberian forests (Costa et al. 1990; Gil and Aránzazu 1993; Morla 1993).

From a theoretical point of view, these results can be interpreted in the context of the mechanistic theory of plant competition and community structure (Tilman 1982, 1988). According to this view, species distributions along a soil moisture gradient are controlled by shade tolerance at the upper limit of the gradient and by drought tolerance at the lower limit. In mesic sites, and for a fixed range of disturbance and initial conditions, the model converges on a monospecific equilibrium dominated by the most shade-tolerant species (holm oak in this case). As rainfall decreases, however, the qualitative behaviour of the model changes as a result of the differential responses in seedling mortality to light. Therefore, this result follows Tilman's (1988) prediction of a generalized trade-off between shade and drought tolerance in plant communities. Plant water balance depends simultaneously on evaporative demand and supply (Cowan 1986; Schulze et al. 1987), hence lower soil moisture levels increase the detrimental effect of radiation on seedling mortality at a rate dependent on species morphology. Accordingly, this effect is likely to be more severe for shade-tolerant species such as holm oak that require a larger photosynthetic apparatus and have a higher transpirative demand.

Mediterranean forests have developed according to a variety of ecological pathways related to disturbances, water balance and ecological strategies (Ruiz de la Torre 1990; Montoya 1993; Espelta et al. 1995) and their long-term dynamics under the prevailing conditions of massive land-use change are uncertain (Romane et al. 1992). There is, therefore, a need for diagnostic tools that scale up the biological and human aspects that operate in these systems at different scales. This integration is more likely to be achieved with the use of simple and biologically interpretable models, rather than by coupling models developed at different levels of biological organization, simply because complex models are difficult to interpret in the light of the mechanisms that bear upon them. The implementation of simple, realistic and yet predictive models of vegetation dynamics for the Mediterranean region will require a clear understanding of how competition and environmental factors regulate species distributions. This understanding seems feasible only under research programs that consider the natural history of these systems and integrate experimentation, fieldwork and modelling.

References

Botkin DB (1993) Forest dynamics: an ecological model. Oxford University Press, Oxford
Botkin DB, Janak JF, Wallis JR (1972) Rationale, limitations, and assumptions of a northeastern forest simulator. IBM J Res Dev 16:106–116
Castell C, Terradas J, Tenhunen JD (1994) Water relations, gas exchange, and growth of resprouts and mature plant shoots of *Arbutus unedo* L. and *Quercus ilex* L. Oecologia 98:201–211
Chaparro J (1996) Distribución potencial del bosque y de sus especies arbóreas en zonas mediterráneas semiáridas: modelos y aplicaciones. PhD Thesis, University of Murcia, Murcia

Costa M, García M, Morla C, Saínz H (1990) La evolución de los bosques de la Península Ibérica: una interpretación basada en datos paleobiogeográficos. Ecología 1:31–58

Cowan IR (1986) Economics of carbon fixation in higher plants. In: Givnish TJ (ed) On the economy of plant form and function. Cambridge University Press, Cambridge

Crawley MJ (ed) (1997) Plant ecology, 2nd edn. Blackwell, Oxford

Czarán T, Bartha S (1992) Spatiotemporal dynamic models of plant populations and communities. Trends Ecol Evol 7:38–42

Edwards AWF (1972) Likelihood. Expanded edn. Cambridge University Press, New York

Espelta JM (1996) Regeneration of holm oak (*Quercus ilex*)-Aleppo pine (*Pinus halepensis*) forests: experimental study of seedling response to light intensity and water availability. PhD Thesis, Autonomous University of Barcelona, Bellaterra

Espelta JM, Riba M, Retana J (1995) Patterns of seedling recruitment in west Mediterranean *Quercus ilex* forests influenced by canopy development. J Veg Sci 6:465–472

Floret C, Galan MJ, Le Floc'h E, Orshan G, Romane F (1989) Growth forms and phenomorphology traits along an environmental gradient: tools for studying vegetation? J Veg Sci 1:71–80

Gil L, Aránzazu MP (1993) Los pinos como especies básicas de la restauración forestal en el medio Mediterráneo. Ecología 7:113–126

González Vázquez E (1947) Selvicultura. Libro primero: Fundamentos naturales y especies forestales. Los bosques ibéricos. Dossat, Madrid

Gracia CA (1992) Inventari ecològic i forestal de Catalunya. CREAF Publications, Barcelona

Grossman WD (1991) Model- and strategy-driven geographical maps for ecological research and management. In: Risser PG (ed) Long-term ecological research. John Wiley, New York, pp 242–254

Horn HS (1975) Markovian properties of forest succession. In: Cody M, Diamond J (eds) Ecology and evolution of communities. Harvard University Press, Cambridge, pp 196–211

Keeley JE (1991) Seed germination and life history syndromes in the California chaparral. Bot Rev 57:81–116

Kobe RK (1996) Interspecific variation in sapling mortality and growth predicts geographic variation in forest composition. Ecol Monogr 66:1–43.

Kruger FJ, Mitchell DT, Jarvis JUM (eds) (1983) Mediterranean-type ecosystems: the role of nutrients. Springer, New York

Landsberg JJ (1986) Physiological ecology of forest production. Academic Press, London

Mac Arthur RH (1972) Geographical ecology. Harper and Row, New York.

Malanson GP, O'Leary JF (1995) The coastal sage scrub-chaparral boundary and response to global climate change. In: Moreno JM, Oechel WC (ed) Global change and Mediterranean-type ecosystems, Springer, Berlin, pp 203–224

Montoya JM (1993) Encinas y encinares. Ediciones Mundi-Prensa, Madrid

Morla C (1993) Significación de los pinares en el paisaje vegetal de la Península Ibérica. In: Silva FJ, Vega G (eds) Congreso Forestal Español. Ponencias y comunicaciones, vol 1. Xunta de Galicia, Lourizán, Pontevedra, pp 361–370

Olazábal S (1916) Composición de las masas arbóreas. Rev Montes 40:29–302

Oliver CD, Larson BC (1990) Forest stand dynamics. McGraw-Hill, New York

Pacala SW (1997) Dynamics of plant communities. In: Crawley MC (ed) Plant ecology, 2nd edn. Blackwell, Oxford, pp 532–555

Pacala SW, Deutschman DH (1995) Details that matter: the spatial distribution of individual trees maintains forest ecosystem function. Oikos 74: 357–365

Pacala SW, Silander JA (1985) Neighborhood models in plant population dynamics. I. Single species models of annuals. Am Nat 125:385–411

Pacala SW, Canham CD, Saponara J, Silander JA, Kobe RK, Ribbens E (1996) Forest models defined by field measurements: estimation, error analysis and dynamics. Ecol Monogr 66:1–43

Peet RK, Loucks OL (1977) A gradient analysis of southern Wisconsin forests. Ecology 58:485–499

Pigott CD, Pigott S (1993) Water as a determinant of the distribution of trees at the boundary of the Mediterranean zone. J Ecol 81:557–566

Prentice IC, Leemans R (1990) Pattern and process and dynamics of forest structure: a simulation approach. J Ecol 78:340–355

Rambal S (1984) Water balance and pattern of root uptake by a *Quercus coccifera* L. evergreen scrub. Oecologia 62:18–25

Retana J, Riba M, Castell C, Espelta JM (1992) Regeneration by sprouting of holm oak (*Quercus ilex*) stands exploited by selection thinning. Vegetatio 99/100:355–364

Retana J, Gracia M, Espelta JM (1996) Caracterización de masas mixtas de pino y encina en el macizo del Montseny (Nordeste de la Península ibérica). Cuad Soc Esp Cienc For 3:167–179

Rodà F, Àvila A, Bonilla D (1990) Precipitation, throughfall, soil solution and streamwater chemistry in a holm oak (*Quercus ilex*) forest. J Hydrol 116:167–183

Romane F (1987) Efficacité de la distribution des formes de croissance des végétaux pour l'analyse de la végétation à l'échelle régionale. PhD Thesis, Université de Droit, d'Economie et des Sciences d'Aix-Marseille, Aix-Marseille

Romane F, Bacillieri R, Bran D, Bouchet MA (1992) Natural degenerate Mediterranean forests: which future? The examples of the holm oak (*Quercus ilex* L.) and chestnut (*Castanea sativa* Mill.) coppice stands. In: Teller A, Mathy P, Jeffers JNR (eds) Responses of forest ecosystems to environmental changes. Elsevier, London, pp 374–380

Ruiz de la Torre J (1990) Distribución y características de las masas forestales españolas. Ecología 1:11–30

Sala A, Tenhunen JD (1994) Site-specific water relations and stomatal response of *Quercus ilex* in a Mediterranean watershed. Tree Physiol 14:601–617

Schulze ED, Robichaux RH, Grace J, Rundel PW Ehleringer JR (1987) Plant water balance. BioScience 37:30–37

Shugart HH (1984) A theory of forest dynamics: the ecological implications of forest succession models. Springer, Berlin

Shugart HH, West DC (1977) Development of an Appalachian deciduous forest succession model and its application to assessment of the impact of the chestnut blight. J Environ Manage 5:161–179

Tenhunen JD, Catarino FM, Lange OL, Oechel WC (eds) (1987) Plant responses to stress. Functional analysis in Mediterranean ecosystems. Springer, Berlin

Ter Braak CJF, Prentice IC (1988) A theory of gradient analysis. Adv Ecol Res 18:271–316

Terradas J, Savé R (1992) The influence of summer and winter stress and water relationships on the distribution of *Quercus ilex* L. Vegetatio 99/100:137–145

Tilman D (1982) Resource competition and community structure. Princeton University Press, Princeton

Tilman D (1988) Plant strategies and the dynamics and structure of plant communities. Princeton University Press, Princeton

Turc L (1958) Le bilan d'eau des sols: relations entre les precipitations, l'evaporation et l'ecoulement. Société Hydrotech de France, Algiers, Paris

Walter H, Lieth H (1960) Klimadiagramm Weltatlas. Veb Gustav Fisher, Jena

Whittaker RH (1956) Vegetation of the Great Smoky Mountains. Ecol Monogr 26:1–80

Whittaker RH (1975) Communities and ecosystems. Macmillan, New York

Part 3   Light Harvesting and Gas Exchange

9 Leaf Traits and Canopy Organization

Santiago Sabaté, Anna Sala and Carlos A. Gracia

9.1 Introduction

Carbon, water and nutrient fluxes through forest canopies depend on the functional and morphological attributes of individual leaves within the canopy, which, in turn depend on the amount and arrangement of foliage. Canopy structure and canopy function are tightly coupled attributes that result in an optimal "functional structure" ultimately determined by abiotic and biotic factors such as incident radiation, water availability, nutrient availability, competitive interactions, herbivory, etc. Holm oak (*Quercus ilex* L.) is a broadleaved evergreen tree characteristic of the western Mediterranean region that forms dense stands in a variety of environments (Chap. 1). Here, we compile information on structural, and functional characteristics of holm oak canopies at two locations in NE Spain, differing in resource availability: the Avic catchment in the Prades mountains and the permanent plot at La Castanya in the Montseny mountains (see Chap. 2). The Avic catchment is a more xeric site with higher annual incident radiation and lower annual precipitation than La Castanya. Our goal is to analyse how much structural and functional variation exists between and within canopies and how this variation relates to environmental conditions.

9.2 Total Leaf Biomass and Leaf Area Index

Variations in total canopy leaf biomass and leaf area index (LAI) of Mediterranean plants have long been related to environmental conditions (Specht and Specht 1989). Reductions in exposed leaf area have been cited as a long-term response (years) to the intensity and duration of the summer dry period (Poole and Miller 1981). In Mediterranean regions, where the typical long dry summer period limits plant growth and production, higher LAIs are found in sites with higher water availability where moderate to high precipi-

Ecological Studies, Vol. 137
Ferran Rodà et al. (eds) Ecology of Mediterranean
Evergreen Oak Forests
© Springer-Verlag, Berlin Heidelberg 1999

Table 9.1. Environmental conditions [mean values of: annual radiation (RAD), annual temperature (T) and annual precipitation (P)], leaf biomass (LB), leaf area index (LAI) and stem area index (SAI, excluding main trunks) of holm oak forests at Avic (Prades), Le Rouquet (southern France) and the permanent plot at La Castanya (Montseny)

	Avic catchment[a]		Le Rouquet[b]	La Castanya[c]
	Ridge site	Valley site		
RAD (GJ m^{-2} year^{-1})	5.7	4.7	5.2	3.4
T (°C)	12.8	13.8	12.4	9
P (mm)	647	647	770	862
LB (Mg ha^{-1})	8.4 ± 0.7[d]	9.3 ± 0.6[d]	7.0	6.1
LAI (m^2 m^{-2})	4.6 ± 0.4	5.3 ± 0.3	4.5	5.7
SAI (m^2 m^{-2})	1.3 ± 0.4	1.3 ± 0.2	–	–

[a] Sala (1992); Sabaté (1993).
[b] Lossaint and Rapp (1978); Cole and Rapp (1981).
[c] Escarré et al. (1987); Gracia (1983).
[d] Mean (± SE) values obtained from destructive sampling of vertical prisms. A lower average figure for the Avic and Teula catchments (6.6 Mg ha^{-1}, see Table 3.3) is obtained by dimensional analysis, which incorporates the effects of crown gaps.

tation is not lost to deep drainage and where deep soils may increase soil water storage. This is the case of the permanent plot La Castanya (located at a valley bottom) and the valley site of the Avic catchment (Table 9.1), where impermeable bedrock substrate prevents deep water drainage and increased soil depth improves soil water retention. At Avic, significant differences in total canopy LAI (including the understory) exist between the ridge and the valley sites (Sala et al. 1994) in spite of the fact that annual precipitation is the same at the two sites (Bellot 1989). Understory species are present in significant amounts (LAI = 0.7 ± 0.3) only at the valley site, where increased soil depth and runoff from upper slopes result in higher water availability. Across the four locations shown in Table 9.1, total leaf biomass on a ground area basis is not correlated with LAI. Leaf biomass of closed-canopy holm oak forests estimated by dimensional analysis applied to forest survey data lies usually in the range 5–7 Mg ha^{-1} (Tables 3.3 and 9.1) and seems to be not very sensitive to the mean annual precipitation of the site.

While spatial variations in leaf production and LAI are well known, temporal changes are much less documented. In Chapter 3, a mean leaf production of 2.3 Mg ha^{-1} year^{-1} is reported for Avic based on multi-year litterfall sampling. However, the new-leaf production of any given year is strongly affected by soil water availability, particularly during the period preceding the spring growth (Table 9.2). Thus, leaf production in a wet year (1988) was much higher than in a dry year (1989). These temporal changes clearly indicate the extent to which growth is limited by reduced water availability (see also Chap. 13). Moreover, during the same period, new leaf production was lower at the ridge site (more xeric) compared with the valley site. Nevertheless, differences between sites may be also increased due to differences in

Table 9.2. New-leaves production at Avic (ridge and valley sites). Means ± SE. Pre-growth rainfall (September–June) from Riudabella meteorological station is also given

Site	Current-leaves production[a] (Mg ha^{-1} year^{-1})		
	1988	1989	Mean
Ridge	2.9 ± 0.3	1.6 ± 0.6	2.2 ± 0.4
Valley	3.4 ± 0.1	2.5 ± 0.1	2.9 ± 0.2
Pre-growth rainfall (mm)	734	328	531

[a] Samples from summer 1988 and summer 1989 (n= 3 at each SITE and DATE). ANOVA probabilities: DATE P = 0.011, SITE P = 0.060.

nutrient availability, as reflected by the lower leaf N content at the ridge site (see below).

The number of standing leaf cohorts, and the amount of leaves in each of them, vary considerably from year to year depending on climatic conditions. In drier sites such as the Avic catchment, it is possible to find up to 4-year-old leaves, although the most abundant leaf age cohorts are the two youngest ones (1- and 2-year-old leaves). The relative amount of leaves of the two youngest cohorts varies considerably between years, with a relative increase in the amount of young leaves during wet years (i.e. years with high precipitation preceding the growth period) and a decrease during dry years (Sabaté 1993; and unpubl. data). Thus, in a wet year (1988) leaf turnover at Avic was 2.7 and 2.8 years at the ridge and valley sites, respectively, but in a dry year (1989) it increased to 5.2 and 3.7 years, respectively (S. Sabaté, unpubl. data). Somewhat lower but comparable values were reported by Lledó (1990) using different methods (Bellot et al. 1992; Lledó et al. 1992). Slow leaf turnover in holm oak, particularly during dry years, is characteristic of slow growing plants with parsimonious nutrient cycling (Chap. 18).

9.3 Vertical Profile Organization

Holm oak canopies are taller in valley bottom sites where water availability and total LAI is higher (Gracia 1983; Sala et al. 1994). At La Castanya and at the valley site of the Avic catchment canopies are between 8 and 12 m high. At the ridge site of Avic, canopies are much shorter, ranging between 3 and 6 m. The vertical distribution of LAI is typically unimodal, with a large proportion of leaf area in the upper meter of the canopy. At Avic, approximately 60% of the total leaf area is found in the uppermost meter of the canopy and more than 81% in the upper 2 m (Fig. 9.1). Despite large differences in canopy height between the ridge and valley sites of the Avic catchment, the vertical organization of the canopy is remarkably similar at both sites (Fig. 9.1),

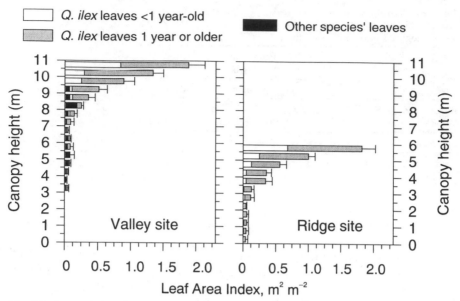

Fig. 9.1. Canopy vertical distribution of projected leaf area index (m² m⁻²) in holm oak forest at two sites of contrasted topographic position within the Avic catchment (Prades). Values are mean ± SE of 15 vertical columns 0.5 × 0.5 m on a side

and it is also similar to that found in holm oak stands at La Castanya (Gracia 1983) and at Le Rouquet (Eckardt et al. 1978). Interannual changes in water availability appear to have no effect on the vertical distribution of LAI (Sala et al. 1994).

According to Horn (1971) mono-layer canopies are adapted to shady environments and are more effective during late successional stages, while multi-layer canopies are more productive under open conditions and tend to appear in early successional stages. In the western Mediterranean Basin holm oak is a late successional species that, under sufficient water availability and in the absence of fire and herbivory, forms dense canopies that remain in a dynamic equilibrium. The tendency to accumulate most of the LAI in the upper canopy has also been related to major activity in carbon uptake occurring during periods of low solar angle (autumn, spring and during fair winter conditions; Sala et al. 1994). During these periods, the desiccation risk from low soil water availability and/or high evaporative demand is reduced, and the annual carbon balance may thus be optimised (Sala et al. 1994).

The proportion of young (less than 1-year-old) to old (1 year or older) leaves generally decreases from top to bottom of the canopy (Fig. 9.1). At Avic, young leaves contribute from 33% (ridge site) to 35% (valley site) of the total leaf area displayed in the uppermost meter of the canopy. Values decrease to 20 and 21% at the two sites, respectively, in the second meter and to 13 and 15% in the third meter. The tendency to accumulate young leaves at the top of the canopy suggests maximum leaf functional activity in higher canopy levels where irradiance does not limit photosynthesis (see below).

Total projected stem area on a ground area basis (SAI; excluding main trunks) was very similar at the two Avic sites (Table 9.1). SAI contributed ca. 19 and 23% of the total phytomass projected area (leaves and stems) at the ridge and valley sites, respectively. Thus, stems contribute significantly to light extinction within the canopy and should be taken into account in modelling efforts to estimate canopy gas exchange.

9.4 Morphological and Functional Leaf Traits

Leaves of holm oak are typically sclerophyllous, with thick cuticles and abundant sclerification. Short spines on the leaf margins are not uncommon. In general, sclerophyllous leaves are considered to be adaptive in dry, nutrient-poor (especially phosphorus-poor) environments due to their increased resistance to water loss, nutrient conservation mechanisms and herbivory protection (Dunn et al. 1976; Rundel 1988; Specht and Rundel 1990; Herms and Mattson 1992; Turner 1994; Feller 1995, 1996). The typical long dry summer period characteristic of Mediterranean areas imposes severe water limitations on growth. Growth is also limited by phosphorus, particularly on calcareous soils (Sardans 1997), and by nitrogen and potassium (Sabaté 1993). Herbivory is another important biotic factor that may strongly influence holm oak growth and production (Picolo and Terradas 1989). Thus, sclerophyllous leaves in holm oak appear to be a response to the combined pressures imposed by limited water and nutrients and by herbivory.

9.4.1 Leaf Size and Leaf Specific Mass

Mean area and dry weight of holm oak leaves are highly variable, ranging from 2 to 10 cm^2, and from 40 to 110 mg, respectively, or even less during extremely dry years (Mayor 1990; Sabaté et al. 1992; Sabaté 1993; Sala et al. 1994; and unpubl. data from Prades and Montseny). Leaf mass per unit of leaf area, or leaf specific mass (LSM), is an important trait strongly dependent on the incident radiation and water availability (Specht and Rundel 1990; Rambal et al. 1996). Because of the high sensitivity of LSM to incident radiation and water availability, LSM varies between sites and, particularly, within closed canopies where strong gradients of radiation occur (Hollinger 1989; Ellsworth and Reich 1993; Rambal et al. 1996). For instance, LSM of holm oak leaves at the top of the canopy is 22 mg cm^{-2} at Avic (Prades, xeric site) compared with 16-17 mg cm^{-2} in the permanent plot at La Castanya (Montseny, mesic site). However, within-canopy variations in LSM in response to incident radiation are much stronger than site-to-site variations resulting from changes in water availability. At Avic, for instance, LSM ranged from 22 mg cm^{-2} at the top of the canopy to 12 mg cm^{-2} at the bottom (Fig. 9.2) with a canopy-integrated mean of 18 mg cm^{-2}. The dependency of LSM on the incident radiation is reflected by a strong negative linear relationship between LSM and cumulative LAI (r = -0.91, P<0.01). Increases in LSM at the top of the canopy are associated with increases in structural compounds and with an increase in the number of layers of the palisade parenchyma (Wagner et al. 1993). As in many other species (Hollinger 1989; Ellsworth and Reich 1993), within-canopy changes in area-based functional traits of holm oak leaves (such as N content and photosynthetic capacity) are mostly a consequence of changes in LSM (Rambal et al. 1996; see below).

Temporal changes in LSM may also occur due to nutrient and carbon storage-mobilization processes in leaves (Sabaté 1993). Typically, LSM increases as leaves age and accumulate structural and storage compounds. However, decreases in LSM in older leaves prior to senescence may also occur (Sala et al. 1994) as nutrients and stored compounds are retranslocated to younger leaves or stems (Sabaté 1993).

In addition to radiation and water availability, nutrient availability may also affect leaf traits. Depending on bedrock type, LSM of holm oak has been shown to respond differently to different nutrients. For instance, in a fertilization experiment (Chap. 13) in the Prades holm oak forest, which lies on schist, LSM decreased and leaf area increased in response to increased nitrogen availability while there was no response to phosphorus addition (Sabaté 1993). In contrast, in another experiment in a different area on calcareous marls, phosphorus fertilization resulted in decreased LSM (Sardans

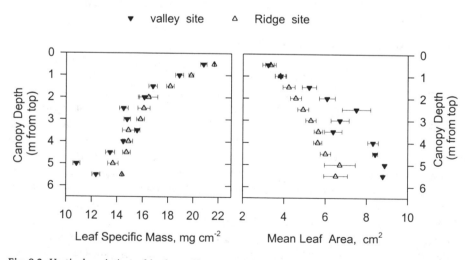

Fig. 9.2. Vertical variation of leaf specific mass (*left*) and mean leaf projected area (*right*) of mature holm oak leaves in the Avic catchment at ridge and valley sites. Values are mean ± SE; *n* = 15; same samples as Fig. 9.1

1997). These different responses may reflect a growth response to the most limiting nutrient in each soil type.

Concurrent with decreases of LSM with increased canopy depth, there are substantial increases in leaf size from top to bottom of the canopy. While leaf size at the top of the canopy is similar at the ridge and valley sites of Avic, leaves below 1 m from the top of the canopy are significantly larger at the valley site than at the ridge site (Fig. 9.2). These differences may be due, in part, to increased leaf growth at the valley site due to greater water availability compared to the ridge site. Stronger radiation extinction within canopies at the valley site (due to greater LAI and the presence of understory) may also contribute to increased leaf size at lower canopy levels at this site. Small leaf size in the upper canopy layers where LAI is greater allows for efficient overall direct light capture while it improves light penetration deeper in the canopy. Radiation capture at the bottom of the canopy, where only 10% of the incident radiation is available (Chap. 11), is facilitated by exposing large surface areas.

In Mediterranean climates, where periods of greatest evaporative demand coincide with periods of reduced soil water availability, changes in LSM and leaf size within the canopy have adaptive value. For instance, the decrease in leaf size at the top of the canopy confers a mechanism to improve thermoregulation and heat dissipation during summer drought (Gates 1980). Improved thermoregulation may significantly contribute to leaf cooling, which, in turn, may prevent tissue mortality due to excessive leaf temperatures. Similarly, higher LSM at the top of the canopy may be advantageous to reduce water loss by transpiration during the summer, because increases in LSM are associated with increased resistance to water loss (Specht 1988; Specht and Specht 1989). However, increased LSM is also associated with increased resistance to diffusion of CO_2, thus posing a potential limitation to carbon uptake. Relatively high leaf stomatal density at the top of the canopy (Table 9.3) may compensate for the negative effects of increased LSM and improve CO_2 uptake during favourable periods. Stomata of holm oak are highly sensitive to atmospheric and soil drought and strong stomatal closure prevents excessive water loss under high evaporative demand and/or low soil moisture, when the risk of desiccation is higher (Sala and Tenhunen 1994; Chaps. 10 and 11).

Table 9.3. Stomatal and hair density in holm oak leaves from top to bottom of the canopy in the permanent plot at La Castanya, Montseny (Savé 1986)

Canopy depth from top (m)	Stomatal density (no. mm^{-2})	Hair density (no. mm^{-2})
0	135.5	108.4
2	108.1	101.8
4	98.5	85.4
6	76.8	12.5

Leaf pubescence in holm oak (hairs are only present in the abaxial face) increases from the bottom to the top of the canopy (Table 9.3). Increases in leaf pubescence may result in increases in the thickness and humidity of the leaf boundary layer, thus reducing the leaf to air water vapour pressure gradient and minimizing water loss by transpiration. Although increased water use efficiency appears to be a direct consequence of increased pubescence, the direct link between leaf pubescence and leaf gas exchange has never been accurately tested.

9.4.2 Chlorophyll Content

Chlorophyll content on a leaf area basis changes considerably within holm oak canopies. In the upper 3 m of the canopy at La Castanya, leaf chlorophyll content is inversely related to the incident photosynthetically active radiation (PAR) (Gracia 1983). Chlorophyll contents range from 600–700 mg m^{-2} at the top of the canopy to a maximum of approximately 1000 mg m^{-2} at 3 m below the top (Fig. 9.3). Below this point chlorophyll decreases in parallel with decreases of incident PAR to values between 400 and 200 mg m^{-2} (Fig. 9.3). Gracia (1983) reported very similar patterns of leaf-area-based chlorophyll content in beech (*Fagus sylvatica*) while Ellsworth and Reich (1993) did not find within-canopy variation in *Acer saccharum*.

At La Castanya, shaded leaves are relatively enriched with chlorophyll b: the chlorophyll a/b ratio decreases from 2.66 at the top of the canopy to 2.24 at 9 m below the top (Gracia 1983). The enrichment in chlorophyll b is due to the extensive stacking of grana in chloroplasts of shaded leaves. These stacks

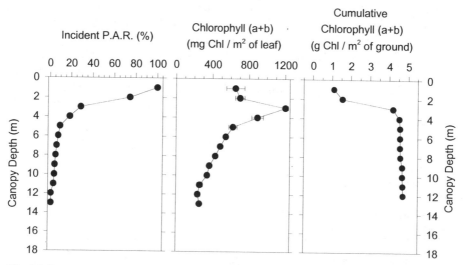

Fig. 9.3. Percentage of PAR (*left*), chlorophyll content on a leaf (*middle*) and ground (*right*) area basis with increased canopy depth at La Castanya (Montseny) holm oak forest

contain a high concentration of the light-harvesting complex II, where most of the chlorophyll b is found (Anderson 1986).

In holm oak, the patterns of leaf-area-based chlorophyll content within the canopy appear to be a balance between the optimal use of resources and the maximization of canopy carbon input. Leaf photosynthetic activity in the top 3 m of the canopy is high due to adequate light and relatively high N content (see below). Within this active portion of the canopy, where more than 90% of the total canopy LAI is found (Fig. 9.1), PAR levels decrease rapidly and leaf chlorophyll concentration increases (Fig. 9.3). This is a typical plastic response to increase light absorption in the shade, where photosystems are no longer light saturated. It appears, then, that at PAR levels above 30% of maximum incident PAR, investments in chlorophyll in leaves located up to 3-m deep into the canopy result in a positive photosynthetic return. However, below 3 m, where PAR levels decrease below 30% of the maximum, investment in chlorophyll is no longer advantageous, probably as a result of excessively low light levels. Most leaves already absorb 80–85% of the available light and a further doubling of chlorophyll concentration only increases absorption by 3 to 6% (Evans 1988).

9.4.3 Nutrient Content

Holm oak leaves are important nutrient sinks and have a relatively high nutrient concentration compared to other tree components (Chap. 18). Seasonal trends of leaf-area-based N and P concentrations of a single age cohort of holm oak leaves at different canopy depths are shown in Fig. 9.4 for the ridge and valley sites of the Avic catchment.

Leaf N content is higher at the valley site than at the ridge site. Higher N content at the valley site may be related to higher total soil nutrient content at the former site due to increased soil depth (Sabaté et al. 1995). N and P content differences between both sites of Avic are larger when concentrations are expressed on a dry weight basis (Sabaté et al. 1995). Presumably leaf tissue at the ridge site (more xeric) has a higher content of structural compounds with little or no N and P than at the valley. Leaf K concentration is higher at the ridge site than at the valley site on both a leaf area and a dry weight basis (Sabaté 1993; Sabaté et al. 1995). As K plays a key role in osmotic processes and stomatal control of water loss (Marschner 1995) it is not surprising to find higher K contents at the more xeric ridge site. This is indeed consistent with the fact that trees at this site exhibit a higher resistance to water stress compared with trees at the valley bottom (Sala et al. 1994).

Leaf-area-based N and P contents decrease significantly from top to bottom of the canopy (Fig. 9.4). However, on a dry weight basis, differences are very small (Sabaté et al. 1995), indicating that changes in area-based nutrient content result from the strong decrease of LSM from the top to the bottom of the canopy. As increases in N content are related to increases in photosynthetic capacity (Field and Mooney 1986; Evans 1989; Reich et al. 1992; Rambal

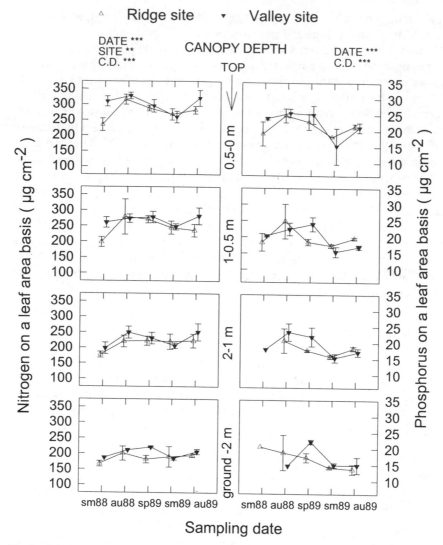

Fig. 9.4. Nitrogen and phosphorus content on a leaf area basis at the Avic catchment (ridge and valley sites) at different canopy depths. Analyzed leaves were pooled samples (by canopy depth) of those shown in Figs. 9.1 and 9.2. Differences between sampling dates (*DATE*), sites (*SITE*) and canopy depth (*C.D.*) are indicated above the *panels* (* $P < 0.05$, * * $P < 0.01$ and *** $P < 0.001$). Values are mean ± SE of leaves produced in spring 1988 and sampled in summer (sm), autumn (au), and spring (sp) of 1988 and 1989; $n = 3$

et al. 1996), high N content in the upper canopy (where LAI is greater) suggests that canopy carbon input is maximized where radiation is not limiting photosynthesis. The decrease in N content with canopy depth is consistent with the decrease in leaf chlorophyll content (Sect. 9.4.2), and lower new leaf production in the lower canopy layers (Fig. 9.1).

At Avic, leaf-area-based N and P contents tend to be lower in summer than in autumn or winter (Fig. 9.4) This is a common temporal trend that has been documented in other holm oak forests (see De Lillis and Fontanella 1992). The lower leaf N and P contents during summer suggest that during the active growth period in late spring, N and P are retranslocated from leaves to new growth and that during the autumn, when growth activity is low and there is no water stress, leaves exhibit N and P replenishment (see also Chap. 18; Chabot and Hicks 1982; Nambiar and Fife 1987; Chapin 1988). Temporal trends suggesting withdrawal and resorption are more apparent in the uppermost canopy layer than at lower canopy layers, where growth activity is limited by light. Escudero et al. (1992) did not find temporal changes of N content in holm oak as resorption processes. However, their results were expressed on a dry weight basis, which makes it difficult to distinguish whether changes in leaf nutrient concentrations are simply due to dilution–concentration effects (i.e. relative increases or decreases of nutrient-poor carbon compounds) or to resorption of mineral nutrients. In particular, resorption of mineral nutrients may not affect foliar nutrient concentrations on a dry weight basis if non-structural carbohydrates are also being resorbed at a similar rate.

In conclusion, holm oak canopies exhibit a typically unimodal leaf area distribution, with most of the leaf area accumulated in the upper canopy. As a consequence of this skewed leaf area distribution and the radiation extinction patterns associated with it, strong gradients in leaf morphological and functional traits are found from the top to the bottom of the canopy. Leaves at the top of the canopy are smaller, have higher stomatal and hair densities, higher LSM and higher area-based nutrient content than leaves at the bottom of the canopy. Holm oak leaves appear to be very plastic depending on the environmental conditions (mainly radiation and water availability) in which they develop. Overall, holm oak exhibits many characteristics of a stress tolerant, slow-growing species with parsimonious carbon and nutrient cycling (Chaps. 13 and 18).

References

Anderson JM (1986) Photoregulation of the composition, function and structure of thylakoid membranes. Annu Rev Plant Physiol 37:93–136

Bellot J (1989) Análisis de los flujos de deposición global, trascolación, escorrentía cortical y deposición seca en el encinar mediterráneo de l'Avic (Sierra de Prades, Tarragona). PhD Thesis, University of Alicante, Alicante

Bellot J, Sánchez JR, Lledó MJ, Martinez P, Escarré A (1992) Litterfall as a measure of primary production in Mediterranean holm-oak forest. Vegetatio 99/100:69–76

Chabot BF, Hicks DJ (1982) The ecology of leaf life spans. Annu Rev Ecol Syst 13:229–259

Chapin III FS (1988) Ecological aspects of plant mineral nutrition. Adv Mineral Nutr 3:161–191

Cole DW, Rapp M (1981) Elemental cycling in forest ecosystems. In: Reichle D (ed) Dynamic properties of forest ecosystems. Cambridge University Press, Cambridge, pp 341–409

De Lillis M, Fontanella A (1992) Comparative phenology and growth in different species of Mediterranean maquis of central Italy. Vegetatio 99/100:83–96

Dunn TA, Shropshire FM, Song LC, Mooney HA (1976) The water factor and convergent evolution in Mediterranean-type vegetation. In: Lange OL, Kappen L, Schulze ED (eds) Water and plant life. Springer, Berlin, pp 492–505

Eckardt FE, Berger A, Méthy M, Heim G, Sauvezon R (1978) Interception de l'énergie rayonnante, échanges de CO_2, régime hydrique et production chez différents types de végétation sous climat méditerranéen. In: Moyse A (ed) Les processus de la production végétale primairie. Gauthier Villars, Paris, pp 1–75

Ellsworth DS, Reich PB (1993) Canopy structure and vertical patterns of photosynthesis and related leaf traits in a deciduous forest. Oecologia 96:169–178

Escarré A, Ferrés Ll, López R, Martín J, Rodà F, Terradas J (1987) Nutrient use strategy by evergreen-oak (Quercus ilex ssp. ilex) in NE Spain. In: Tenhunen JD, Catarino FM, Lange OL, Oechel WC (eds) Plant responses to stress. Functional analysis in Mediterranean ecosystems. Springer, Berlin, pp 429–435

Escudero A, del Arco JM, Garrido MV (1992) The efficiency of nitrogen retranslocation from leaf biomass in Quercus ilex ecosystems. Vegetatio 99/100:225–237

Evans JR (1988) Acclimation by the thylakoid membranes to growth irradiance and the partitioning of nitrogen between soluble and thylakoid proteins. Aust J Plant Physiol 15:93–106

Evans JR (1989) Photosynthesis and nitrogen relationships in leaves of C_3 plants. Oecologia 78: 9–19

Feller IC (1995) Effects of nutrient enrichment on growth and herbivory of dwarf red mangrove (Rhizophora mangle). Ecol Monogr 65:477–505

Feller IC (1996) Effects of nutrient enrichment on leaf anatomy of dwarf Rhizophora mangle L. (red mangrove). Biotropica 28:13–22

Field C, Mooney HA (1986) The photosynthesis-nitrogen relationship in wild plants. In: Givnish T (ed) On the economy of plant form and function. Cambridge University Press, London, pp 25–55

Gates DM (1980) Biophysical ecology. Springer, Berlin

Gracia CA (1983) La clorofila en los encinares del Montseny. PhD Thesis, University of Barcelona, Barcelona

Hollinger DY (1989) Canopy organization and foliage photosynthetic capacity in broad-leaved evergreen montane forest. Funct Ecol 3:53–62

Horn HS (1971) The adaptive geometry of trees. Princeton University Press, Princeton

Herms DA, Mattson WJ (1992) The dilemma of plants to grow or to defend. Q Rev Biol 67: 283–335

Lledó MJ (1990) Compartimentos y flujos biogeoquímicos en una cuenca de encinar del Monte Poblet. PhD Thesis, University of Alicante, Alicante

Lledó MJ, Sánchez JR, Bellot J, Boronat J, Ibáñez JJ, Escarré A (1992) Structure, biomass and production of a resprouted holm-oak (Quercus ilex L.) forest in NE Spain. Vegetatio 99/100: 51–59

Lossaint P, Rapp M (1978) La fôret méditerranéenne de chênes verts (Quercus ilex L.). In: Lamotte M, Bourlière F (eds) Problémes d'écologie: structure et fonctionnement des écosystèmes terrestres. Masson, Paris, pp 129–185

Marschner (1995) Nutrition of higher plants, 2nd edn. Academic Press, London

Mayor X (1990) El paper dels nutrients com a factors limitants de la producció primària de l'alzinar de la conca del Torrent de la Mina (Montseny). MSc Thesis, Autonomous University of Barcelona, Bellaterra

Nambiar EKS, Fife DN (1987) Growth and nutrient retranslocation in needles of radiata pine in relation to nitrogen supply. Ann Bot 60:147–156

Picolo R, Terradas J (1989) Aspects of crown reconstruction and leaf morphology in Quercus ilex L. and Quercus suber L. after defoliation by Lymantria dispar L. Acta Oecol Oecol Plant 10:69–78

Poole DK, Miller DC (1981) The distribution of plant water-stress and vegetation characteristics in southern California chaparral. Am Mid Nat 105:32–43

Rambal S, Damesin C, Joffre R, Méthy M, Lo Seen D (1996) Optimization of carbon gain in canopies of Mediterranean evergreen oaks. Ann Sci For 53:547–560

Reich PB, Walters MB, Ellsworth DS (1992) Leaf life-span in relation to leaf, plant, and stand characteristics among diverse ecosystems. Ecol Monogr 63:365–392

Rundel PW (1988) Leaf structure and nutrition in Mediterranean-climate sclerophylls. In: Specht RL (ed) Mediterranean-type ecosystems. A data source book. Kluwer, Dordrecht, pp 157–167

Sabaté S (1993) Canopy structure and nutrient content in a *Quercus ilex* L. forest of Prades mountains: effects of natural and experimental manipulation of growth conditions. PhD Thesis, University of Barcelona, Barcelona

Sabaté S, Calvet S, Gracia CA (1992) Preliminary results of a fertilization–irrigation experiment in a *Quercus ilex* L. forest in relation to leaves and twigs characteristics. Vegetatio 99/100: 283–287

Sabaté S, Sala A, Gracia CA (1995) Nutrient content in *Quercus ilex* canopies: seasonal and spatial variation within a catchment. Plant Soil 168/169:297–304

Sala A (1992) Water relations, canopy structure, and canopy gas exchange in a *Quercus ilex* forest: variation in time and space. PhD Thesis, University of Barcelona, Barcelona

Sala A, Tenhunen JD (1994) Site-specific water relations and stomatal response of *Quercus ilex* L. in a Mediterranean watershed. Tree Physiol 14:601–617

Sala A, Sabaté S, Gracia C, Tenhunen JD (1994) Canopy structure within a *Quercus ilex* forested watershed: variations due to location, phenological development and water availability. Trees 8:254–261

Sardans J (1997) Resposta de quatre espècies llenyoses mediterrànies a diferent disponibilitat d'aigua i nutrients. PhD Thesis, Autonomous University of Barcelona, Bellaterra

Savé R (1986) Ecofisiologia de les relacions hídriques de l'alzina al Montseny (La Castanya). PhD Thesis, Autonomous University of Barcelona, Bellaterra

Specht RL (1988) Vegetation, nutrition and climate –examples of integration. (2) Climatic control of ecomorphological characters and species richness in Mediterranean ecosystems of Australia. In: Specht RL (ed) Mediterranean-type ecosystems. A data source book. Kluwer, Dordrecht, pp 149–155

Specht RL, Rundel PW (1990) Sclerophylly and foliar nutrient status of Mediterranean-climate plant communities in southern Australia. Aust J Bot 38:459–474

Specht RL, Specht A (1989) Canopy structure in *Eucalyptus*-dominated communities in Australia along climatic gradients. Oecol Plant 10:191–213

Turner IM (1994) Sclerophylly: primarily protective? Funct Ecol 8:669–675

Wagner J, Pelaez Menendez S, Larcher W (1993) Bioclima e potenziale di produttività di *Quercus ilex* L. al limite settentrionale dell'areale di distribuzione. Parte III. adattamento morfologico e funzionale delle foglie alle radiazioni luminose. Studi Trentini Sci Nat 68:37–51

10 Gas Exchange and Water Relations

Robert Savé, Carles Castell and Jaume Terradas

10.1 Introduction

Mediterranean environments are often characterized by a double stress: summer drought and winter cold (Mitrakos 1980; Miller 1981; Terradas and Savé 1992). Summer drought results from the coincidence of low summer precipitation with high temperature, high irradiance, and high water vapour pressure deficit (Di Castri and Mooney 1973); it has been traditionally recognized as the main climate constraint characterizing Mediterranean-type ecosystems. However, some degree of stress can be also due to winter cold, which may be determinant in montane and/or continental sites. This is reflected in the seasonal patterns of plant activity. Photosynthetic activity is typically relatively high in spring, decreases strongly in summer due to drought (Oechel et al. 1981, Tenhunen et al. 1990), increases again after the first autumn rains, and decreases or ceases during the winter months. Plant growth and leaf transpiration may follow similar patterns.

Effects of drought and cold stresses on the water relations of holm oak (*Quercus ilex* L.) have been long studied (Larcher 1960, 1981; Kyriakopoulos and Larcher 1976; Kyriakopoulos and Richter 1977; Lossaint and Rapp 1978; Savé 1986; Comín et al. 1987; Rambal 1992; Castell et al. 1994; Sala and Tenhunen 1994, 1996). Compared to temperate deciduous trees, holm oak follows a conservative strategy with low rates of gas exchange per unit of leaf area. Stomatal regulation is considered to be efficient in this species (Larcher 1960). However, under moderate water stress, holm oak maintains higher stomatal conductance for longer periods than deciduous oaks (Acherar and Rambal 1992) and than shallow-rooted Mediterranean woody species. So, transpiration continues and low xylem water potentials develop. High capacity to adjust stomatal conductance and to maintain low water potentials, combined with deep roots (Chap. 4) allowing access to deep soil water, are the clues to holm oak responses to water availability.

In this chapter we analyze the ways in which holm oak copes with this double stress. We first deal briefly with the adaptive meaning of sclerophylly.

Ecological Studies, Vol. 137
Ferran Rodà et al. (eds) Ecology of Mediterranean
Evergreen Oak Forests
© Springer-Verlag, Berlin Heidelberg 1999

We then summarize two of our own studies in NE Spain that illustrate two revealing situations: water relations of holm oak in a mesic site, and the eco-physiology of stool sprouts. We go on to address some of the major mechanisms involved, and finally consider the ecosystem and biogeographical implications of the observed behaviour.

10.2 Sclerophylly and Water Stress

The double-stress climate is associated with some recurrent or convergent plant characteristics in the five Mediterranean-climate regions of the world. Among these traits, sclerophylly has received much attention (e.g. for holm oak: Ilijanic and Gracanin 1972; Larcher 1960; Aussenac and Valette 1982; Salleo et al. 1997). Sclerohyllous leaves, such as those of holm oak, are useful to minimize water losses and to resist those losses that do occur. Holm oak leaves can tolerate leaf water saturation deficits of up to 37% (Larcher 1960), which are lethal to most mesophytic trees. However, sclerophylly is not a universal feature in Mediterranean-climate plants, neither is it restricted to them. In Mediterranean-climate areas, sclerophylly is found in isomorphic evergreen plants, such as holm oak, but, when the intensity and length of summer drought increase, heteromorphic evergreen and even summer-deciduous plants replace the isomorphic sclerophylls (Margaris 1981). Then, the advantages of sclerophyllous leaves are probably restricted to a limited portion of these Mediterranean-climate areas. In addition, the main advantage of sclerophylly for holm oak probably does not lie in reducing water losses since its stomata remain open much longer than in deciduous oaks (Salleo and Lo Gullo 1990). So, holm oak is not a drought-avoider but a drought-tolerant species. Furthermore, sclerophylly can also be an evolutionary response to low nutrient availability (Loveless 1961) and to herbivory. In fact, sclerophylly can result from selection for increased leaf longevity under situations of resource shortage (Turner 1994). Once acquired, sclerophylly can serve a protective function useful in facing a variety of abiotic and biotic stresses.

10.3 Transpiration and Water Status of Leaves in a Mesic Site

Our first study in the mesic permanent plot at La Castanya in the Montseny experimental area (Chap. 2) yielded information on the patterns of water use of holm oak growing at relatively high soil water availability (Savé 1986). This plot lies at the base of a steep NW-facing slope where the trees benefit from a deep soil, lateral soil water inputs, relatively high rainfall, and reduced evaporative demand through topographic shading. Even under these favourable conditions, results show that water-loss behaviour in holm oak is intermediate between those of xerophytic and mesophytic plants. Daily leaf-transpiration rates, measured in this plot by the quick-weigh method, are above those of conifers, but well below those of some Mediterranean sclerophyllous broadleaved trees (e.g. *Pistacia lentiscus, Laurus nobilis* or *Phillyrea latifolia*; Ilijanic and Gracanin 1972), and nearly 50% of those found on average in a beech forest at Montseny (Savé 1986). Transpiration rates of holm oak fluctuate seasonally, being on average 44% lower at La Castanya in winter than in summer. Transpiration declines significantly when soil temperature decreases in autumn. Cuticular transpiration rates at La Castanya are on average between 2 and 8% of maximum stomatal transpiration. This confirms the significance of the thick cuticles of sclerophyllous leaves in reducing water loss. In contrast, cuticular transpiration rates are much higher in mesophytic, deciduous leaves: e.g. 34% in *Quercus humilis* and 32% in *Fagus sylvatica* (Larcher 1960; Rabella 1991).

Stomatal conductances measured with a steady state porometer by Savé (1986) in the permanent plot at La Castanya are within the range given by Berger et al. (1977) for holm oak ($0.02-0.2$ cm s^{-1}). Total stomatal closure was not frequent in this site, where we generally found only midday stomatal regulation. This may be due to the favourable water balance in this mesic plot. In southern France, holm oak shows an earlier decline of transpiration in the morning, and longer periods of stomatal closure (Berger et al. 1977) than those we observed in our permanent plot.

10.4 Within-Canopy Variations

Variations along the vertical canopy profile in leaf morphology and leaf age structure (Chap. 9) and in canopy microclimate promote significant changes in leaf transpiration rates. On average, sun leaves in the permanent plot at La Castanya show transpiration rates 19% higher than shade leaves. Maximum daytime conductance is usually found at the top of the canopy (10 m above the ground; Savé 1986). However, in summer stomatal conductance decreases at midday in sun leaves, and a peak of conductance appears at 8 m (Fig. 10.1). These differences arise in part because stomatal regulation does not proceed simultaneously over the whole canopy, resulting in large variations in stomatal conductance, water content and gas exchange along vertical profiles, mainly when the canopy is under high irradiance (Rabella et al. 1983). In contrast, much smaller within-canopy differences have been found in cork oak (*Quercus suber* L.), a closely related species but having a sparser crown (Oliveira et al. 1992; Oliveira 1995).

The above patterns must be interpreted in terms of the morphological and functional traits of sun leaves that increase their resistance to stress. Furthermore, sun leaves maintain shaded canopy leaves under a lower stress

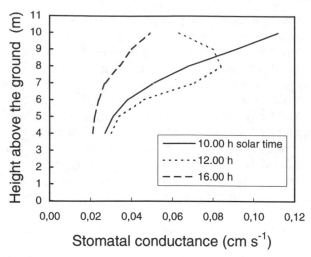

Fig. 10.1. Generalized patterns of stomatal conductance of holm oak leaves from several canopy levels on clear summer days, in the permanent plot at La Castanya (Montseny)

level. Thus, at midday, when evaporative demand is highest, hydroactive stomatal regulation and low light interception caused by the vertical orientation of upper leaves (Morris 1989; Burriel et al. 1993) promote a decrease in stomatal conductance and transpiration rates in the top layer. These midday depressions in transpiration and assimilation rates should not be interpreted as photoinhibition, because recovery of gaseous exchange occurs in the afternoon (Correia et al. 1990). A secondary peak in transpiration may occur in the subtop layer, and the tree can continue to photosynthesize. Thus, top leaves, due to their xeromorphic characteristics, act as an epidermic surface that protects the rest of the canopy (Savé 1986).

Leaf water saturation deficit during the day is almost always highest in the topmost leaves. Canopy-averaged values at La Castanya are usually around 13.5%. According to Larcher (1960, 1980), leaf water saturation deficit of 8% is optimal for net photosynthesis in holm oak, 12.5% corresponds to the highest water use efficiency, 17% is the limit for absolute stomatal closure, and 37% is the sublethal level.

Xylem water potential decreases with increasing height in the canopy. Absolute minima observed by Savé (1986) at La Castanya during five summer days (which included the hottest day for 100 years, 7 July 1982) were only –2.2 MPa at the top of the canopy (10 m), and –1.5 MPa at the bottom of the canopy (4 m). Mean maximum values were –0.70 ± 0.27 MPa for sun levels (8 and 10 m) and –0.56 ± 0.23 MPa for shade levels (4 and 6 m). Mean minima were –1.6 3± 0.21 MPa and –1.16 ± 0.11 MPa for sun and shade levels, respectively. Much lower extreme values, around –4 MPa, have been found for holm oak in the south-facing slopes at La Castanya (Savé, unpubl. data), at Prades (Sala and Tenhunen 1996) and in other areas (Rambal and Debussche

1995). In holm oak, the critical predawn xylem potential for stomatal closure in summer is –3.0 to –3.5 MPa (Aussenac and Valette 1982).

10.5 Ecophysiology of Stool Sprouts

Holm oak forests in NE Spain are frequently thinned or burned, and holm oak resprouts vigorously after disturbance (Chap. 5). During much of the between-disturbance period, holm oak forests are made up of young stool sprouts, whose ecophysiological responses may differ from those of adult trees. In particular, the much larger root to shoot ratio of sprouts, which rely upon the pre-existing root system of the stool, results in a much larger availability of soil resources per unit of leaf area in sprouts compared with mature trees. This can greatly modify the ecophysiological behaviour of holm oak. To test this, water relations and gas exchange in mature holm oak and in young sprouts produced by experimentally clipped stools were studied at Collserola by Castell et al. (1994). This site lies in the coastal range near Barcelona, and it has a drier and warmer climate than the permanent plot at Montseny. Shoot elongation of mature holm oak occurs mainly in spring, and stops in mid-August, while few individuals show a new flush in autumn. This pattern seems closely related to the decrease in predawn water potential during summer drought (Fig. 10.2). In contrast, midday water potential changes very little, remaining above –3.0 MPa, the potential below which extensive cavitation occurs (Lo Gullo and Salleo 1993; see Sect. 10.8). Similar patterns of gas exchange in summer have been found for *Quercus suber* in Portugal under comparable climate conditions (Oliveira 1995).

In summer, mature holm oaks at Collserola maintain a relatively high xylem water potential due to a 50% reduction in stomatal conductance from spring values. At the same time, net photosynthesis is reduced by ca. 85% (Fig. 10.2). These differences suggest some inhibitory effects of high light intensities at the chloroplast level. After the first autumn rains, xylem water potential, leaf conductance and net photosynthesis in mature trees quickly recover to the spring levels (Fig. 10.2). Although the rates of net photosynthesis are fairly high in autumn, there is no or low shoot elongation at this time, probably because production is dedicated to ripening fruits, as happens in other species (Bazzaz et al. 1987). Winter cold does not seem to affect markedly holm oak ecophysiology in this area (mean December temperature is 7 °C), and net photosynthesis is quite high in sunny winter days.

Seasonal patterns of water relations, gas exchange and growth have also been evaluated for young sprouts. In spring, the elongation rates of sprouts are much higher than those observed in mature plants, then they decrease in August, and vigorously increase again in autumn. High sprout growth rates are closely related to their greater relative water availability, since the pre-existing root system gives them a high root to shoot ratio. High water poten-

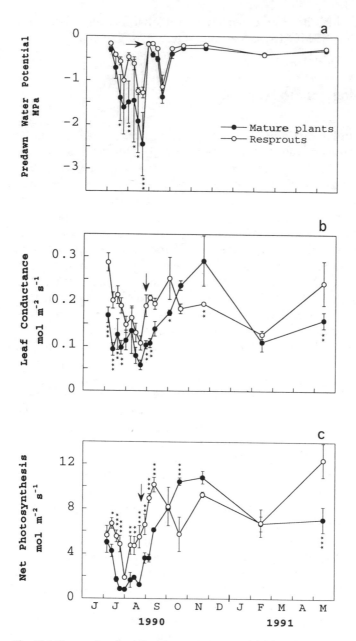

Fig. 10.2. Seasonal course of predawn water potential (a) and mean values during the light saturation phase of the day for leaf conductance (b) and net photosynthesis (c) of mature plant and
sprout leaves of holm oak from June 1990 to May 1991. Conductance and photosynthesis were
measured with a portable ADC IRGA-porometer. Mean values for sprouts versus mature plant
shoots were compared by repeated measures analysis of variance (significant differences on a
particular date are indicated as * $P < 0.05$, ** $P < 0.01$, *** $P<0.001$; $n = 3$). *Arrow* indicates the
first autumn rain greater than 10 mm

tial in holm oak sprouts allows them to maintain high leaf conductance and high photosynthesis rate during the summer, with a less conservative use of water than mature plants (Fig. 10.2). However, the observed increase in leaf conductance and net photosynthesis may not be fully explained just by the increased leaf water potential, and enhanced nutrient availability might be involved in this response (Fleck et al. 1996). Growth of sprouts can also be fostered by the mobilization of carbohydrates from burls and large roots. The morphological and physiological traits exhibited by young holm oak sprouts are involved in the fast recovery of the leaf area index and canopy carbon and water fluxes after disturbance (Leonardi and Rapp 1990; Chap. 23).

Seasonal courses of gas exchange in potted holm oak seedlings (Espelta 1996; Chap. 7) are similar to those found in mature plants and very different from those observed in sprouts, thus confirming the role of the root to shoot ratio in determining the water status and gas exchange rates in holm oak.

10.6 Drought-Tolerance Mechanisms

Pressure-volume curves (Tyree and Richter 1981, 1982) reveal that sun leaves are more drought-tolerant than shade leaves. Osmotic potential at full turgor ($\Psi_{\pi100}$), osmotic potential at zero turgor ($\Psi_{\pi0}$) and volumetric modulus of elasticity (ε) of 1-year-old leaves are -1.49, -1.82 and 7.31 Mpa, respectively, in the upper canopy and -0.98, -1.35 and 6.64 MPa, respectively, in the lower canopy of the permanent plot at La Castanya (Terradas and Savé 1992). These data agree with the results reviewed by Morgan (1984), and indicate the different capacity of sun and shade leaves for osmotic adjustment. However, the relatively high values of $\Psi_{\pi0}$ and the relatively low values of ε seem to indicate that holm oak shows higher capacity for elastic than osmotic adjustment as a tolerance mechanism to drought stress (Salleo et al. 1997). A moderate osmotic adjustment under drought conditions (Kyriakopoulos and Richter 1991) may also contribute to drought tolerance, but this is only possible by maintaining gas exchange at relatively low levels.

10.7 Effects of Temperature on Plant Water Relations

Since water availability is a key factor in holm oak performance, many ecological factors can affect holm oak through their effects on water relations. We will try here to analyze the indirect effects of temperature through changes in ecophysiological parameters related to water relations. Sun and shade leaves show significant differences in their responses to high temperatures, reaching the turgor loss point at 33 and 26 °C, respectively. In sun leaves, transpiration rate increases with increasing temperatures up to 33 °C, while stomatal resistance does not show significant changes. Above ca. 35 °C, the transpiration rate decreases while the stomatal resistance increases sharply. Shade leaves show a similar behaviour, although the change of tendency in the transpiration rate and in the stomatal resistance occurs at air temperatures around 25 °C (Savé 1986).

In winter, the xylem water potential can reach very low values in holm oak. Predawn values of less than −4.0 MPa were measured at Prades in a ridge-top plot and near −2.0 MPa in a valley-bottom plot (Sala et al. 1990). During the day, the xylem potential increases to between −1.0 and −2.2 MPa. At Montseny, the minimal values of winter xylem potential of holm oak decrease along an altitude gradient from 350 to 1130 m a.s.l. (Savé et al. 1982, 1988). Electrolyte leakage and cuticular transpiration rate of the trees growing at the highest altitude are 150 and 90% higher, respectively, than those of trees growing at the lowest elevation. These changes agree with the deleterious effects of frost drought stress described by Tranquillini (1982) on plant growth and development.

10.8 Xylem Embolism

Xylem embolism has been recognized as a serious cause of lowered plant productivity when plants are submitted to abiotic (drought, cold, freeze) or biotic (tracheomycoses, fungus, nematodes) stress (Gartner 1995; Sperry 1995). Observations made at Prades and elsewhere in the Catalan pre-coastal range during the extreme drought of 1994 suggest that, in comparison with other Mediterranean tree species, holm oak could be quite sensitive to embolism, since it suffered heavy crown withering, while *Phillyrea latifolia* and *Pinus halepensis* fared much better. Plants with smaller xylem conduits develop cavitation at lower water potentials (Tyree and Dixon 1986). Thus, despite their lower efficiency in water conduction, short and narrow xylem conduits would be advantageous to plants exposed to xylem embolism (Zimmerman 1983; Lo Gullo et al. 1995). Compared to Euro-Siberian deciduous oaks, holm oak has small xylem conduits: in 1-year-old internodes, 35% of the conduits have diameters < 20 μm, and only 20% have diameters > 40 μm (Lo Gullo and Salleo 1993). However, there is evidence that vulnerability to embolism is determined by the pit-membrane pore diameter rather than by the conduit diameter (Zimmerman 1983). This agrees with the "air-seeding" hypothesis (Sperry and Tyree 1990), which proposes that gaseous emboli would enter xylem conduits through the pit-membrane pores, whenever the pressure differences across them are high enough to make the radius of the air-water meniscus smaller than that of the pore.

Embolism can occur in summer under drought stress. When holm oak plants dried under a mild water stress [a value of about 0.9 in the ratio (leaf water potential)/(turgor loss point water potential)], only the conduits wider than 30 μm were embolised and the resultant losses in hydraulic conductivity were about 30%. At this stage, recovery was easy with a moderate water supply. With increasing water stress, losses in hydraulic conductivity reached 85% of the initial value, and the percentage of recovery fell dramatically (Lo Gullo and Salleo 1993). In the permanent plot at La Castanya, the ratio (leaf water potential)/(turgor loss point water potential) was usually close to 0.9, indicating a low probability of embolism due to drought stress. At higher altitudes (above 1000 m a.s.l.) this ratio increased to 1.9–2.5, which can explain the deleterious effects of drought stress on oak growth and survival (Tyree and Dixon 1986).

Xylem embolism can occur in winter as well. Holm oak plants exposed to an air temperature of −2.5 °C showed a 50% loss of hydraulic conductivity, a loss that increased to 94% at −11 °C (Sakai and Larcher 1987). The recovery rate was negatively correlated with the intensity and duration of exposure to cold. The above temperatures are not uncommon throughout the geographical range of holm oak (Terradas and Savé 1992) and they can limit the growth and survival of holm oak at high altitudes.

10.9 Canopy Transpiration and Ecosystem Water Budgets

Annual water consumption by entire forests is difficult to determine directly. From water budgets in our small experimental catchments at Montseny and Prades, total evapotranspiration in dense holm oak forests is estimated to average 500–600 mm year^{-1} (Chap. 19). This estimate not only refers to canopy transpiration but also includes interception by the canopy (Chap. 15), interception by the forest floor, and direct evaporation from the soil. The model described in Chapter 11 predicts a canopy transpiration of 450–460 mm for 1989 (a dry year) in two plots of contrasted topographic position within the Avic catchment at Prades. On a site with high soil water availability, such as the mesic permanent plot at La Castanya, integration of leaf transpiration rates yielded a canopy transpiration of 628 mm in 1980 (Savé 1986), again a dry year. Despite being based on different methods, sites and years, these estimated water requirements agree with the geographical distribution of holm oak forests, which are dominant in humid and sub-humid Mediterranean climates. Nevertheless, in Mediterranean areas there is a great interannual variability of precipitation, and even in sites whose average rainfall lies above the water requirements of holm oak forests, years with much lower water availability are frequent.

10.10 Biogeographical Implications

The ability of holm oak to maintain relatively low but sustained rates of foliar water loss is an important trait to understand its success in a range of
Mediterranean environments. The water requirements and the degree of
drought tolerance of holm oak preclude, however, the existence of dense
holm oak forests in areas having either a mean annual rainfall lower than
400–450 mm or extended periods with a predawn xylem potential below
−3.0 MPa. This helps us to understand the above-mentioned limitation of
holm oak forests to humid and sub-humid Mediterranean environments.

Most of the extant holm oak forests in NE Spain are found in areas characterized by humid mesoMediterranean conditions with cold winters, especially near the upper limit of their altitudinal range. A very cold winter can
be expected every 10–12 years in northern or mountain sites within the
Mediterranean region (Larcher 1981). This is relevant because the lowest xylem water potentials in montane holm oak forests are often found in winter
(Sala et al. 1988, 1990). Winter transpiration rates and xylem water potential
decline along altitudinal gradients, as we observed at Montseny. We can expect winter water deficits to develop if there is a reduced water uptake when
the evaporative demand is high, due to strong, dry winds, or high radiation.
Our results show a very low xylem potential (−3.0 MPa) in winter at the highest altitude along the gradient (1130 m a.s.l.). Below values of −2.2 MPa, water
loss is exclusively cuticular. Because cuticle thickness increases as the growing season becomes longer and warmer, we might expect cuticular transpiration to increase in colder environments (Tranquillini 1976; Savé et al. 1988)
and so trees growing at high altitudes would have a less efficient control of
water losses. Water stress is thus involved in limiting holm oak distribution
not only towards dry climates but also towards cold climates.

References

Acherar M, Rambal S (1992) Comparative water relations of four Mediterranean oak species.
Vegetatio 99/100:177–184
Aussenac G, Vallette JC (1982) Comportement hydrique estival de *Cedrus arlantica Manettl,
Quercus ilex* L. et *Quercus pubescens* Will. et de divers pins dans le Mont Ventoux. Ann Sci
For 39:41–62
Bazzaz FA, Chiariello NR, Coley PD, Pitelka LF (1987) Allocating resources to reproduction and
defense. BioScience 37:58–67
Berger A, Eckardt FE, Méthy M, Heim G, Sauvezon R (1977) Interception de l'énergie rayon-
nante, échange de CO_2, régime hydrique et production chez différents types de végétation
sous climat méditerranéen. In: Moyse A (ed) Les processus de la production végétale primaire. Gauthier Villars, Paris, pp 1–15
Burriel JA, Calvet S, Sala A, Gracia CA (1993) Ángulo foliar en *Quercus ilex:* modulación por el
ambiente, y contribución a la economía hídrica de la planta. In: Silva FJ, Vega G (eds) Congr
Forestal Español, Ponencias y Comunicaciones, vol 1. Xunta de Galicia, Lourizán, Pontevedra, pp 225–232

Castell C, Terradas J, Tenhunen JD (1994) Water relations, gas exchange, and growth of resprouts and mature plant shoots of *Arbutus unedo* L. and *Quercus ilex* L. Oecologia 98:201–211

Comín M P, Escarré A, Gracia CA, Lledó MJ, Rabella, R, Savé R, Terradas J (1987) Water use by *Quercus ilex* L. in forests near Barcelona, Spain. In: Tenhunen JD, Catarino FM, Lange OL, Oechel WC (eds) Plant response to stress: functional analysis in Mediterranean ecosystems. Springer, Berlin, pp 259–266

Correia MJ, Chaves MMC, Pereira JS (1990) Afternoon depression in photosynthesis in grapevine leaves. Evidence for a high light stress effect. J Exp Bot 41:417–426

Di Castri F, Mooney HA (eds) (1973) Mediterranean-type ecosystems. Springer, New York

Espelta JM (1996) La regeneració de boscos d'alzina (*Quercus ilex* L.) i pi blanc (*Pinus halepensis* Mill.): estudi experimental de la resposta de les plàntules a la intensitat de llum i a la disponibilitat d'aigua. PhD Thesis, Autonomous University of Barcelona, Bellaterra

Fleck I, Grau D, Sanjosé M, Vidal D (1996) Carbon isotope discrimination in *Quercus ilex* resprouts after fire and tree fell. Oecologia 105:286–292

Gartner BL (1995) Patterns of xylem variation within a tree and their hydraulic and mechanical consequences. In: Gartner BL (ed) Plant stems. Physiology and functional morphology. Academic Press, San Diego, pp 125–149

Ilijanic L, Gracanin M (1972) Zum Wasserhaushalt einiger mediterraner Pflanzen. Ber Dtsch Bot Ges 85:329–339

Kyriakopoulos E, Larcher W (1976) Saugspannungsdiagramm fur austrocknende Blater von *Quercus ilex* L. Z Pflanzenphysiol 77:268–271

Kyriakopoulos E, Richter H (1977) A comparison of methods for the determination of water status in *Quercus ilex* L. Z Pflanzenphysiol 82:14–27

Kyriakopoulos E, Richter H (1991) Desiccation tolerance and osmotic parameters in detached leaves of *Quercus ilex*. Acta Oecol 12:357–367

Larcher W (1960) Transpiration and photosynthesis of detached leaves and shoots of *Quercus pubescens* and *Quercus ilex* during desiccation under standard conditions. Bull Res Counc Isr 8D:213–224

Larcher W (1980) Physiological plant ecology. Springer, Berlin

Larcher W (1981) Low temperature effects on Mediterranean sclerophylls: an unconventional viewpoint. In: Margaris NS, Mooney HA (eds) Components of productivity of Mediterranean-climate regions. Dr W Junk Publishers, The Hague, pp 259–266

Leonardi S, Rapp M (1990) Production de phytomasse et utilisation des bioéléments lors de la reconstitution d'un taillis de chêne vert. Acta Oecol 11:819–834

Lo Gullo MA, Salleo S (1993) Different vulnerabilities of *Quercus ilex* L. to freeze- and summer drought-induced xylem embolism: an ecological interpretation. Plant Cell Environ 16:511–516

Lo Gullo MA, Salleo S, Piaceri EC, Rosso R (1995) Relations between vulnerability to xylem embolism and xylem conduit dimensions in young trees of *Quercus cerris*. Plant Cell Environ 18:661–669

Lossaint P, Rapp M (1978) La forêt méditerranéenne de chênes verts. In: Lamotte M, Bourlière F (eds) Problèmes d'écologie: structure et fonctionnement des écosystèmes terrestres. Masson, Paris, pp 129–185

Loveless AR (1961) A nutritional interpretation of sclerophylly based on differences in the chemical composition of sclerophyllous and mesophytic leaves. Ann Bot 25:168–184

Margaris SN (1981) Adaptative strategies in plants dominating Mediterranean-type ecosystems. In: Di Castri F, Goodall DW, Specht RL (eds) Mediterranean-type shrublands. Elsevier, New York, pp 309–314

Miller PC (1981) Conceptual basis and organization of research. In: Miller PC (ed) Resource use by chaparral and matorral. Springer, New York

Mitrakos (1980) A theory for Mediterranean plant-life. Oecol Plant 1:245–252

Morgan JM (1984) Osmoregulation and water stress in higher plants. Annu Rev Plant Physiol 35:299–319

Morris JT (1989) Modelling light distribution within the canopy of the marsh grass *Spartina alterniflora* as a function of canopy biomass and solar angle. Agric For Meteorol 46:349–361

Oechel WC, Lawrence W, Mustafa J, Martínez J (1981) Energy and carbon acquisition. In: Miller PC (ed) Resource use by chaparral and matorral. Springer, Berlin, pp 151–184

Oliveira G (1995) Autecologia do sobreiro *(Quercus suber* L.) em montados portugueses. PhD Thesis, University of Lisboa, Lisboa

Oliveira G, Correira OA, Martins-Louçao MA, Catarino FM (1992) Water relations of cork-oak *(Quercus suber* L.) under natural conditions. Vegetatio 99/100:199–208

Rabella R (1991) Ecofisiologia de les relacions hídriques del faig al Montseny. PhD Thesis, Autonomous University of Barcelona, Bellaterra

Rabella R, Savé R, Terradas J (1983) Conducta hídrica vertical del encinar montano de La Castanya (Montseny). V Reunión de la Sociedad Española de Fisiología Vegetal, Murcia

Rambal S (1992) *Quercus ilex* facing water stress: a functional equilibrium hypothesis. Vegetatio 99/100:147–153

Rambal S, Debussche G (1995) Water balance of Mediterranean ecosystems under a changing climate. In: Moreno JM, Oechel WC (eds) Global change and Mediterranean-type ecosystems. Springer, New York, pp 386–407

Sakai A, Larcher W (1987) Frost survival of plants. Springer, Berlin

Sala A, Tenhunen JD (1994) Site-specific water relations and stomatal response of *Quercus ilex* L. in a Mediterranean watershed. Tree Physiol 14:601–617

Sala A, Tenhunen JD (1996) Simulations of canopy net photosynthesis and transpiration in *Quercus ilex* L. under the influence of seasonal drought. Agric For Meteorol 78:203–222

Sala A, Pícolo R, Piñol J (1988) Efectos del frío en las relaciones hídricas de *Quercus ilex* en la sierra de Prades (Tarragona). Options Méditer 3:57–62

Sala A, Burriel JA, Tenhunen JD (1990) Spatial and temporal controls on transpiration within a watershed dominated by *Quercus ilex*. Proc on *Quercus ilex* L. ecosystems: function, dynamics and management, Montpellier-Barcelona, September 1990

Salleo S, Lo Gullo MA (1990) Sclerophylly and plant water relations in three Mediterranean *Quercus* species. Ann Bot 65:259–270

Salleo S, Nardini A, Lo Gullo MA (1997) Is sclerophylly of Mediterranean evergreens an adaptation to drought? New Phytol 135:603–612

Savé R (1986) Ecofisiologia de les relacions hídriques de l'alzina al Montseny. PhD Thesis, Autonomous University de Barcelona, Bellatera

Savé R, Rabella R, Gascón E, Terradas J (1982) Transpiration and diffusion resistance of leaves of *Quercus ilex* L. at La Castanya (Montseny, Catalonia, NE Spain). USDA For Serv Gen Tech Rep PSW–58:632

Savé R, Rabella R, Terradas J (1988) Effects of low temperature on *Quercus ilex* ssp. *ilex* water relations. In: Di Castri F, Floret Ch, Rambal S, Roy J (eds) Time scales and water stress. Proc 5th Int Conf on Mediterranean ecosystems. International Union of Biological Sciences, Paris, pp 1103–1105

Sperry JS (1995) Limitations on stem water transport and their consequences. In: Gartner BL (ed) Plant stems. Physiology and functional morphology. Academic Press, San Diego, pp 105–124

Sperry JS, Tyree MT (1990) Water-stress-induced xylem embolism in three species of conifers. Plant Cell Environ 13:427–436

Tenhunen J, Sala A, Harley PC, Dougherty RL, Reynolds JF (1990) Factors influencing carbon fixation and water use by Mediterranean sclerophyll shrubs during summer drought. Oecologia 82:381–393

Terradas J, Savé R (1992) The influence of summer and winter stress and water relationships on the distribution of *Quercus ilex* L. Vegetatio 99/100:137–145

Tranquillini W (1976) Water relations and alpine timberline. In: Lange OL, Kappen L, Schulze ED (eds) Water and plant life. Springer, Berlin, pp 473–491

Tranquillini W (1982) Frost-drought and its ecological significance. In: Lange OL, Nobel PS, Osmond CR, Ziegler H (eds) Physiological plant ecology. Encyclopaedia of plant physiology, vol 11. Springer, Berlin, pp 379–400

Turner IM (1994) Sclerophylly: primarily protective? Funct Ecol 8:669–675

Tyree MT, Dixon MA (1986) Water stress induced cavitation and embolism in some woody plants. Physiol Plant 66:397–405

Tyree MT, Richter H (1981) Alternative methods of analyzing water potential isotherms: some cautions and clarifications. I. The impact on non-linearity and of some experimental errors. J Exp Bot 32:643–653

Tyree MT, Richter H (1982) Alternative methods of analyzing water potential isotherms: some cautions and clarifications. II. Curvilinearity in water potential isotherms. Can J Bot 60: 911–916

Zimmerman MH (1983) Xylem structure and the ascent of sap. Springer, Berlin

11 Modelling Canopy Gas Exchange During Summer Drought

Anna Sala

11.1 Introduction

The structure and function of woody Mediterranean vegetation is strongly influenced by the periodical occurrence of a summer dry period. A hierarchy of structural and physiological adjustments (e.g. changes in leaf area, rooting depth, and stomatal opening), each operating at different time scales (from years to minutes) have been shown to be effective mechanisms by which Mediterranean sclerophylls cope with limited water availability during these drought periods (Rambal 1993). In holm oak (*Quercus ilex* L.) canopies, leaf area index (LAI) varies between approximately 2.5 and 5.5 $m^2 m^{-2}$ depending on long-term site water availability (Table 11.1). Higher LAIs occur in areas with moderate to high annual precipitation, where deep soils with minimal deep drainage ensure high soil water availability (e.g. valley bottom of the Avic catchment; Sala et al. 1994; Table 11.1). Lower LAI values occur in areas where annual precipitation may be high but soil water availability is low due to deep infiltration of water (e.g. limestone substrate) and low soil water retention (e.g. Puéchabon, France; Rambal et al. 1996; Table 11.1). Closely related to changes in leaf area are within-canopy changes in leaf structure and

Table 11.1. Leaf area index (LAI) of dense holm oak stands in the Mediterranean Basin

LAI ($m^2 m^{-2}$)	Location	Annual precipitation (mm)	Reference
5.3–4.6	Avic, Spain	658	Sala et al. (1994)
4.9–4.5	Castelporziano, Italy	670	Gratani and Fiorentino (1988) Gratani (1993)
4.4	Le Rouquet, France	755	Eckardt et al. (1978)
4.0	Montpellier–Camp Redon, France	667	Rambal et al. (1996)
3.9	Bosco Mesola, Italy	614	Pitacco et al. (1992)
2.9	Puéchabon, France	809	Rambal et al. (1996)

Ecological Studies, Vol. 137
Ferran Rodà et al. (eds) Ecology of Mediterranean
Evergreen Oak Forests
© Springer-Verlag, Berlin Heidelberg 1999

function, which improve the light environment within the canopy and canopy carbon balance (Sala et al. 1994; Rambal et al. 1996; Sala and Tenhunen 1996; Chap. 9). Sensitive stomatal control of water loss in holm oak is also a well documented short-term response (hours to days) by which canopy water use is reduced and the damaging effects of transient or prolonged drought are avoided (e.g. Tenhunen et al. 1987; Terradas and Savé 1992; Tetriach 1993; Sala and Tenhunen 1994).

Here, I will use a process-based simulation model that incorporates physiological behaviour and vegetation structure (Tenhunen et al. 1990; Harley and Tenhunen 1991; Reynolds et al. 1992; Sala and Tenhunen 1996) to examine the relative effects of structural and physiological adjustments to drought on the annual water loss and carbon assimilation of holm oak canopies. This type of analysis, performed under varying intensities of summer drought, will provide a better understanding of the complex interactions between canopy structure and function of Mediterranean sclerophylls such as holm oak. It will also provide better insight into the trade-off between optimization of water use vs. carbon uptake in Mediterranean sclerophylls, a topic that needs further investigation (Rambal 1993). I will first provide a brief description of the model and the main results obtained from its application in the holm oak forest of the Avic catchment (Prades Mountains, NE Spain).

11.2 Model Description

A full description of the model used and its parameterization is provided in Sala and Tenhunen (1996). The model incorporates a canopy model of light interception and microclimate (Caldwell et al. 1986; Tenhunen et al. 1990; Reynolds et al. 1992) which estimates detailed microclimatic data within the canopy. Microclimatic conditions and leaf photosynthetic properties in different canopy layers are used to calculate photosynthesis rates following a mechanistically based model of C_3 leaf photosynthesis (Farquhar and Von Caemmerer 1982; Tenhunen et al. 1990; Harley and Tenhunen 1991). Stomatal conductance is then calculated as a function of photosynthesis rates with the empirical model described by Ball et al. (1987; see below). Transpiration rates are calculated from stomatal conductance. Initially, the model assumes leaf temperature equal to air temperature. Since leaf temperature depends on the transpiration rate, final equilibrium leaf temperature is solved by an energy balance approach with successive within- and between-canopy layer iterations.

Inputs to the model include 1-h time steps of the driving variables (global short-wave radiation, air temperature, relative humidity and wind velocity above the canopy, and soil surface temperature) and structural, optical, and physiological properties of leaves in each canopy layer. Outputs of the model are diurnal courses of leaf conductance, transpiration and net photosynthesis for sun and shade leaves in different canopy layers. Total daily canopy carbon assimilation and transpiration are calculated from the hourly contributions of all leaves. While the model considers light interception by stems, values of canopy carbon uptake include only the gas exchange of leaves and ignore respiration by twigs and stems.

11.3 Model Implementation at the Avic Catchment

Simulations of canopy gas exchange in holm oak were carried out at two sites (ridge top and valley bottom) along a slope and microclimate gradient at the Avic catchment (Prades Mountains, NE Spain; Sala et al. 1994; Sala and Tenhunen 1996; see Chap. 2 for site description).

Canopy structural characteristics model inputs (leaf area index, stem area index, leaf size, leaf inclination angle and stem inclination angle) were derived from measurements described in Chapter 9 and Sala et al. (1994). The degree of leaf clustering was chosen to allow the best agreement between predicted light values and measured data in different canopy levels (Eckardt et al. 1978; A. Sala et al. unpubl.). Leaf optical properties were derived from reported values in the literature (Eckardt et al. 1978) and were assumed to be constant throughout the canopy.

Parameters describing leaf photosynthetic characteristics in leaves in the uppermost canopy layer were set as reported by Tenhunen et al. (1990) for the closely related scrub oak species *Quercus coccifera*. These parameters were changed as a function of cumulative leaf area index such that photosynthetic capacity of the leaves at different canopy layers was linearly related to the cumulative LAI and to leaf nitrogen (Field 1983; Sabaté et al. 1995; Chap. 9). The model assumes that photosynthetic capacity in each canopy layer does not change during the year (Kaiser 1987; Epron and Dreyer 1993). CO_2 partial pressure was set equal to 350 ppm.

11.3.1 Incorporating the Effects of Seasonal Water Deficits on Stomatal Regulation

According to the model of Ball et al. (1987), a scaling factor (g_F) relates stomatal conductance to a "stomatal index". This is, in turn, directly proportional to the net assimilation rate and the relative humidity within the leaf boundary layer and inversely proportional to the CO_2 partial pressure within the boundary layer. By only changing g_F Tenhunen et al. (1990) were able to incorporate the integrated effects of long-term water stress on stomatal regulation and produce realistic simulations of gas exchange during the year. Sala and Tenhunen (1996) optimized the value of g_F to allow the closest fit

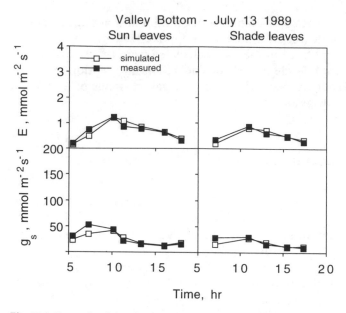

Fig. 11.1. Example of simulated and measured diurnal courses of stomatal conductance (g_s) and transpiration rates (E) in sun ($g_F = 6.4$) and shade ($g_F = 6.2$) leaves on a clear summer day at the valley bottom site of the Avic catchment. For each day, the value of g_F was determined from least squares between measured and simulated data. (Modified from Sala and Tenhunen 1996 with permission)

between simulated and measured diurnal courses of stomatal conductance in sun and shade leaves over the course of the year (Fig. 11.1). They found that when pre-dawn water potential (Ψ_{pd}) was below −1 MPa, the best g_F determined for each day was linearly related to the corresponding Ψ_{pd} measured during that day ($g_F = 18.3 + 4.2\ \Psi_{pd}$; $r^2 = 0.88$; $P < 0.01$). This relationship reflects a functional dependency of g_F on Ψ_{pd} of holm oak. At the Avic catchment, Ψ_{pd} is related to catchment discharge and is a good indicator of long-term soil water availability (Sala and Tenhunen 1994). Thus, the relationship between g_F on Ψ_{pd} indicates that stomatal sensitivity to photosynthesis rates, relative humidity and CO_2 concentration under natural conditions changed gradually over the course of the year and that these changes were coupled to long-term water availability. These results reinforced once more that the parameter g_F reflects an integrated stomatal response (such as changes in stomatal patchiness, Beyschlag et al. 1992; and/or root signals, Tenhunen et al. 1994) to the degree of water stress experienced by the plant. While there is no mechanistic definition of g_F, and the use of photosynthesis rates as a predictor of stomatal conductance is based on correlative behaviour, the use of g_F is extremely useful as a modelling tool because it provides a good phenomenological description of the effects of decreased soil water availability on stomatal regulation at the leaf and at the canopy level (e.g. Tenhunen et al. 1990, 1994; Sala and Tenhunen 1996).

11.3.2 Model Performance

The use of g_F as a link between long-term water availability and stomatal regulation allowed effective predictions of diurnal and seasonal patterns of gas exchange of sun and shade leaves for different days during the year (Fig. 11.2; cf. Eckardt et al. 1978; Sala 1992; Gratani 1993, 1996; Tetriach 1993; Castell et al. 1994; Sala and Tenhunen 1994). Gas exchange rates were maximum in late spring and early summer when environmental conditions were favourable. In June, stomatal conductance was constant during the day in response to low relative humidity. While simulated transpiration rates were

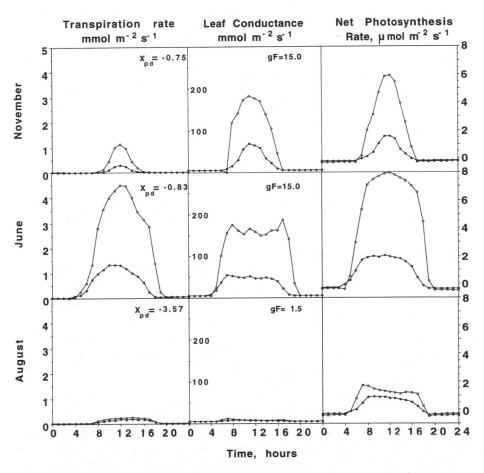

Fig. 11.2. Simulated diurnal courses of gas exchange of leaves located at 0.12 m (sun leaves, *open symbols*) and 1 m from the top of the canopy (shade leaves, *solid symbols*) during clear days at the valley-bottom site of the Avic catchment. Also indicated are the pre-dawn xylem water potential in MPa (X_{pd}) and the value of g_F (see text)

very low during cold months due to decreased evaporative demand, stomatal conductance and photosynthesis rates were relatively high. Significant carbon input during fair winter conditions in Mediterranean sclerophylls was also measured by Larcher and Tisi (1990). At the whole canopy level, simulated peak canopy carbon input during clear autumn days ranged from 11.3 to 15 μmol m^{-2} s^{-1}, values comparable with those reported by Valentini et al. (1991).

Simulations of monthly transpiration and carbon input in holm oak canopies at the ridge top and valley bottom sites of the Avic catchment (Fig. 11.3) were done by selecting representative days from detailed meteorological records available during 1989 at each site (Sala and Tenhunen 1996). The value of g_F for each month was determined from annual courses of Ψ_{pd} at each site and the linear dependency between g_F and Ψ_{pd}. Because of reduced light intensity, the contribution of shade leaves deep in the canopy to canopy water loss and carbon input is very small; thus canopy totals at the ridge top (LAI = 4.6 m^2 m^{-2}) and valley bottom (LAI = 5.3 m^2 m^{-2}) during periods of adequate water availability were very similar. Stands at the valley bottom used more water during the critical period in late spring and early summer, which led to severe soil water depletion and summer water stress at the valley bottom site (particularly during August; Sala and Tenhunen 1994). A decline in measured catchment streamflow following increases in precipitation suggests that the basin was not fully recharged and that soil water may have been severely depleted (e.g. August–September). Severe soil water depletion occurred during late spring and early summer (May–June) when simulated canopy transpiration was at a maximum and precipitation was low. The result was severe water stress during August and September when trees exhibited very conservative gas exchange behaviour (Fig. 11.3; Sala and Tenhunen 1994). Periods of high precipitation and low canopy transpiration allowed the soil to recharge and an increase in measured streamflow was measured.

Estimated total annual canopy transpiration was 464 mm year^{-1} at the valley bottom site and 453 mm year^{-1} at the ridge top. This represents 85 and 87% of the annual precipitation during 1989 (534 mm) at the ridge top and valley bottom sites, respectively. These values differ less than 10% from the ones reported by Piñol (1990) by using the input/output hydrological balance method. While this is not a validation, these results suggest, once more, that model predictions of water use by holm oak canopies are realistic and consistent with measured hydrological parameters.

The results summarized here (see also Sala and Tenhunen 1996) demonstrate that the modelling approach used at the Avic catchment (which is based on detailed field studies) provides a realistic description of diurnal and seasonal patterns of leaf and canopy response to water availability. By using the same approach we can now address specific questions regarding the relative effect of physiological and structural adjustments to drought on the water loss and carbon input of holm oak canopies.

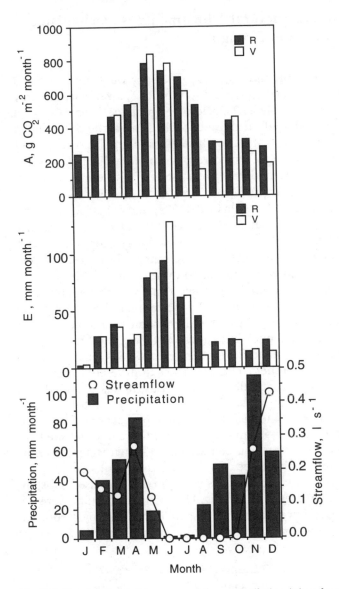

Fig. 11.3. Simulated monthly canopy carbon assimilation (*A*) and transpiration (*E*) on a ground area basis at the ridge-top (R) and valley-bottom (V) sites of the Avic catchment. *Bottom*: monthly precipitation (*bars*) and streamflow (*open symbols*) measured at Avic during 1989. (From Sala and Tenhunen 1996 with permission)

11.4 Physiological vs. Structural Adjustments as Mechanisms to Cope with Summer Drought

Annual water use, carbon input and water use efficiency were simulated in a series of canopies with LAI varying from 2 to 6 m^2 m^{-2}, a range commonly found in holm oak stands (Table 11.1). Micrometeorological, physiological and structural inputs were based on those described in Sala and Tenhunen (1996). For each LAI, simulations were conducted assuming three degrees of water availability during summer, each corresponding to a distinct seasonal pattern of Ψ_{pd}: no water stress, moderate water stress and severe water stress (Fig. 11.4). Canopy structural characteristics for each LAI were varied according to the patterns described in Sala et al. (1994) and Sabaté et al. (Chap. 9). For instance, it was assumed that canopy height increased with successive increments of LAI and that the size of leaves in the lower canopy increased while leaf inclination angles at the bottom of the canopy decreased. The degree of leaf clustering was also varied throughout the canopy to allow realistic light attenuation patterns within the canopy. Thus, at each LAI, intensity of photosynthetically active radiation at the bottom of the canopy was enough to keep leaves at or slightly above their light compensation point. The proportion of LAI to stem area index in each canopy layer and for the entire canopy was maintained according to field measurements (Sala et al. 1994). As described earlier, leaf photosynthetic characteristics in different canopy layers were changed according to cumulative LAI. Thus, it was assumed that leaves at the bottom of a canopy with a LAI of 6 m^2 m^{-2} exhibited a higher degree of shade-adaptation than leaves at the bottom of a canopy with a LAI of 2 m^2 m^{-2}. The value of g_F used as input for representative days selected per month was determined from seasonal courses of pre-dawn water potential (Fig. 11.4) which represent each degree of water availability considered.

At each level of water stress, annual transpiration rates increased substantially with successive increases in LAI (Fig. 11.5). Similar results were reported by Rambal (1993). However, the increase in annual canopy carbon input with increases of LAI was much more curvilinear, with large increases at

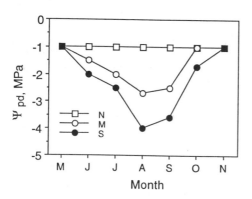

Fig. 11.4. Seasonal courses of pre-dawn water potential (Ψ_{pd}) representing three degrees of water availability during the summer. N, M and S indicate no, moderate, and severe water stress, respectively

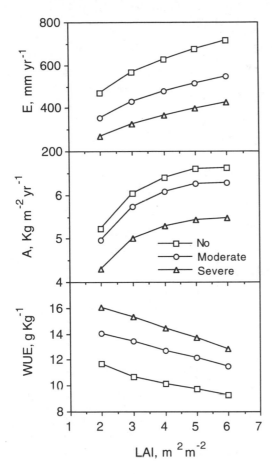

Fig. 11.5. Annual canopy transpiration (E), carbon input (A), and water use efficiency (WUE) simulated in holm oak canopies of increasing LAI. Simulations were performed assuming three degrees of water stress during summer (no water stress, moderate and severe)

lower LAI and almost no change at higher LAI. Even though leaf inclination angles and degree of leaf clustering was changed with successive increases in LAI to reproduce realistic patterns of light extinction within the canopy, self-shading is eventually unavoidable and leaves in lower portions of canopies with LAI of 4 m^2 m^{-2} or greater contribute very little to the total canopy carbon input. The critical LAI values above which further increases in LAI no longer result in an increase of canopy carbon gain appear to be approximately at 5 m^2 m^{-2} (Fig. 11.5). These are, in fact, the upper limits of LAI commonly encountered in holm oak canopies (Table 11.1).

The importance of changes in leaf arrangement (leaf angle and/or clustering degree) to improve the light environment in canopies of high LAI is illustrated when comparing simulated annual canopy carbon input, water use and WUE in canopies with leaves randomly arranged (not clustered) and with leaves clustered such that measured canopy light penetration is compatible with measured values (see above). Differences in canopy carbon input

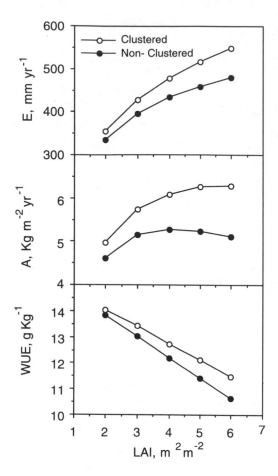

Fig. 11.6. Annual canopy transpiration (*E*), carbon input (*A*), and water use efficiency (*WUE*) simulated in holm oak canopies under moderate summer water stress (see Fig. 11.4). Simulations were performed assuming clustered and non-clustered leaves

and water use were small at lower LAIs but increased at higher LAI (Fig. 11.6). Simulated annual carbon input in canopies with non-clustered leaves peaked at a LAI of 4 m^2 m^{-2} and decreased thereafter. In contrast, values in canopies with clustered leaves reached a plateau at LAI of 5 m^2 m^{-2}. Non-clustered leaf arrangement had a greater effect on canopy carbon input than on water use. As a result, decreases of WUE with increases in LAI were slightly more pronounced in canopies with random leaves than in canopies with clustered leaves. These results indicate that at LAIs at or above 4 m^2 m^{-2}, small changes in leaf arrangement (in this modelling example imposed only via changes in the degree of leaf clustering) may have small but significant effects on the long-term balance between carbon input and water use in holm oak forests. Although variation in leaf clustering in holm oak canopies has not been measured, leaf angle in holm oak has been shown to change within and between canopies, and even seasonally (Burriel et al. 1993; Tappeiner et al. 1993). However, while these changes occur and have potential adaptive value, results in Figs. 11.5 and 11.6 indicate as well that there is a limit beyond which changes in leaf arrangement cannot compensate for increases in self-shading associated with increases in LAI.

At all levels of water stress, a reduction of canopy LAI from 6 to 2 $m^2 m^{-2}$ resulted in an annual water saving of approximately 35% (Fig. 11.5). However, at any LAI, larger water savings (up to 41%) were simulated by stomatal regulation of water loss (imposed via the parameter g_F). These results indicate that short-term functional responses (stomatal control) characteristic of holm oak and other Mediterranean sclerophylls may be more effective than long-term structural responses (changes in leaf area) in reducing annual water loss (Tenhunen et al. 1990). Thus, even if stands exhibited relatively high LAI, stomatal regulation of water loss would serve to prevent excessive water loss during periods of summer drought. However, the results also indicate that LAI values greater than 5 $m^2 m^{-2}$ do not result in significant increments of annual carbon input (Caldwell et al. 1986). Furthermore, WUE decreases approximately 5% with each successive increase in LAI. Thus, increases in LAI up to 5 $m^2 m^{-2}$ are associated with increases in carbon input but also with decreases of WUE. This is because as LAI increases light penetration within the canopy decreases and the contribution of leaves lower in the canopy to the total canopy carbon input decreases.

11.5 Conclusions

The model used here allowed effective predictions of diurnal and seasonal courses of gas exchange in holm oak. The effects of summer drought on stomatal regulation were successfully introduced by changing a single model parameter which can be estimated from pre-dawn water potential. Because of the mechanistic nature of the model, a direct linkage with a hydrological model (e.g. Beven and Kirkby 1979) would allow us to estimate spatial and seasonal patterns in canopy water use and to assess the impacts of climate change, at least in the short term (Reynolds et al. 1992), on the hydrology of the entire catchment.

The analysis of the relative effects of structural vs. functional adjustments as mechanisms to cope with reduced site water availability suggests a delicate trade-off between water loss and carbon input for Mediterranean sclerophylls. This trade-off may shift depending on biotic and abiotic interactions. Stands supporting LAIs below 3 $m^2 m^{-2}$ are found in dry sites where, according to model simulations, strong reductions of carbon input are compensated for by substantial increases in WUE. Under conditions of reduced water availability, optimization of water use may be crucial to survive and coexist with other xerophytic Mediterranean sclerophylls characteristic of dry areas such as *Quercus coccifera*. In sites where the water balance is more favourable, LAIs up to 5 $m^2 m^{-2}$ may be found, suggesting that maximizing canopy carbon input (even if at the expense of increased water use and reduced WUE) may allow holm oak to outgrow potential competitors.

References

Ball JT, Woodrow IE, Berry JA (1987) A model predicting stomatal conductance and its contribution to the control of photosynthesis under different environmental conditions. In: Binggins IJ (ed) Progress in photosynthesis research, vol IV 5. Martinus Nijhoff, Dordrecht, pp 221–224

Beven K, Kirkby MJ (1979) A physically based, variable contributing area model of basin hydrology. Hydrol Sci Bull 24:43–69

Beyschlag W, Pfanz H, Ryel RJ (1992) Stomatal patchiness in Mediterranean evergreen sclerophylls: phenomenology and consequences for the interpretation of midday depression in photosynthesis and transpiration. Planta 187:546–553

Burriel JA, Calvet S, Sala A, Gracia CA (1993) Ángulo foliar en Quercus ilex: modulación por el ambiente, y contribución a la economía hídrica de la planta. In: Silva FJ, Vega G (eds) Congr Forestal Español. Ponencias y Comunicaciones, vol 1. Xunta de Galicia, Lourizán, Pontevedra, pp 225–232

Caldwell MM, Meister HP, Tenhunen JD, Lange OL (1986) Canopy structure, light microclimate and leaf gas exchange of Quercus coccifera L. in a Portuguese macchia: measurements in different canopy layers and simulations with a canopy model. Trees 1:25–41

Castell C, Terradas J, Tenhunen JD (1994) Water relations, gas exchange, and growth of resprouts and mature plant shoots of Arbutus unedo L. and Quercus ilex L. Oecologia 98:201–211

Eckardt FE, Berger A, Méthy M, Heim G, Sauvezon R (1978) Interception de l'ènergie rayonnante, èchanges de CO_2, règime hydrique et production chez différents types de vègètation sous climat mèditerranèen. In: Moyse A (ed) Les processus de la production vègètale primairie. Gauthier Villars, Paris, pp 1–75

Epron D, Dreyer E (1993) Photosynthesis of oak leaves under water stress: maintenance of high photochemical efficiency of photosystem II and occurrence of non-uniform CO_2 assimilation. Tree Physiol 13:107–117

Farquhar GD, Von Caemmerer S (1982) Modeling of photosynthetic response to environment. In: Lange OL, Nobel PS, Osmond CB, Ziegler H (eds) Encyclopedia of plant physiology, vol 12B, Physiological Plant Ecology II, Water relations and carbon assimilation. Springer, Berlin, pp 549–587

Field C (1983) Allocating leaf nitrogen for the maximization of carbon gain: leaf age as a control on allocation program. Oecologia 56:341–347

Gratani L (1993) Response to microclimate of morphological leaf attributes, photosynthetic and water relations of evergreen sclerophyllous shrub species. Photosynthetica 29:573–582

Gratani L (1996) Leaf and shoot growth dynamics of Quercus ilex L. Acta Oecol 17:17–27

Gratani L, Fiorentino E (1988) Leaf area index for Quercus ilex L. high maquis. Photosynthetica 22:458

Harley PC, Tenhunen JD (1991) Modeling the photosynthetic response of C_3 leaves to environmental factors. In: Boote KJ, Loomis RS (eds) Modeling crop photosynthesis – from biochemistry to canopy. ASA, Madison, Wisconsin, pp 17–39

Kaiser W (1987) Effects of water deficit on photosynthetic capacity. Physiol Plant 71:142–149

Larcher W, Tisi F (1990) Bioclima invernale e rendimento carbonico di Quercus ilex al limite settentrionale delle leccete prealpine. Atti Accad Naz Lincei Mem Lincee Sci Fis Nat Ser IX (1) Fac I:1–22

Piñol J (1990) Hidrologia i biogeoquímica de conques forestades de les Muntanyes de Prades. PhD Thesis, University of Barcelona, Barcelona

Pitacco A, Gallinaro N, Giulivo C (1992) Evaluation of actual evapotranspiration of a Quercus ilex L. stand by the Bowen ratio-energy budget method. Vegetatio 99–100:163–168

Rambal S (1993) The differential role of mechanisms for drought resistance in a Mediterranean evergreen shrub: a simulation approach. Plant Cell Environ 16: 35–44

Rambal S, Damesin C, Joffre R, Méthy M, Lo Seen D (1996) Optimization of carbon gain in canopies of Mediterranean evergreen oaks. Ann Sci For 53:547–560

Reynolds JF, Chen JL, Harley PC, Hilbert DW, Dougherty RL, Tenhunen JD (1992) Modeling the effects of elevated CO_2 on plants: extrapolating leaf response to a canopy. Agric For Meteorol 61:69–94

Sabaté S, Sala A, Gracia CA (1995) Nutrient content in *Quercus ilex* canopies: seasonal and spatial variation within a catchment. Plant Soil 168–169:297–304

Sala A (1992) Water relations, canopy structure, and canopy gas exchange in a *Quercus ilex* forest: variations in time and space. PhDThesis, University of Barcelona, Barcelona

Sala A, Tenhunen JD (1994) Site-specific water relations and stomatal response of *Quercus ilex* L. in a Mediterranean watershed. Tree Physiol 14:601–617

Sala A, Tenhunen JD (1996) Simulations of canopy net photosynthesis and transpiration in *Quercus ilex* L. under the influence of seasonal drought. Agric For Meteorol 78:203–222

Sala A, Sabaté S, Gracia CA, Tenhunen JD (1994) Canopy structure within a *Quercus ilex* forested watershed: variations due to location, phenological development and water availability. Trees 8:254–261

Tappeiner U, Jiménez MS, Morales D, Larcher W (1993) Bioclima e potenziale di produttività di *Quercus ilex* L. al limite settentrionale dell'areale di distribuzione. Parte II. Struttura della chioma ed energia radiante. Studi Trentini Sci Nat 68:19–35

Tenhunen JD, Pearcy RW, Lange OL (1987) Diurnal variation in leaf conductance and gas exchange in natural environments. In: Zeiger E, Farquhar G, Cowan I (eds) Stomatal function. Stanford University Press, Stanford, pp 323–351

Tenhunen JD, Sala A, Harley PC, Dougherty RL, Reynolds JF (1990) Factors influencing carbon fixation and water use by Mediterranean sclerophyll shrubs during summer drought. Oecologia 82:381–393

Tenhunen JD, Hanano R, Abril M, Weiler EW, Hartung W (1994) Above- and belowground environmental influences on leaf conductance of *Ceanothus thyrsiflorus* growing in a chaparral environment: drought response and the role of abscisic acid. Oecologia 99:306–314

Terradas J, Savé R (1992) The influence of summer and winter stress and the water relationships on the distribution of *Quercus ilex* L. Vegetatio 99/100:163–168

Tetriach M (1993) Photosynthesis and transpiration of evergreen Mediterranean and deciduous trees in an ecotone during a growing season. Acta Oecol 14:341–360

Valentini R, Scaracia Mugnozza GE, De Angelis P, Bimbi R (1991) An experimental test of the eddy correlation technique over a Mediterranean macchia canopy. Plant Cell Environ 14: 987–994

12 GOTILWA: An Integrated Model of Water Dynamics and Forest Growth

Carlos A. Gracia, Estíbaliz Tello, Santiago Sabaté and Juan Bellot

12.1 GOTILWA: an Overview

GOTILWA is a simulation model of forest growth. Its name, GOTILWA, is an acronym for Growth Of Trees Is Limited by WAter. The name itself defines the main characteristic of the model. Water is, very often, the limiting factor for plant growth (Piñol et al. 1991; Sala 1992; Chap. 13) and thus it constitutes a key factor in the model (Tello et al. 1994). In a standard simulation, daily climatic data are analyzed. From the interaction between daily rainfall and the forest structure, the amount of intercepted water by the canopy layer, throughfall and stemflow are estimated. This effective rainfall increases the water stored in the soil which is used by the trees. The proportion of sapwood to heartwood, the leaf area of each tree and, consequently, the leaf area index (LAI) of the forest are all highly dependent on water availability in the model.

Carbon uptake by trees is computed by using water use efficiency (WUE) and the water transpired by each tree on a monthly basis. The pool of carbon gained leads to an increase, primarily, in the mobile carbon stored in the plant. A fraction of this carbon compensates the maintenance respiration, while the remaining carbon, if any, is available for storage and for sustaining growth.

Associated with the formation of new biomass is an additional metabolic cost which constitutes the growth respiration. The difference between carbon fixation and maintenance plus growth respiration is net primary production (NPP). The carbon balance of trees determines the processes of leaf formation and leaf fall, tree ring formation, the rate of change of sapwood into heartwood and, consequently, the changes in tree structure within each size class. These changes will affect in turn the hydrological fluxes which will subsequently take place in what might be described as a feedback process (Sala and Tenhunen 1994).

Ecological Studies, Vol. 137
Ferran Rodà et al. (eds) Ecology of Mediterranean
Evergreen Oak Forests
© Springer-Verlag, Berlin Heidelberg 1999

12.1.1 Input Data

The input data of GOTILWA comprise three different sets (Table 12.1). A first data set is the climate data: daily values of maximum and minimum temperature (°C), rainfall (mm day^{-1}), duration of rain (min day^{-1}), solar irradiance (MJ day^{-1}) and potential evapotranspiration (E_o, mm day^{-1}). Daily E_o is used to calculate the evaporative coefficient (Specht 1972) and the actual evapotranspiration (E_a). If E_o is not available, GOTILWA estimates the daily values using the method of Samani and Heargreaves (1985), based on the use of maximum and minimum temperature and solar irradiance above the atmosphere, as described by McKenney and Rosenberg (1993).

Table 12.1. Input variables used by GOTILWA to simulate forest growth

Climatic variables	
Maximum daily temperature	°C
Minimum daily temperature	°C
Precipitation	mm day^{-1}
Irradiance	MJ day^{-1}
Duration of rainfall	min day^{-1}
Potential evapotranspiration	mm day^{-1}
Catchment or stand variables	
Altitude at highest point	m a.s.l.
Altitude of discharge point	m a.s.l.
Horizontal distance between these points	m
Length of discharge front	m
Soil depth at highest point	m
Soil depth at discharge point	m
Maximum soil water storage capacity	m
Minimum soil water storage capacity	m
Soil hydraulic conductivity	m day^{-1}
Tree variables and functions	
Throughfall function	$T = a\,R + b\,D_r + c$
Interception function	$I = a\,LA\,R\,e^{c\,R}$
Stemflow function	$S = [\,1 - e^{a\,LA}\,]\ [\,b\,(R - I)^c\,]$
Aboveground biomass function	$B = a\,D^b$
Ratio belowground/aboveground biomass	Unitless
Fraction cross-sectional area of sapwood	Unitless
Leaf area per unit of sapwood	m^2 cm^{-2}
Leaf specific mass	mg cm^{-2}
Fine root biomass/leaf biomass	Unitless
Water use efficiency at maximum PAR	mmol CO_2 mol^{-1} H_2O
Water use efficiency at minimum PAR	mmol CO_2 mol^{-1} H_2O
Mobile carbon stored in leaves	g / 100 g
Mobile carbon stored in woody tissues	g / 100 g
Fraction of incident PAR reaching the soil	%
Fraction of respiring sapwood	Value per one

T, throughfall; I, interception; S, stemflow; R, rainfall; D_r, duration of rainfall; LA, leaf area; D, tree dbh.

A second set of input data is related to the functions used to describe the hydrological fluxes through the canopy (throughfall, stemflow and interception) and through the soil (drainage). For each daily event of rainfall GOTILWA estimates the canopy fluxes as functions of the amount of rainfall and the leaf area of each tree size class. The drainage is calculated by Darcy's law, from topographic and soil parameters.

The third set of data is formed by the variables used to define the structure and function of the trees. Mean diameter (D) is the basic parameter used to characterize the tree size. Initial aboveground biomass of a tree is estimated using the allometric relationship:

$$B_i = a \cdot D_i^b$$

where B_i is the aboveground biomass of tree i, the diameter of which is D_i, and a and b are empirical constants.

The biomass of each fraction at any given time is the result of the balance between carbon fixation and respiration. The amount of water transpired during a period of time determines the fraction of active xylem (sapwood) and, thus, the leaf area of the tree. Since the ratio leaf area/sapwood area should remain constant, any change in water availability results in changes in the LAI and changes in the sapwood/heartwood ratio.

12.1.2 Temporal Resolution

Interception, throughfall and stemflow are calculated daily from the amount of rainfall data and the forest structure (density and size of the trees). The soil water content is used to estimate both daily streamflow and actual evapotranspiration. The daily Ea is integrated on a monthly basis and then the subsequent physiological processes of trees, namely carbon uptake, respiration, carbon allocation, litterfall, and growth, are estimated on this monthly basis.

12.1.3 Spatial Resolution

Two different spatial approaches can be used. In one of them, GOTILWA works at the stand level integrating the values of the output variables on a hectare area basis. In the second one, GOTILWA analyzes a catchment as a unit integrating the differences in forest structure, and the main topographical and pedological features. In this case a classification of the whole catchment in categories as stands of uniform characteristics should be provided. GOTILWA can subsequently analyze each of these uniform categories and integrate the results to provide the output at the catchment level.

12.2 Main Features and Initial Hypotheses

The flow chart of the GOTILWA model is shown in Fig. 12.1. The main features and initial hypotheses of the model are:

1. Transpiration is one of the main driving forces of the physiological processes of forest growth.
2. Available soil water is transpired by trees as much as possible. If E_0 is lower than the amount of available water, transpiration reaches the E_0 value. The remaining water can be lost as streamflow according to Darcy's law.
3. Carbon uptake is calculated from the amount of water transpired using the water use efficiency (WUE) of the trees. A different value of WUE can be defined for dominant and suppressed trees.
4. The carbon fixed is transformed to gross primary production (GPP) using the carbon content of dry matter.
5. A part of the GPP is used to compensate the respiratory cost of all the living tissues. The remaining primary production is used in the formation of new biomass which involves a growth respiratory cost.
6. All the physiological processes such as leaf shedding, leaf formation, etc. are temperature dependent. GOTILWA uses the value of Q_{10} based on the monthly average temperature.
7. Translocation of mobile carbon from leaves takes place prior to leaf abscission. The amount of carbon retranslocated depends on the amount of mobile carbon stored in leaves. Leaf specific mass (LSM) of shed leaves decreases according to this translocation (Sabaté 1993).
8. Leaf area supported by an individual tree is proportional to the sapwood cross-sectional area in a constant value. Changes in leaf area are translated into changes in the sapwood cross-sectional area.
9. A set of priority rules is used in respiration and carbon allocation processes. If the fixed carbon exceeds the cost of respiration, it is used to increase the pool of mobile carbon stored, first of all, in leaves and, after that, in woody tissues. The remaining carbon, if any, is used to build new biomass.
10. During periods in which GPP does not compensate the maintenance respiration, the carbon stored in leaves and in stem sapwood is used to compensate the leaf and stem respiration while carbon stored in coarse roots is used to compensate the respiration of fine roots.
11. If the mobile carbon stored does not compensate the respiratory cost, leaf shedding reduces the amount of leaf tissue and, thus, reduces respiration until both values, respiration and carbon availability, compensate each other.
12. If this compensation point cannot be reached, even when all the leaves are shed, the tree metabolism is unbalanced and it dies.

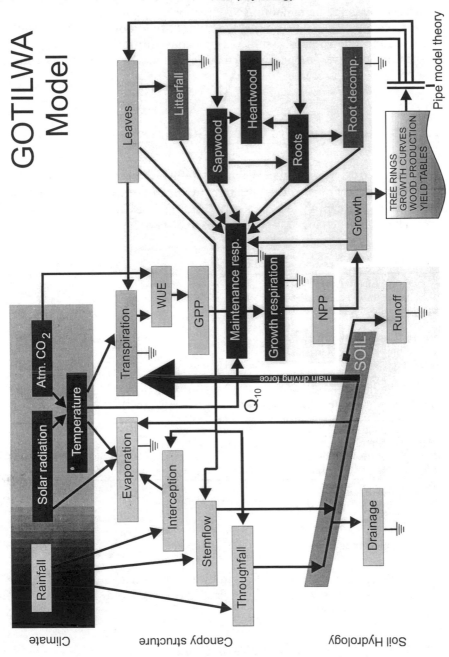

Fig. 12.1. Flowchart showing main paths and processes in the GOTILWA model. The main driving force in the model is transpiration resulting from hydrological flows occurring in canopy and soil

13. During favourable periods in which GPP exceeds respiration and carbon storage demand, the remaining NPP is allocated to new tissues.
14. Due to a scarcity of data on the dynamics of fine roots, the proportion of leaves and fine roots is assumed to be constant; thus, the formation or shedding of leaves implies the formation or mortality of the proportional amount of fine root mass.

12.3 Constant Parameters Used by GOTILWA

GOTILWA uses a set of constant values to define different processes such as the respiration rate of the different components of the tree, the light extinction through the canopy, etc. Table 12.2 summarizes these values. Any of these values can be changed to fit different tree species.

Table 12.2. Constant values used by GOTILWA

PAR to global radiation ratio	0.42
μEinsteins per watt in solar radiation	4.12
Atmospheric CO_2, base concentration (ppm)	350
Q_{10}	3
Reference temperature for Q_{10} (°C)	20
Energy equivalence of OM (cal g^{-1})	4700
OM to carbon ratio	2.10
N content (g) per 100 g of dry matter (leaves)	1.20
Respiration rate of structural components of leaves (cal g^{-1} day^{-1})	33.30
Respiration rate of non-structural components of leaves (cal g^{-1} day^{-1})	55.50
Respiration rate of living components of wood (cal g^{-1} day^{-1})	35.00
Grams of leaf tissue formed by 1 g of carbohydrates	0.68
Maximum transpiration rate (mmol H_2O m^{-2} leaf s^{-1})	5.00
Extinction coefficient of direct PAR	0.81
Sap flow before cells collapse (L cm^{-2})	375

OM, organic matter.

12.3.1 Forest Structure

In GOTILWA, the forest is characterized by an initial distribution of diameter (D) classes. All trees within a size class are considered structurally and physiologically identical.

12.3.2 Interception, Stemflow and Throughfall

Interception and stemflow are estimated daily for each individual tree as a function of its leaf area and the amount of rainfall. They are integrated by adding the values obtained for all the trees in the forest. Throughfall is calculated from the amount of rainfall by the equation derived by Bellot (1989). The functions of these fluxes are:

$$I = 0.1225 \cdot LA \cdot 0.9965 \cdot R \cdot e^{c \cdot R}$$

$$S = \left[1 - e^{-0.013 \cdot LA}\right] \cdot \left[1.0 \cdot (R - I)^{1.191}\right]$$

$$T = 0.8711 \cdot R - 0.2363 \cdot D_r - 0.494,$$

where T, S and I account for throughfall, stemflow and interception, R is the amount of rainfall, D_r is the duration of this rainfall event and LA is the leaf area of each particular tree.

12.3.3 Actual Evapotranspiration

Actual evapotranspiration (Ea) is estimated using the evaporative coefficient modified for application on a daily basis. Basically, E_a is the result of three components: the available soil water (W), the available energy to evaporate this water (E_o) and the ability of plants to promote transpiration, the so-called evaporative coefficient (k). An increase in any one of these variables increases the final Ea. The following equation summarizes this relation:

$$Ea = k \cdot Eo \cdot W.$$

12.3.4 Streamflow

The streamflow leaving a catchment or a stand can be estimated by applying Darcy's law to the saturated soil area.

$$S_i = \frac{K_H \cdot A_s \cdot \frac{\partial H}{\partial X}}{S},$$

where S_i is the streamflow (mm) during the day i, K_H is the soil hydraulic conductivity (m day^{-1}), A_S is the saturated discharge area (m^2), $\delta H/\delta X$ is the hydraulic gradient (m m^{-1}) and S is the stand or catchment area (m^2). The value of A_S is $L \cdot Z_S$, Z_S being the saturated soil depth at the discharge point which is linearly related to the soil water content and L is the discharge length.

12.3.5 Actual Transpiration

The total actual transpiration is distributed among trees belonging to different diameter classes. The amount of water transpired by each tree is a function of the incident photosynthetically active radiation (PAR) reaching its canopy. The incident PAR on the canopy of a particular tree is estimated assuming that the canopies of all the larger trees are at a higher level. These trees intercept the available PAR according to a light extinction function. Thus, incident PAR on a tree canopy of diameter i is estimated as:

$$PAR_i = 100 \cdot e^{-\mu \cdot \sum_{j-i+1}^{n} LA_j \cdot N_j} \, ,$$

where μ is the PAR extinction coefficient, LA_j is the leaf area of a tree size j, N_j is the number of trees in the diameter class j and $j = i + 1$ to n account for all the trees larger than the trees of class i.

Transpiration is described as a Michaelis-Menten function of the incident PAR on the canopy. If $TR_{*,m}$ is the water transpired during a month by the forest, one tree of size i transpires:

$$TR_{i,m} = TR_{*,m} \cdot \dfrac{LA_i \cdot \dfrac{PAR_i}{K_m + PAR_i}}{\sum_{i=1}^{n} LA_i \cdot N_i \cdot \dfrac{PAR_i}{K_m + PAR_i}} \, ,$$

where LA_i is the leaf area of tree size i and K_m is the Michaelis–Menten constant (the percentage of PAR at which transpiration reaches 50% of the maximum transpiration rate).

12.3.6 Water Use Efficiency

The monthly carbon uptake of each tree is estimated from the water transpired by the tree and the mean value of water use efficiency (WUE, in mmol CO_2/mol H_2O). A different WUE for dominant (largest D classes) and suppressed trees (smallest D classes) can be defined. The WUE for the remaining classes is estimated by interpolating between both values according to the PAR reaching the canopy of each tree size class.

12.3.7 Gross Primary Production

The gross primary production (GPP) of a tree is estimated by combining the amount of transpired water and the WUE of the tree. The millimoles of absorbed CO_2 are estimated as:

$$U_{CO_2} = WUE_i \cdot TR_i \dfrac{1000}{18} \cdot \dfrac{atmCO_2}{350} \cdot \dfrac{\%N_{leaves}}{1.2\%N_{leaves}} \, ,$$

where TR_i is the water transpired by a tree of size i; the ratio 1000/18 is a factor to convert liters of water into moles; the ratio $atmCO_2/350$ corrects for the concentration of atmospheric CO_2, taking as a base concentration the present value of 350 ppm; the ratio% $N_{leaves}/1.2\%$ N_{leaves} adjusts the carbon uptake, considering 1.2 g N in 100 g dry matter of leaves as the base value for

350 ppm atmospheric CO_2. These corrections are only used in climate change simulations.

Absorbed CO_2 is converted into organic matter (OM) using:

$$GPP_i = U_{CO_2} \cdot 0.012 \cdot 2.1 \, ,$$

where U_{CO_2} is the carbon uptake (mmol), the factor 0.012 transforms the moles of carbon into grams, and 2.1 is the proportion of OM to C which is considered as a constant value (see Table 12.2)

12.3.8 Leaf Respiration

The respiration of any component of the plant is temperature-dependent. The role of temperature is introduced in GOTILWA using the Q_{10} function estimated as:

$$Q_{10,t} = Q_{10,20}^{\frac{T_m - 20}{10}} \, ,$$

where T_m is the mean temperature of the month. The final respiration rate depends on the fraction of mobile carbon present in the leaf. This fraction has a base respiration rate of 55.5 cal g^{-1} DM day^{-1} while the structural components have a base respiration rate of 33.3 cal g^{-1} DM day^{-1}, so the respiration of the leaf tissues is:

$$R_l = 55.5 \cdot C_m \cdot Q_{10,t} + 33.3 \cdot (1 - C_m) \cdot Q_{10,t} \, ,$$

where R_l is the leaf respiration rate, C_m is the fraction of mobile carbon in leaves and $Q_{10,t}$ is the value of Q_{10} at temperature t.

12.3.9 Wood Respiration

Similarly, the living woody tissues depend on the Q_{10} value and the base respiration rate, which has been estimated as 35 cal g^{-1} DM $year^{-1}$. The fraction of living xylem is a constant fraction of sapwood.

12.3.10 Fine Root Respiration

Fine root dynamics seems to parallel leaf dynamics in most ecosystems in which both have been analyzed (Kummerow and Ellis 1989). GOTILWA assumes that fine root biomass is a constant fraction of leaf biomass. Any variations in leaf area or leaf biomass can be translated into a proportional variation in fine root biomass.

12.3.11 Growth Respiration

Owing to biosynthesis costs and to the transport of carbon between organs and across cell membranes, the formation of new biomass in plants from net carbon uptake has a respiration cost. On average, 1 g of carbohydrates gives about 0.68 g of new tissues and the difference is consumed during the process of growth respiration.

12.3.12 Net Primary Production and Carbon Allocation

GPP minus total respiration (the sum of leaf, wood, fine root and growth respiration) is the net primary production (NPP). NPP is allocated to the different parts of the plant following a set of allocation rules. The plant uses a fraction of this NPP to form new leaves and fine roots to compensate their turnover. The remaining NPP is allocated to the pool of mobile carbon in leaves and woody tissues. If the NPP exceeds the carbon storage demand it is used in the formation of leaves and fine roots, in the case that leaf area is lower than the leaf area that can be supported by the existing sapwood area. The remaining NPP, if any, is invested in the production of new leaves and new wood in such a proportion that the new sapwood area and the new leaf area fit the constant ratio.

12.4 Model Validation

Actual weather data recorded at Poblet Meteorological Station, close to the Prades experimental area (Chap. 2), between 1981 and 1995 were used to simulate the growth of the holm oak (*Quercus ilex* L.) forest in Prades. To calibrate the model, a series of field data collected between 1981 and 1995 for tree density, increase in diameter, litterfall, tree ring width, LAI, basal area, aboveground biomass (Djema et al. 1994) and streamflow have been used even if all the series do not cover the entire period (Lledó et al. 1992).

A thinning experiment began in Prades in 1992 (Chap. 23). This experiment applied four thinning intensities replicated in three plots (0, control; 55 ± 2, 72 ± 3 and 79 ± 4% of basal area). The model validation used data from field measurements obtained from this experiment between 1992 and 1995 (Albeza et al. 1995), long-term series of litterfall recorded between 1982 and 1989 (Bellot et al. 1992) and weather data recorded during the same period. Figure 12.2 shows the agreement between observed and simulated tree ring increments in the control, minimum and maximum thinning plots as well as in the observed and simulated monthly litterfall.

Other field-measured variables such as mean diameter, LAI, basal area or aboveground biomass have been compared with the results obtained in the

MODEL VALIDATION

MONTHLY LITTERFALL

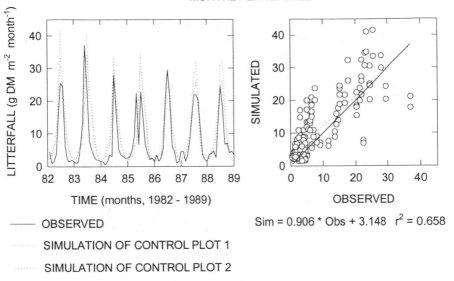

OBSERVED

SIMULATION OF CONTROL PLOT 1

SIMULATION OF CONTROL PLOT 2

Sim = 0.906 * Obs + 3.148 $r^2 = 0.658$

TREE RING WIDTH (mm), 1992 - 94

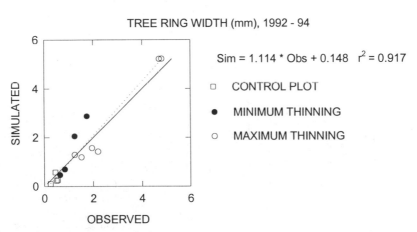

Sim = 1.114 * Obs + 0.148 $r^2 = 0.917$

□ CONTROL PLOT

● MINIMUM THINNING

○ MAXIMUM THINNING

Fig. 12.2. Validation of litterfall and tree ring increment of GOTILWA model. Simulated versus observed values of monthly litterfall from 1982 to 1989 and tree ring width recorded from 1992 until 1994. Both variables were measured in the Prades experimental area. *Continuous straight lines* are equality lines; *dashed straight lines* are linear regressions

simulations. In all cases, the simulations reproduce both the annual values and the observed annual patterns. Figure 12.3 shows the annual pattern of transpiration, leaf formation, tree ring formation and litterfall. Transpiration shows a clear pattern with two maximum values in spring and autumn,

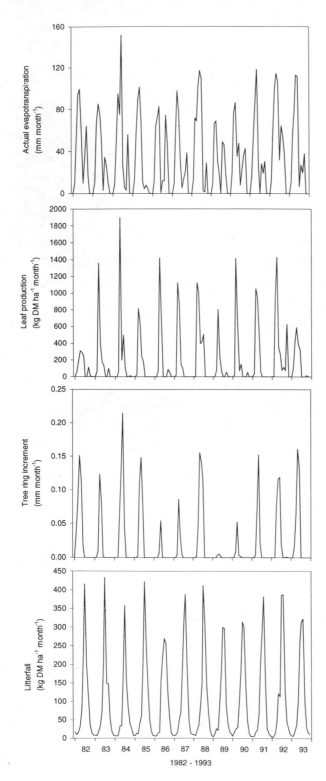

Fig. 12.3. Monthly evapotranspiration, leaf production, tree ring formation, and litterfall during the period 1982–1993 simulated using actual climate data. Leaf and tree ring formation show two main periods in spring and autumn, according to the transpiration pattern which is characteristic of the Mediterranean-type climate. Litterfall shows a single peak in summer when water stress is higher

characteristic of the Mediterranean climate areas. The annual pattern of other variables is also in agreement with field observations.

12.5 Model Application

GOTILWA has been used to explore how the thinning intensities applied to the Prades holm oak forest modify the effects of an increase in both atmospheric CO_2 and temperature and a reduction in water availability as predicted by most general circulation models (GCM) for the Mediterranean area (Rotmans et al. 1994). Two climate scenarios have been used in this analysis: reference (present climate, no change) and climate change in which we assume a linear increase of 150 ppm CO_2, a temperature increase of 3 °C, and a decrease of 15% in rainfall over the next 50 years.

Results after 50 years of simulation show that both climate change and thinning could have opposite effects on some main variables. This is the case, for example, for actual evapotranspiration (E_a). Thinning promotes a reduction of LAI (Fig. 12.4g) and might affect transpiration. E_a increases with increasing thinning intensity, 6% under the present climate scenario but only 2% under the climate change scenario. The slight differences shown by E_a are not reflected in the GPP (Fig. 12.4a). Despite the reduction of E_a, an increase in atmospheric CO_2 of 43% promotes a significant increase of 35% in GPP in all thinning treatments. The difference is due to the counterbalance of water and temperature stress. This result agrees with the prediction of Mooney et al. (1991) that in areas where water limits productivity, increased CO_2 will boost water use efficiency and is similar to the predictions obtained from the SPUR model applied to semi-arid catchments (Skiles and Hanson 1994).

Maintenance respiration involves an important proportion of GPP (Fig. 12.4b). This fraction decreases from 59% in the control to 48% in the maximum thinning treatment under the present climate. Under climate change scenario, the amount of GPP invested in maintenance respiration is much higher (72% in control and 64% in maximum thinning treatment). These values show that thinning promotes a proportional decrease of maintenance costs due to the reduction of total biomass. Climate change promotes an increase (about 62%) in maintenance respiration due mainly to the effect of increased temperature on the Q_{10} factor. In spite of the increasing GPP, the NPP does not significantly change due to this higher respiratory cost. Both climate change and thinning affect in a similar way litterfall (Fig. 12.4e) and leaf production (Fig. 12.4f). Thinned plots recover the former LAI under both climate scenarios after a few years. Increased temperature increases leaf shedding (71, 93 and 104% for control, minimum and maximum thinning treatments, respectively) and promotes an important increase in leaf production (57, 78 and 97% for control, minimum and maximum thinning, respectively).

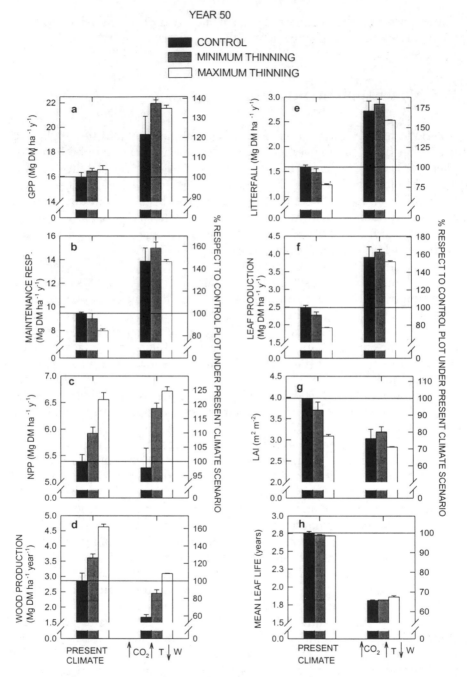

Fig. 12.4. Final values of modelled gross primary production (*GPP*, **a**), maintenance respiration (**b**), net primary production (*NPP*, **c**), wood production (**d**), litterfall (**e**), leaf production (**f**), leaf area index (*LAI*, **g**) and mean leaf life (**h**) in control and thinned plots after 50-year simulation under both climate scenarios (*T* increased temperature; *W* decreased precipitation). *Right axes* indicate relative value as a percentage with respect to control treatment under present climate scenario

Allocation of carbon to leaf and wood production differs with thinning intensity and climate scenario. Although leaf production does not show a clear pattern according to thinning intensity (Fig. 12.4f), the fraction of NPP allocated to leaf production decreases as thinning intensity increases under both climate scenarios (from 46 to 29% in the present climate, and from 74 to 56% with climate change). Thus, the resulting fraction of NPP invested in wood production increases as thinning intensity increases. The effect of climate change on leaf turnover has important consequences for the final structure of the forest. Leaf maintenance costs are so high under the climate change scenario (9.44 and 7.79 Mg DM ha^{-1} year^{-1} in control and maximum thinned plots, respectively) that LAI is reduced when compared to the present climate scenario (Fig. 12.4g). The reduction of LAI is more important in the control treatment (from 3.98 to 3.03 m^2 m^{-2}) than in thinning treatments (from 3.09 to 2.83 m^2 m^{-2} in maximum thinned plots) because the proportional respiratory costs are lower in thinning treatments. Even more important is the reduction of mean leaf life (Fig. 12.4h) from 2.7 years under the present climate scenario to 1.8 years under the climate change scenario due to the higher leaf turnover promoted by climate change. Changes in leaf turnover are also reflected in the final mean stem diameter. Thinning slightly decreases the proportion of NPP invested in leaf production and allows wood production to be increased.

The total biomass after 50 years of simulation shows roughly the same value under the present climate scenario (465 Mg DM ha^{-1}) while under the climate change scenario the control treatment shows a lower value than the thinning treatment (219 and 366 Mg DM ha^{-1}, respectively). Climate change promotes a lower final value of total biomass in all treatments. The higher maintenance respiration and leaf turnover while NPP remains constant leads to the mortality of a significant number of trees. The density of control treatment is 7661 stems ha^{-1} at the end of the present-climate simulation while it is 5074 stems ha^{-1} under the climate change scenario. The reduction of density is lower in the case of minimum thinning treatment (from 4775 to 4330 stems ha^{-1}), and there is no mortality in the maximum thinning treatment (2038 trees ha^{-1}).

Thinning effects tend to diminish after 50 years of simulation and most of the variables tend to converge to their former values, as shown by GPP (Fig. 12.4a) and total biomass (data not shown). The consequences of thinning differ in some major variables, such as the reduction of the proportion of GPP invested in maintenance respiration and the percentage of NPP used in leaf turnover.

Climate change promotes the opposite effects on the former variables. Even if the GPP increased by about 29%, climate change would promote a drastic reduction in LAI and total biomass because the temperature rise involves a marked increase in respiratory cost and leaf turnover. This is an important point to take into account if the results are compared with other predictions of the effects of climate change, where the total biomass generally increases due to the rise in CO_2 (Mooney et al. 1991; Skiles and Hanson 1994; Medlyn and Dewar 1996). In holm oak forests water and temperature stresses are limiting factors and their effects can overcome the fertilization effect of CO_2 increase.

Holm oak forests are highly sensitive to the increasing temperature due to climate change and, even if WUE increases, most of the increased carbon gain is accumulated in the soil due to higher litterfall and mortality (for control and minimum thinning treatments). This situation would be less critical for thinning treatments, because there is a higher water availability and the respiration increase is proportionally lower than that in control treatments. Thinning promotes a slight attenuation of the negative effects of climate change, and could be a useful tool for improving the response to the potential hydrological stress induced by climate change.

References

Albeza E, Sabaté S, Escarré A, Gracia CA (1995) A long-term thinning experiment on a *Quercus ilex* L. forest: changes in leaf characteristics and dynamics in response to different treatment intensities. In: Jenkins A, Ferrier RC, Kirby C (eds) Ecosystem manipulation experiments. Ecosystems Research Report 20, Commission of the European Communities, Luxembourg, pp 200–208

Bellot J (1989) Análisis de los flujos de la deposición global, trascolación, escorrentía cortical y deposición seca en el encinar mediterráneo de L'Avic (Sierra de Prades, Tarragona). PhD Thesis, University of Alicante, Alicante

Bellot J, Sánchez JR, Lledó MJ, Martínez P, Escarré A (1992) Litterfall as a measure of primary production in Mediterranean holm-oak forest. Vegetatio 99/100:69–76

Djema A, López B, Sabaté S, Gracia C (1994) A long-term thinning experiment on a *Quercus ilex* L. forest: changes in root biomass and nutrient contents. Not Biol (Santiago de Chile) 2(3):21

Kummerow J, Ellis BA (1989) Structure and function in chaparral shrubs. In: Keeley SC (ed) The California chaparral. Paradigms reexamined. Natural History Museum of Los Angeles County, Los Angeles, pp 141–150

Lledó MJ, Sánchez JR, Bellot J, Boronat J, Ibáñez JJ, Escarré A (1992) Structure biomass and production of a resprouted holm-oak (*Quercus ilex*) forest in NE Spain. Vegetatio 99/100:51–59

McKenney MS, Rosenberg NJ (1993) Sensitivity of some potential evapotranspiration estimation methods to climate change. Agric For Meteorol 64:81–110

Medlyn BE, Dewar RC (1996) A model of long-term response of carbon allocation and productivity of forests to increased CO_2 concentration and nitrogen deposition. Global Change Biol 2:367–376

Mooney HA, Drake BG, Luxmore RJ, Oechel WC, Pitelka LF (1991) Predicting ecosystem responses to elevated CO_2 concentration. BioScience 41:96–104

Piñol J, Lledó MJ, Escarré A (1991) Hydrological balance of two Mediterranean forested catchments (Prades, northeast Spain). Hydrol Sci J 36:95–107

Rotmans J, Hulme M, Downing TE (1994) Climate change implications for Europe. An application of the ESCAPE model. Global Environ Change 4:97–124

Sabaté S (1993) Canopy structure and nutrient content in a *Quercus ilex* L. forest of Prades mountains: effect of natural and experimental manipulation of growth conditions. PhD Thesis, University of Barcelona, Barcelona

Sala A (1992) Water relations, canopy structure, and canopy gas exchange in a *Quercus ilex* forest: variations in time and space. PhD Thesis, University of Barcelona, Barcelona

Sala A, Tenhunen JD (1994) Site-specific water relations and stomatal response of *Quercus ilex* in a Mediterranean watershed. Tree Physiol 14:601–617

Samani ZA, Heargreaves GH (1985) Water requirements, drought and extreme rainfall manual for the United States International Irrigation Center. Utah State University, Logan

Skiles JW, Hanson JD (1994) Responses of arid and semiarid watersheds to increasing carbon dioxide and climate change as shown by simulation studies. Climatic Change, 26:377–397

Specht RL (1972) The vegetation of South Australia. AB James Publisher, Adelaida

Tello E, Sabaté S, Bellot J, Gracia C (1994) Modelling the responses of Mediterranean forest to climate change: the role of canopy in water fluxes. Not Biol (Santiago de Chile) 2(3):55

Part 4   Nutrient Cycling at the Stand Level

13 Water and Nutrient Limitations to Primary Production

Ferran Rodà, Xavier Mayor, Santiago Sabaté and Victoria Diego

13.1 Introduction

Primary production is typically low in holm oak (*Quercus ilex* L.) forests (Chap. 3). Water availability is likely to limit primary production in most Mediterranean terrestrial ecosystems given that summer drought is a characteristic trait of Mediterranean-type climates (di Castri 1981). Despite its deep roots (Chap. 4) and its relative tolerance to foliar water deficits, such water limitation also applies to holm oak (Chaps. 10 and 11).

Nutrient availability can be also involved in the low production of Mediterranean forests, either through the inherently low fertility of some soils or through recurrent nutrient export resulting from a centuries-long history of harvesting and fire. Nutrient limitations to plant growth are widespread in Mediterranean-type shrublands in Australia, South Africa, and California (Hellmers et al. 1955; Oechel et al. 1981; Specht 1981; McMaster et al. 1982; Witkowski et al. 1990). From these and other studies, and from the general close link between sclerophylly and oligotrophic soils (Loveless 1961), the view emerged that low nutrient availability shaped the evolution and functioning of Mediterranean-type ecosystems (e.g. Kruger et al. 1983). However, the soils of the Mediterranean Basin are usually more eutrophic than those of Mediterranean-type shrublands in Australia, South Africa, and California.

Despite substantial information on nutrient distribution and cycling in holm oak forests (Chap. 18), we did not know whether, or to what extent, these forests were nutrient limited. We therefore conducted a fertilization and irrigation experiment to assess (1) the relative importance of water, nitrogen, and phosphorus as limiting factors for tree growth and primary production in holm oak forests, and (2) the responses to increased resource availability at the tree, stand, and ecosystem levels.

In this chapter, we first summarize the experiment layout, and then explore how the aboveground net primary production and its components were affected during the first 3 years of the experiment. We go on to discuss the mechanisms of the observed responses and, finally, incorporate what we

Ecological Studies, Vol. 137
Ferran Rodà et al. (eds) Ecology of Mediterranean
Evergreen Oak Forests
© Springer-Verlag, Berlin Heidelberg 1999

learnt in the follow-up of the experiment during the fourth year, which was exceptional in having a high spring–summer rainfall.

13.2 Experiment Layout

The experiment was performed in the Prades experimental area (NE Spain; Chap. 2), in a stand of closed-canopy holm oak, which shows the typically high density of small-sized stems developed by resprouting after earlier coppicing. This stand structure is highly relevant to the present experiment, since the ability of individual stems to respond to added water or nutrients might be limited in a crowded canopy by the low amounts of light intercepted by their narrow crowns.

The stand occupied a mid-slope position on a steep, southeast-facing slope at an altitude of 900 m. The soil was a shallow, stony, clay–loam Xerochrept on phyllites and metamorphic sandstones. The tree layer was 6 m high. At the start of the experiment, the density of stems of all species having a diameter at 50 cm from the ground of ≥ 2 cm averaged 18 275 stems ha^{-1}. Mean basal area at 50 cm from the ground was 47.9 m^2 ha^{-1}. Aboveground biomass of the tree layer averaged 128 Mg ha^{-1} (range 81–177 Mg ha^{-1}), a typical figure for closed-canopy holm oak forests in NE Spain (Chap. 3). Holm oak accounted on average for 91% of total aboveground biomass.

Details of the experimental design and methods can be found in Mayor and Rodà (1994) and Sabaté and Gracia (1994). Briefly, we used a factorial design with eight treatments combining three factors (irrigation, N fertilization, and P fertilization), each at two levels: with and without experimental addition. We used three replicated plots per treatment. Weekly drip irrigation (ca. 20 mm $week^{-1}$) was applied during the warm seasons of 1989 through 1991. Fertilization was conducted in March 1989 in a single dose. N-fertilized plots received 250 kg N ha^{-1} as ammonium nitrate. P-fertilized plots received 125 kg P ha^{-1} as calcium superphosphate.

13.3 Aboveground Net Primary Production

Aboveground net primary production (ANPP) was estimated as the sum of aboveground biomass increment and fine litterfall. Stem diameters were measured before the start of the experiment, and again 3 years later. These data and allometric regressions derived for this forest (Lledó 1990; S. Sabaté et al. unpubl. results) allowed us to compute the aboveground biomass. Fine litterfall was collected monthly. Here, we focus on the main results obtained in the first 3 years after treatment application began (1989–1991). Overall, this was a dry period, and rainfall was below average during the springs of 1989 and 1990.

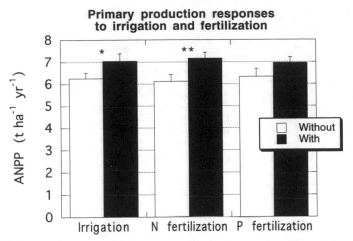

Fig. 13.1. Aboveground net primary production (ANPP) in holm oak forest at Prades during the first 3 years of the irrigation/fertilization experiment. Each column is the mean (+ 1 SE) of the 12 plots receiving/not receiving application of the specified factors

In control (i.e. untreated) plots, ANPP averaged only 5.1 Mg ha^{-1} year^{-1}, a typical figure for holm oak forests under dry conditions (Chap. 3). Irrigation and N fertilization each significantly increased ANPP (Fig. 13.1), while P fertilization did not. This demonstrates that ANPP in this forest is simultaneously limited by water and N availability. The effects of irrigation and of N fertilization were of similar (and moderate) magnitude: each factor increased ANPP on average by 0.8–1.1 Mg ha^{-1} year^{-1}. There were no significant interactions between the experimental factors, perhaps because the power of the experiment to detect interactions was rather low. Even so, the highest mean ANPP (8.1 Mg ha^{-1} year^{-1}) was observed in plots receiving irrigation plus NP-fertilization (INP treatment).

13.4 Components of Primary Production

We measured or estimated the two major components of ANPP: net increment of aboveground biomass and litterfall. Mean biomass increment ranged from 1.1 Mg ha^{-1} year^{-1} in control plots to 3.0 Mg ha^{-1} year^{-1} in INP plots. Mean fine litterfall ranged from 3.9 Mg ha^{-1} year^{-1} in control plots to 5.2 Mg ha^{-1} year^{-1} in INP plots.

Biomass increment and litterfall were affected by the experimental treatments in contrasting ways. Adding water to this holm oak forest significantly increased biomass increment but not litterfall; adding nitrogen had the opposite effect (Table 13.1). Therefore, increasing the availabilities of these two resources (water and nitrogen) had a similar overall effect on ANPP, but by

Table 13.1. Net increment of aboveground biomass and total fine litterfall (mean ± SE) in n experimental plots during the first 3 years after treatment application began. Lower part of table shows the result of pooling together all plots sharing the same experimental factor

Treatment	n	Biomass increment	Litterfall ($Mg\ ha^{-1} year^{-1}$)	Leaf litterfall
Control	3	1.14 ± 0.20	3.95 ± 0.14	2.43 ± 0.07
N fertilization (N)	3	1.76 ± 0.46	4.86 ± 0.16	2.58 ± 0.18
P fertilization (P)	3	2.48 ± 0.48	4.09 ± 0.12	2.31 ± 0.04
NP	3	1.92 ± 0.38	4.75 ± 0.60	2.61 ± 0.29
Irrigation (I)	3	2.43 ± 0.74	3.94 ± 0.35	2.13 ± 0.38
IN	3	2.47 ± 0.19	4.78 ± 0.42	2.13 ± 0.36
IP	3	2.51 ± 0.30	3.91 ± 0.25	2.04 ± 0.14
INP	3	2.96 ± 0.31	5.17 ± 0.23	2.60 ± 0.19
Grouped treatments				
Irrigation				
Without	12	1.82 ± 0.22	4.41 ± 0.18	2.48 ± 0.08
With	12	2.59 ± 0.20**	4.45 ± 0.21	2.23 ± 0.14*
N fertilization				
Without	12	2.14 ± 0.27	3.97 ± 0.10	2.23 ± 0.10
With	12	2.28 ± 0.21	4.89 ± 0.17***	2.48 ± 0.13*
P fertilization				
Without	12	1.95 ± 0.26	4.38 ± 0.18	2.32 ± 0.13
With	12	2.47 ± 0.19	4.48 ± 0.21	2.39 ± 0.11

*, $P < 0.1$; **, $P < 0.01$; ***, $P < 0.001$.

means of very different mechanisms, as we will discuss later. Phosphorus fertilization had no effect on the amount of litterfall, and its effect on biomass increment was on the verge of significance. In any case, P fertilization had at most a much smaller effect on ANPP components than irrigation and N fertilization did.

13.4.1 Biomass Increment

Given the way we computed stand biomass, any increments in the estimated aboveground biomass must result from increments in the diameter growth rates of trees, which were analyzed by Mayor and Rodà (1994) and Mayor et al. (1994). The mean diameter increment in control plots averaged 0.31 (SE 0.07) mm year^{-1} (Mayor et al. 1994). This low growth rate testifies to the fact that holm oak is indeed a slow growing species, particularly under dry conditions and in high density stands of resprout origin. Of course, the diameter growth of individual stems was highly dependent on the position of the stem within the canopy. For dominant stems (i.e. those receiving direct sunlight) mean diameter growth was 0.56 mm year^{-1}, while suppressed stems barely grew at all.

Irrigation significantly increased stem growth, but fertilization with either N or P did not. As fertilization and factor interactions had no detectable effects on stem growth, we pooled plots, taking into account irrigation categories only: for all non-irrigated plots taken together ($n = 12$), mean diameter increment was 0.33 ± 0.05 (SE) mm year^{-1}, whilst for irrigated plots ($n = 12$) it was 0.57 ± 0.05 (SE) mm year^{-1}. Water addition during the warm season thus increased the stem diameter increment on average by 66%.

13.4.2 Litterfall

The litterfall rate measured in our control plots (3.9 Mg ha^{-1} year^{-1}) was similar to those recorded in many other holm oak forests, but, surprisingly, it was much higher than in a nearby north-facing plot in the Avic catchment (Bellot et al. 1992; Chap. 3). In our control plots, holm oak leaf fall averaged 2.1 Mg ha^{-1} year^{-1} and accounted for 54% of fine litterfall.

The enhancing effect of N fertilization on fine litterfall (Table 13.1) was distributed among the different litterfall components. In particular, average fruit fall increased by 54% and holm oak leaf fall by 11%. Despite not having an effect on total fine litterfall, irrigation did increase fruit fall on average by 48%, an effect similar to that of nitrogen. Surprisingly, irrigation slightly decreased holm oak leaf fall, by 10% when averaged over the 3 years and over all plots receiving one of the irrigation treatments.

13.5 Mechanisms of Response

A detailed mechanistic analysis of the observed responses was not the objective of our experiment. Nonetheless, our studies conducted on the experimental plots shed light on some of the mechanisms underlying the effects of increased resource availability on primary production.

13.5.1 Improved Water Status

Weekly irrigation during the warm season was generally instrumental in increasing the soil water content of the forest floor and upper mineral soil of irrigated plots (V. Diego, unpubl. data). Mature sun-leaves from the topmost branches produced during the second growing season after treatment application began had a significantly lower ^{13}C content in irrigated than in control plots (F. Rodà et al. unpubl. data). This shows that, averaging over the lifespan of the leaf, irrigated trees had a lower water use efficiency than control trees (Farquhar et al. 1989) as a result of their improved water status. In this water-limited forest, where almost the entire annual precipitation is evapotranspired (Piñol et al. 1995; Chap. 19), the supplementary water added via irrigation was probably used to increase transpiration, either lengthening the periods of active gas exchange at the leaf surface and/or increasing the mean stomatal conductance. Both responses would allow a higher rate of net carbon fixation, eventually increasing stem diameter growth in irrigated trees, as we observed.

N fertilization did not significantly change the ^{13}C content of holm oak leaves. Thus, and although improved N nutrition often leads to higher water use efficiency, the increased N supply did not have such an effect in our holm oak plots.

13.5.2 Increased Photosynthetic Capacity

N fertilization significantly increased the N concentration in topmost sun leaves produced during the second growing season after fertilization, from a mean of ca. 1.3% in non N-fertilized plots to ca. 1.5% in N-fertilized plots (Sabaté and Gracia 1994). Irrigation and P fertilization did not affect the foliar N concentration. Given the general relationship between foliar N content and photosynthetic capacity (Field and Mooney 1986), it is likely that our N-fertilized holm oaks had an increased rate of net carbon fixation, leading to a higher ANPP. Such an effect was not involved in the ANPP response to irrigation. At the canopy level, increased carbon gain resulted probably also from the increased proportion of current-year leaves making up the leaf population in N-fertilized plots and, to a lesser extent, in irrigated plots (Fig. 13.2b,c).

13.5.3 Increased Light Interception

Light interception capacity is closely linked to primary production (Landsberg 1986; Cannell 1989). By the second summer after treatment application began, the total leaf area displayed per unit of cross-sectional area of topmost branches had increased as a result of N fertilization (Fig. 13.2a). This implies that N-fertilized canopies had a greater capacity to intercept light. At this time, irrigation and P fertilization had no effect on total leaf area supported per branch.

In April 1992, 3 years after the start of the experiment, light penetration was measured below the canopy of each plot. Irrigation and N fertilization each lowered light penetration, in both cases from average values of 91 µE PAR m^{-2} s^{-1} in plots not receiving one of the above factors to 65 µE m^{-2} s^{-1} in those receiving one of them. The effects of water and N were additive, and P fertilization had no effect on light penetration.

Thus, both N fertilization and irrigation seemed to increase the leaf area index (LAI) and the light interception capacity of the stand, though the effects of N were detected 1 year earlier than those of irrigation. An increase in LAI after N fertilization of forests is commonly reported in the literature (e.g. Binkley and Reid 1984; Vose et al. 1994), but we were surprised to find such a

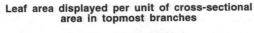

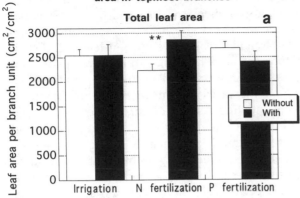

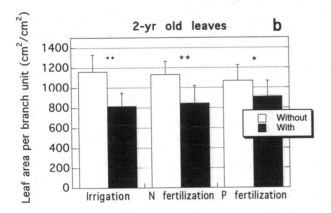

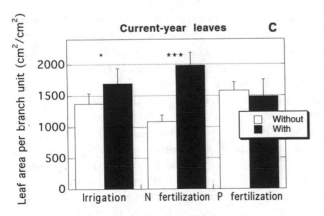

Fig. 13.2. Leaf area displayed per unit of cross-sectional area in topmost branches sampled in August 1990 (i.e. the second summer after treatment application began). a Total leaf area; b area of 2-year-old leaves; c area of already mature current-year leaves. Area of 1-year-old leaves was negligible in all treatments because 1989 was a severely dry year. Each column is the mean (+1 SE) of the 12 plots receiving/not receiving application of the specified factors

high increase in this holm oak forest, where the canopy was already closed
before fertilization and where water availability strongly limits tree growth.

The effect of the experimental treatments on LAI and light interception
was, at least in part, due to the increased size of leaves and stems of new
shoots. We measured the size (weight, area, and length) of leaves and stems
of shoot cohorts borne on topmost branches sampled on several occasions
(Sabaté 1993). For organs produced in the second spring after treatment application began, N fertilization increased the mean weight of individual
leaves and the mean length of individual stems of current-year shoots. For
this same cohort, irrigation increased the mean weight of leaves and standing fruits (sampled in August) and the mean weight and length of stems of
current-year shoots. These responses, added to possible (but unmeasured)
increases in the number of new leaves produced, accounted for the observed
increases in leaf area supported per branch unit (Fig. 13.2a) and in light interception in plots receiving either irrigation or N fertilization. A further
possible mechanism leading to increased LAI could be the extended retention of older leaves. In fact, the opposite was observed: when the leaf area
supported per unit of cross-sectional area of branch was split into leaf cohorts, all experimental factors showed accelerated shedding of the older cohort (2-year-old leaves; Fig. 13.2b). Thus, neither supplemental water, nitrogen, nor phosphorus increased the retention of older leaves. Total leaf area
borne on topmost branches in N-fertilized plots therefore increased despite
accelerated shedding of old leaves. For irrigated plots, total leaf area remained constant because the accelerated shedding of old leaves and the increased production of new leaves were of the same magnitude (Fig. 13.2c).

The above results show that, in irrigated or N-fertilized plots, the canopies
were not in steady state over the first 3 years of the experiment, but rather
they accumulated leaf biomass during this period. This was probably a major
factor behind the increased ANPP induced by both experimental factors.
However, a non-steady state leaf biomass implies that we underestimated the
net biomass increment because we used the same allometric regressions for
estimating the biomass at the start of the experiment and 3 years later. This
procedure was probably suitable for estimating the increment in woody biomass since, given the very low rates of diameter and height growth we found
even in treated plots, it was unlikely that allometric differences in woody
biomass between treated and untreated trees appeared in only 3 years. On
the other hand, for faster-aggrading tree components such as leaves, using
pre-treatment allometric regressions surely underestimated the net biomass
increment and the ANPP of, particularly, N-fertilized plots (Fig. 13.2c). Yet,
this simply reinforces our findings on the effect of N fertilization on canopy
dynamics and litterfall. The net effect of incorporating the increase in
standing leaf biomass in our estimates would be to strengthen the effect of N
fertilization. In particular, ANPP of this holm oak forest is somewhat more
responsive to N fertilization than Fig. 13.1 suggests.

13.6 Follow-up to the Experiment

Our experiment was designed to run for 3 years, and irrigation was discontinued after the third summer (1991). However, in March 1992 fertilizers were applied again in the same plots and doses as in 1989, in order to maintain a high nutrient availability for other studies. The fourth year after the first treatments (1992) happened to be one of exceptionally high spring–summer rainfall. We took advantage of this opportunity to assess the tree growth and litterfall responses to a natural reduction in summer water stress, and to test whether such responses were affected by irrigation during the previous 3 years and by fertilization. We were particularly interested in comparing the effects of high rainfall during the warm season to those of experimental irrigation, and testing whether a high light-interception capacity (as induced by previous irrigation and by N fertilization) had a carry-on positive effect on stem diameter growth. So, stem diameters of all trees were re-measured at the end of this growing season (December 1992), and monthly litterfall collections were continued through 1992.

In control plots, mean stem diameter growth of holm oak during 1992 was 0.67 mm year^{-1} ($\pm$ 0.19 SE among plot means), a growth rate doubling that of the 3 preceding years (Mayor et al. 1994). Diameter growth in control plots in 1992 was higher than that achieved during 1989–1991 in the irrigated plots (mean of the I and IP treatments: 0.54 mm year^{-1}). These two observations clearly indicate that 1992 was a much better year for holm oak growth than the 3 previous years, and that soil water availability during the growing season was probably increased more by the abundant spring–summer rainfall of 1992 than it had been by irrigation in the preceding years. Annual rainfall in 1992 was 855 mm, while mean rainfall plus irrigation in 1989–1991 amounted to only 735 mm. The inability of forest irrigation to equal the effect of natural precipitation has been also observed in experiments in other forest types (Lucier and Hinckley 1982; Gower et al. 1992).

When all holm oak stems (both dominant and suppressed) were included in the analysis, none of the three experimental factors (previous irrigation, N fertilization, and P fertilization) significantly affected stem diameter growth during 1992. For dominant stems, N fertilization significantly increased mean stem diameter growth in 1992 by 42%, compared to non-fertilized stems, while P fertilization and previous irrigation did not. Thus, increased light interception capacity by itself (as shown by the previously irrigated plots) was not conducive to increased stem diameter growth in a year of relatively high water availability. N fertilization increased the diameter growth of dominant trees in 1992, an effect which had not taken place during the 3 previous years. Such a different response could arise from (1) fertilizer being applied a second time in March 1992, (2) a time-delayed response to the first fertilization, or (3) the effects of improved N nutrition or increased light interception capacity coupled with a high soil water availability during the growth season of 1992.

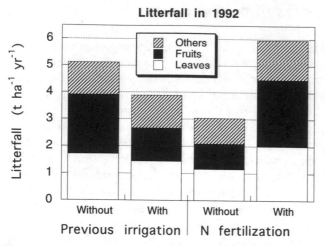

Fig. 13.3. Litterfall in 1992, the fourth year after treatment application began. Fertilization was repeated in 1992 in the same way as in 1989. No irrigation was applied in 1992. Data are shown for holm oak leaves, holm oak fruits, and others (twigs, holm oak catkins, leaves of other species, and unidentified litter). Each column is the mean of the 12 plots receiving/not receiving application of the specified factors

In 1992, litterfall was increased by N fertilization, as it had also been in the 3 previous years, though the effect was much stronger in 1992 (Fig. 13.3). Total fine litterfall in 1992 doubled in response to N fertilization, from an average of 3.0 Mg ha⁻¹ year⁻¹ in non-fertilized plots to 6.0 in fertilized ones. Leaf fall almost doubled, and there was massive fruit production, which accounted on average for 42% of the annual litterfall in N-fertilized plots.

Irrigation during the 3 previous years induced a sharp drop in litterfall in 1992 (Fig. 13.3). The reduction was moderate for holm oak leaves and greater for fruits. During the 3 previous years, irrigation had slightly reduced leaf fall but had increased fruit fall. For control plots, reduction of leaf fall was also seen as a result of the naturally increased water availability in 1992: in these plots, mean fall of holm oak leaves was 2.1 (SE 0.07) Mg ha⁻¹ year⁻¹ in 1989–1991, and only 1.4 (SE 0.22) Mg ha⁻¹ year⁻¹ in 1992. In common with the preceding years, phosphorus application had no effect on total litterfall or major litterfall components in 1992.

13.7 Concluding Remarks

Holm oak is a slow-growing species, often seen as employing a conservative strategy, i.e. low responsiveness to environmental changes and, particularly, parsimonious use of available resources. While this may be the case, our experiment shows that (1) holm oak has a substantial capacity to respond to increased availabilities of soil resources, (2) it does so relatively quickly, and (3) the high stem densities and the probable old age of the stools do not prevent holm oak from taking advantage of increased resource availabilities.

Holm oak responded quite differently to water and to nitrogen additions. Increased soil water availability during the warm season greatly increased the diameter growth rate of tree stems, both in the case of irrigation and in the case of a year of abundant spring–summer rainfall. Surprisingly, increased water availability led to a decrease in leaf fall, even on a multi-year average. On the other hand, N fertilization did not increase the stem diameter growth, at least during the 3 years after the first application, but it did have a strong effect on canopy dynamics. This suggests that holm oak preferentially allocates the increased soil N to leaf and twig production, thus improving its light interception capacity. Phosphorus addition had remarkably few effects on this stand, perhaps because of the already relatively high phosphorus availability in this circumneutral soil.

Our experiment demonstrates that water and nitrogen simultaneously limit ANPP in the widespread high-density holm oak stands of resprout origin. In managing holm oak stands, therefore, it is essential to preserve the water-holding capacity of the soil, and to minimize nitrogen losses from the site.

References

Bellot J, Sánchez JR, Lledó MJ, Martínez P, Escarré A (1992) Litterfall as a measure of primary production in Mediterranean holm oak forest. Vegetatio 99/100:69–76

Binkley D, Reid P (1984) Long-term responses of stem growth and leaf area to thinning and fertilization in a Douglas-fir plantation. Can J For Res 14:656–660.

Cannell MGR (1989) Physiological basis of wood production: a review. Scand J For Res 4: 459–490

di Castri F (1981) Mediterranean-type shrublands of the world. In: di Castri F, Goodall DW, Specht RL (eds) Mediterranean-type shrublands. Elsevier, Amsterdam, pp 1–52

Farquhar GD, Hubick KT, Condon AG, Richards RA (1989) Carbon isotope fractionation and plant water-use efficiency. In: Rundel PW, Ehleringer JR, Nagy KA (eds) Stable isotopes in ecological research. Springer, New York, pp 21–40

Field C, Mooney HA (1986) The photosynthesis–nitrogen relationship in wild plants. In: Givnish TJ (ed) On the economy of plant form and function. Cambridge University Press, Cambridge, pp 25–55

Gower ST, Vogt KA, Grier CC (1992) Carbon dynamics of Rocky Mountain Douglas-fir: influence of water and nutrient availability. Ecol Monogr 62:43–65

Hellmers H, Bonner JF, Kelleher JM (1955) Soil fertility: a watershed management problem in the Sant Gabriel mountains of southern California. Soil Sci 80:189–197

Kruger FJ, Mitchell DT, Jarvis JUM (eds) (1983) Mediterranean-type ecosystems. The role of nutrients. Springer, Berlin

Landsberg JJ (1986) Physiological ecology of forest production. Academic Press, London

Lledó MJ (1990) Compartimentos y flujos biogeoquímicos en una cuenca de encinar del Monte Poblet. PhD Thesis, University of Alicante, Alicante

Loveless AR (1961) A nutritional interpretation of sclerophylly based on differences in the chemical composition of sclerophyllous and mesophytic leaves. Ann Bot 25:168–184

Lucier AA, Hinckley TM (1982) Phenology, growth, and water relations of irrigated and non-irrigated black walnut. For Ecol Manage 4:127–142

Mayor X, Rodà F (1994) Effects of irrigation and fertilization on stem diameter growth in a Mediterranean holm oak forest. For Ecol Manage 68:119–126

Mayor X, Belmonte R, Rodrigo A, Rodà F, Piñol J (1994) Crecimiento diametral de la encina (*Quercus ilex* L.) en un año de abundante precipitación estival: efecto de la irrigación previa y de la fertilización. Orsis 9:13–23

McMaster GS, Jow WM, Kummerow J (1982) Response of *Adenostoma fasciculatum* and *Ceanothus greggii* chaparral to nutrient additions. J Ecol 70:745–756

Oechel WC, Lowell W, Jarrell W (1981) Nutrient and environmental controls on carbon flux in Mediterranean shrubs from California. In: Margaris NS, Mooney HA (eds) Components of productivity of Mediterranean-climate regions. Dr W Junk Publishers, The Hague, pp 53–59

Piñol J, Terradas J, Àvila A, Rodà F (1995) Using catchments of contrasting hydrological conditions to explore climate change effects on water and nutrient flows in Mediterranean forests. In: Moreno JM, Oechel WC (eds) Global change and Mediterranean-type ecosystems. Springer, New York, pp 371–385

Sabaté S (1993) Canopy structure and nutrient content in a *Quercus ilex* L. forest of Prades mountains: effect of natural and experimental manipulation of growth conditions. PhD Thesis, University of Barcelona, Barcelona

Sabaté S, Gracia C (1994) Canopy nutrient content of a *Quercus ilex* L. forest: fertilization and irrigation effects. For Ecol Manage 68:31–37

Specht RL (1981) Primary production in Mediterranean-climate ecosystems regenerating after fire. In: Di Castri F, Goodall DW, Specht RL (eds) Mediterranean-type shrublands. Elsevier, Amsterdam, pp 257–267

Tanner EVJ, Kapos V, Franco W (1992) Nitrogen and phosphorus fertilization effects on Venezuelan montane forest trunk growth and litterfall. Ecology 73:78–86

Vose JM, Dougherty PM, Long JN, Smith FW, Gholz HL Curran PJ (1994) Factors influencing the amount and distribution of leaf area of pine stands. Ecol Bull (Copenhagen) 43:102–114

Witkowski ETF, Mitchell DT, Stock WD (1990) Response of a cape fynbos ecosystem to nutrient additions: shoot growth and nutrient contents of a proteoid (*Leucospermum parile*) and an ericoid (*Phylica cephalantha*) evergreen shrub. Acta Oecol 11: 311–326

14 Precipitation Chemistry and Air Pollution

Antoni Escarré, Adoración Carratalá, Anna Àvila, Juan Bellot,
Josep Piñol and Millán Millán

14.1 Introduction

Interest in precipitation chemistry greatly expanded as a result of the increasing rain acidity over northern Europe and northeastern North America
in the early 1970s. The links between air pollution, acid rain and forest health
have generated many thousands of publications in the last 25 years (e.g.
Drablos and Tollan 1980; Hutchinson and Havas 1980; Ulrich and Pankrath
1983). Precipitation chemistry is indeed relevant to monitor air pollution,
particularly in places distant from source areas, since the principal atmospheric pollutants, sulphur dioxide (SO_2) and nitrogen oxides (NO_x), can be
incorporated as sulphate and nitrate ions in bulk deposition. Changes in the
emission of these gases are reflected in variations in nitrate and sulphate in
precipitation.

Precipitation chemistry is also relevant in ecosystem nutrient cycles (e.g.
Johnson and Lindberg 1992; Likens and Bormann 1995). Nutrient inputs in
precipitation can make a major contribution to total ecosystem inputs and
even to total nutrient capital of the ecosystem in nutrient-poor sites. Nutrients in precipitation can proceed from a variety of natural and anthropogenic sources, including air pollution. For example, excessive atmospheric
deposition of nitrogen is affecting terrestrial and freshwater ecosystems in
many European countries (Aber 1992).

Bulk deposition is the input flux most often used in nutrient cycling studies because it is relatively easy to measure. Research on bulk precipitation
chemistry in the holm oak (*Quercus ilex* L.) forests at Montseny and Prades
(NE Spain) began in 1978 and 1980, respectively. The characteristics of the
precipitation chemistry and the associated nutrient inputs to the ecosystem
are quite well established for these sites (Rodà 1983; Àvila 1988, 1996; Bellot
1989; Bellot and Escarré 1989; Piñol 1990; Rodà et al. 1993).

In this chapter we first summarize the characteristics of precipitation
chemistry at our two holm oak sites, and infer the sources of delivered elements from ion relationships in precipitation. We then take a broader geo-

Ecological Studies, Vol. 137
Ferran Rodà et al. (eds) Ecology of Mediterranean
Evergreen Oak Forests
© Springer-Verlag, Berlin Heidelberg 1999

graphical view using a precipitation network in eastern Spain, and a long temporal view analyzing decadal trends in precipitation chemistry at our holm oak sites. Using back-trajectory analysis and mesoscale meteorology, we next attempt to outline the extent of the airshed ecosystem and to identify the mesoscale determinants of pollutant transport in this area. We finish with an assessment of the relevance of precipitation inputs for the nutrient cycles in holm oak ecosystems.

14.2 Bulk Precipitation Chemistry at Prades and Montseny

A continuous event/weekly sampling schedule of bulk precipitation chemistry started at Prades and Montseny in 1981 and 1983, respectively, as part of catchment monitoring programmes. At Prades, bulk precipitation was sampled in l'Avic catchment until December 1983 and at the outlet of la Teula catchment thereafter (Chap. 2). At Montseny, it was sampled at the outlet of the TM9 catchment (Chap. 2).

Volume-weighted mean (VWM) concentrations for 13-year (Prades) and 11-year (Montseny) periods show that the average bulk precipitation chemistry was similar at both sites (Table 14.1). Except for H^+, ion concentrations were slightly lower at Montseny than at Prades, probably reflecting the higher annual precipitation at Montseny. Precipitation chemistry at both sites is characterized by a moderate marine influence, moderate N concentrations, and positive mean alkalinity, with SO_4^{2-} being mostly balanced by Ca^{2+}. This contrasts with the most frequent situation in central and northern Europe and eastern North America where precipitation is acidic (i.e. has negative alkalinity) and H^+ is the major counterpart to SO_4^{2-} and NO_3^- (Table 14.2).

Table 14.1. Mean chemistry of bulk precipitation at Montseny and Prades. Figures are arithmetic means of annual volume-weighted mean (VWM) concentrations for an 11-year period at Montseny and a 13-year period at Prades

Site		Precipitation	H^+	Alk.	Na^+	K^+	Ca^{2+}	Mg^{2+}	NH_4^+	NO_3^-	SO_4^{2-}	Cl^-
		(mm year^{-1})					(μeq L^{-1})					
Montseny[a]	Mean	876	14.3[b]	13.5	23	4.1	57	9.8	23.1	21	46	29
	CV (%)	23	36	183	21	34	41	26	14	14	18	24
Prades[c]	Mean	551	11.4[b]	23.4	26	8.6	59	10.7	29.1	23	56	30
	CV (%)	27	58	66.4	36	42	36	83	38.1	33	33	25

CV, coefficient of variation; Alk., alkalinity.
[a] Sampling period at Montseny: August 1983–July 1994.
[b] Computed from pH of each sample, i.e. without taking into account neutralization by alkaline samples. If this is included, resultant VWM H^+ concentrations are much lower: 0.36 μeq L^{-1} at Montseny, and 0.21 at Prades.
[c] Sampling period at Prades: November 1981–November 1994.

Table 14.2. Volume-weighted mean precipitation chemistry in 7 localities in eastern Spain along the Mediterranean seaboard (1–7) and in 7 localities from the European Network (EMEP) (8–14). Underlined values in pH and H$^+$ columns correspond to localities were alkaline samples are frequent or dominant; these values, computed from pH of each sample, do not take into account neutralization by alkaline samples, and underestimate pH (and overestimate H$^+$ concentration) of the average precipitation solution. Column labelled pH$_{eq}$ gives the correct average air-equilibrated pH in this case, computed from volume-weighted mean alkalinity

Locality[b]	Rainfall (mm year^{-1})	pH$_{eq}$	pH	H$^+$	Alkal.	Na$^+$	K$^+$	Ca^{2+}	Mg^{2+}	NH$_4^+$	NO$_3^-$	SO$_4^{2-}$	Cl$^-$
								(μeq L^{-1})					
1 Pyrenees (B)	1640	7.16	5.30	5.00	73	23	9	94	8	22	18	51	20
2 Roquetes (W)	428	–	5.84	1.45	–	69	7	142	25	53	46	48	50
3 Morella (B)	678	7.42	5.66	2.19	133	24	12	127	16	42	24	75	34
4 Sogorb (B)	470	7.50	5.60	2.53	175	49	11	202	23	41	28	106	59
5 Ontinyent (B)	780	7.17	4.78	16.58	73	62	9	101	29	53	35	95	79
6 Elx (B)	186	7.57	6.23	0.60	186	60	10	201	38	51	38	99	75
7 Filabres (B)	675	7.40	5.71	0.51	125	37	15	123	34	32	19	58	34
8 Deuselbach (W)	691	–	4.64	22.91	–	22	4.8	16	6.4	33	30	23	25
9 Iraty (W)	668	–	5.07	8.51	–	13	1	21	8	57	22	23	13
10 Aliartos (W)	–	–	4.78	16.60	–	111	12	204	–	158	47	159	128
11 Ispra (W)	1233	–	4.34	45.71	–	7.3	1.5	17	4	71	62	35	10
12 Witteveen (W)	844	–	4.83	14.79	–	113	–	19	25	139	51	87	132
13 Braganza (W)	420	–	5.41	3.89	–	12	3.5	34	11	18	10	25	11
14 Kamernicki (W)	527	–	5.12	7.59	–	53	19	82	26	87	36	75	69

(B) = bulk precipitation. (W) = wet-only precipitation.

[a] (8) Germany, (9) France, (10) Greece, (11) Italy, (12) Norway, (13) Portugal, (14) former Yugoslavia.

[b] References: 1 Camarero and Catalan (1993); 3–5 Carratalá et al. (1996); 6 Carratalá (1993); 7 Domingo (1991); 2 and 8–14 Pedersen et al. (1992).

Acidic precipitation events (pH < 4.5) occurred both at Montseny and Prades. They accounted, respectively, for 20 and 14% of the number of samples. Alkaline rains were more frequent, with 34% (Montseny) and 48% (Prades) of the samples having a pH > 5.6. The pH of individual events (or weekly samples) ranged from 4 to 8 at both sites. Such co-occurrence of acidic and alkaline rains within a site precludes the use of the conventional method of computing mean precipitation pH from VWM H$^+$ concentrations because H$^+$ is not conservative under these conditions. Instead, alkalinity is the conservative property to average (Liljestrand 1985; Young et al. 1988). The resultant VWM alkalinities (Table 14.1) would correspond to an air-equilibrated pH of 6.4 at Montseny and 6.7 at Prades, assuming that bicarbonate accounts for most of the alkalinity. Despite the occurrence of some acidic events, the overall picture that emerges from these data is that acidic precipitation has not been a relevant environmental stress in either holm oak ecosystem during the last 20 years.

14.3 Geographical Patterns

Precipitation chemistry at our two holm oak sites in NE Spain can be compared with that at seven other localities spread along the Mediterranean seaboard of Spain (Table 14.2). These localities span a gradient of increasing regional aridity from the Catalan Pyrenees in NE Spain to Filabres (Almería, SE Spain), though the respective annual rainfalls reported in Table 14.2 partially mask this regional trend due to local orographic effects. Precipitation at all sites has positive VWM alkalinity and high Ca^{2+} concentrations (Table 14.2). Ion concentrations are generally lower in the wettest Pyrenees site and in our two holm oak forest sites (Table 14.1) than at the other sites, which are regionally drier and are surrounded by less vegetated or more agricultural landscapes.

The alkaline character of precipitation all along eastern Spain results from the incorporation of carbonate-rich aerosols into the rain, of local or remote origin. Local sources are minor in forested areas, such as our experimental sites at Prades and Montseny, but they increase with increasing aridity. Nonetheless, a major remote source in the circum-Mediterranean area is the long-range transport of dust particles from North Africa (Löye-Pilot et al. 1986; Glavas 1988; Caboi et al. 1992; Àvila et al. 1996, 1997; Carratalá et al. 1996). This northward transport produces episodes of dust-laden rains (known as red rains), which have high alkalinity and high Ca^{2+} concentrations (Löye-Pilot et al. 1986; Rodà et al. 1993). The important neutralizing role of African red rains at our sites is revealed by the positive relationship between the percentage of annual precipitation occurring as red rains and the annual mean pH (Fig. 14.1). Note that years without important contributions of red rains present acidic mean pHs.

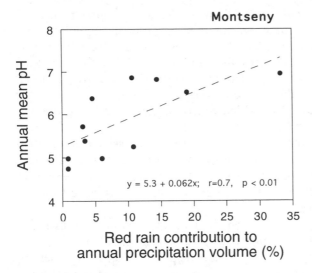

Fig. 14.1. Relationship between annual mean equivalent air-equilibrated pH (computed from annual VWM alkalinity) and percentage of the annual precipitation contributed by red rains at Montseny, 1983–1994

A more detailed evaluation of geographical patterns of bulk precipitation chemistry in eastern Spain was obtained with the network of the Valencian Autonomous Community, which consisted of 27 sampling stations spread over 25 000 km² (Carratalá 1993). In this network, the deposition fluxes of marine-related elements (Cl⁻, Na⁺ and Mg²⁺) were negatively correlated with distance from the sea, as is usually the case. The concentrations of NO₃⁻ also decreased significantly inland (and NO₃⁻ deposition marginally so), probably reflecting the accumulation of human activities along the coast. These patterns did not simply mirror that of annual precipitation, since the latter was uncorrelated with distance from the sea. The annual VWM H⁺ concentration was weakly but positively correlated with the annual precipitation of each locality (r = 0.51, P < 0.01), suggesting that alkaline inputs through dry deposition or washout were increasingly diluted and/or that the source strength of alkaline material decreased with increasing annual precipitation.

14.4 Sources of Ions in Precipitation

A method widely used to distinguish the origin of ions in precipitation is based on the hypothesis that all the Na⁺ or Cl⁻ in precipitation derive from sea-salt aerosols. Hence, it is possible to estimate enrichment ratios for the different ions by comparing the ratios of the ions to Na⁺ (or Cl⁻) in rainwater with those in seawater. At both Prades and Montseny, the ratio Na/Cl was consistently close to the marine ratio (Rodà 1983; Bellot 1989; Piñol 1990). The other ions were all enriched to different degrees, with continental sources accounting for 88–98% of K⁺, Ca²⁺ and SO₄²⁻, and about half of Mg²⁺ in bulk precipitation at both sites.

Another approach to define the origin of the ions in precipitation is through correlation analysis between the ions. Single or multiple linear regressions between SO_4^{2-} and NO_3^- and H^+ have been used to explain the precipitation acidity (Gorham 1955; Cogbill and Likens 1974; Gorham et al. 1984). At Prades and Montseny, correlations between SO_4^{2-} and NO_3^- with H^+ were not significant, while these anions were significantly correlated with Ca^{2+} and NH_4^+ (Àvila 1996), ions which contribute to neutralizing the precipitation acidity. The neutralization capacity of the atmosphere at our sites can also be inferred from the increase in acidity in successive rain events, a fact that suggests the important role of washout in providing alkaline species which become depleted in successive rains (Bellot and Escarré 1989; Carratalá 1993). This finding substantiates the correlational evidence from the Valencian network (Sect. 14.3).

A more general approach relies on multivariate methods as principal component analysis. Different data sets of precipitation chemistry from this Mediterranean area usually produce a first principal component representing the degree of mineralization. Rains of high ion concentrations, mostly corresponding to African red rains, are well separated by this component. The second principal component reflects the marine influence since it is mainly correlated with Na^+ and Cl^-. The third component is associated with the rain acidity. Similar results were obtained at Mont Lozère (southern France), except that the third component was defined by the acid anions SO_4^{2-} and NO_3^- and pH was not significantly correlated with the other elements (Durand et al. 1992).

14.5 The Airshed

Precipitation chemistry depends strongly on the acquisition of gaseous and particulate materials by moving air masses, and on the subsequent transformations of these materials. From an ecosystem perspective, the regions where these airborne materials originate define the loose boundaries of the airshed for a given ecosystem. In this section we first trace the movements of air masses reaching holm oak forests in NE Spain in order to identify the origin of elements delivered by bulk precipitation. We then analyze the particular meteorological conditions prevailing over Iberia in summer that are involved in the buildup of secondary air pollutants above the Spanish Mediterranean coast.

14.5.1 Back Trajectories and Origin of Air Masses

The back trajectory of an air mass arriving at a given locality at a certain time can be computed by tracing its movement through pressure fields or isentropic surfaces in the atmosphere. Using the Valencian precipitation net-

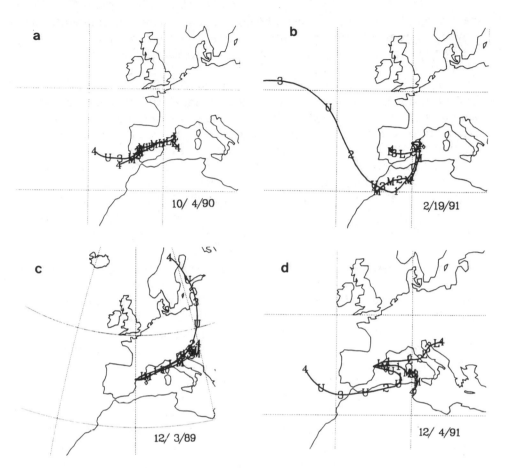

Fig. 14.2. Examples of computed back trajectories at three levels (L 1000 hPa; M 850 hPa; U 700 hPa) for episodes of **a** oceanic, **b** African, **c** continental, and **d** mixed continental and African origins

work referred to above, back trajectories were calculated for 23 episodes of several precipitation days chosen to minimize local influences (Carratalá 1993; Carratalá et al. 1996). Three main routes of transport were found: (1) oceanic, (2) continental, and (3) African. Continental rains produced the greatest amount of precipitation (62%), followed by oceanic (20%) and African rains (18%).

Oceanic trajectories were associated with fronts crossing the Iberian Peninsula from W to E (coming from the Atlantic Ocean), and were characterized by rains of pH close to neutral and low ionic content (Fig. 14.2a). Continental fronts were characterized by high pressures over south Scandinavia or eastern Europe and low pressures around the strait of Gibraltar, with continental cold air from eastern Europe being channelled over the Mediterranean (Fig. 14.2c,d). In this situation, SO_2 and other pollutants emitted from power plants in eastern and central Europe would be transported south-westwards for distances greater than 1500 km, reaching the Mediterranean area through the Triest Gap (Camuffo et al. 1991). The most acid rains reported in the Valencian network were associated with this continental flow.

African air masses were associated with a synoptic situation of high pressure over central Europe and low in SW Spain, producing southern air flows from Africa to reach eastern Spain (Fig. 14.2b). This African influence has been widely reported for the western Mediterranean and, as mentioned in Sect. 14.2, they are of special importance to Spain due to its proximity to the Sahara. Rains produced in these conditions had a very high alkalinity and high concentrations of most ions. At Montseny, the African origin of such air masses has been confirmed by back-trajectory analysis and by the mineralogical composition of the dust contained in red rains (Àvila et al. 1997).

14.5.2 The Iberian Low System, Stratification and Recirculation Processes

The summer meteorology in the western Mediterranean basin differs considerably from that in central and northern Europe (Millán et al. 1992, 1996, 1997). The high coastal mountain ranges surrounding the warm Mediterranean Basin, the region's high insolation and the inland soil aridity favour the formation of convective processes ranging from local, through meso to synoptic scales with marked diurnal cycles. In summer, a thermal low is formed almost every day and is reinforced in the middle hours of the day as insolation increases. At night, the low-pressure system weakens, producing a net flux to the sea. This oscillating movement of air masses favours the incorporation of many anthropogenic and biogenic pollutants into the air masses, remaining there and recirculating for several days (Millán et al. 1996). A similar process has been described for the Italian Peninsula, Greece (Lalas et al. 1983), Los Angeles Bay (USA) (Blumenthal et al. 1978), and Japan (Finlayson-Pitts and Pitts 1986). Recent experimental evidence supports the hypothesis that organized links exist between the circulation systems at different spatial and time scales, in which the aged pollutants emitted in central Europe could enter the central and western Mediterranean and contribute to the pollution climate there (Millán et al. 1996). As much as 60% of the photo-oxidant concentration levels at the coast may result from aged air masses transported in layers from "upstream" source areas. This considerably broadens the spatial extent of ecosystem boundaries compared to the conventional water-divide criterion used in small-catchment studies.

14.5.3 Secondary Pollutants and Their Potential Effects

The recirculatory nature of the coastal flows under strong insolation converts these flows into "large natural photo-chemical reactors" where most of the NO_x emissions and other precursors are transformed into oxidants, acidic compounds, and aerosols. As a result, rural sites inland are exposed to maritime air-masses containing O_3 concentrations around 60 to 80 ppb during the sea-breeze period of the day, and the EU air quality standards for vegetation protection may be exceeded. Furthermore, this situation can persist for several months on the Spanish Mediterranean coast.

The mean annual concentrations of SO_2 and NO_x at the rural stations of the Valencian network (10 and 20 ppb, respectively) are far below those permitted by legislation. However, high concentrations up to 40 and 29 ppb, respectively, were reported with the entry of the sea breeze (Millán et al. 1992; Millán and Sanz 1993). Also, SO_2 peaks of hundreds of ppb during up to several hours were found in forested sites under the influence of major point sources. The high levels of O_3 and the possible synergy between O_3 and other pollutants, such as SO_2 and NO_x, have to be considered. In fact, there is evidence of damage to cultivated plants sensitive to O_3 in the area (Reinert et al. 1991). Similar evidence has been found in natural Mediterranean vegetation only for *Pinus halepensis* (Gimeno et al. 1992) and *Pinus sylvestris* (Sanz et al. 1995). Although the genus *Quercus* seems quite insensitive to O_3, the species *Q. alba* and *Q. gambelis* were found to be sensitive in the USA. Adaptations to drought, like epicuticular waxes and the closure of stomata at noon, minimize the exchange of gases and could be a cause of the lack of symptomatology because the maximum concentration of contaminants occurs when stomata are closed. However, within the Mediterranean range of holm oak, many plant species are active in the humid and warm season (spring) when stomata should be open at moments of high O_3 concentrations.

14.6 Long-Term Trends in Precipitation Chemistry

Similarly to the widespread trend of decreasing S deposition in precipitation reported for the northern hemisphere (Hedin et al. 1987, 1994; Berge 1988; Buishand et al. 1988) SO_4^{2-} concentration and deposition in bulk precipitation has decreased significantly at Prades and Montseny since the early 1980s (Fig. 14.3; Àvila 1996). Linear regressions of VWM SO_4^{2-} concentration on year were used to smooth the interannual variability. Estimates from these regressions yielded a reduction of 36% in the annual SO_4^{2-} concentration at Montseny between 1983 and 1994, and a reduction of 39% in SO_4^{2-} deposition. The corresponding figures for Prades were 60 and 59% between 1981 and 1994. Paralleling the SO_4^{2-} decrease, the frequency of acidic rains also decreased at Montseny (Àvila 1996). The general declining trend for S deposition in Europe and North America has been attributed to the reduction of S emissions resulting from abatement strategies. Between 1980 and 1992, SO_2 emissions decreased around 37% in Europe and 30% in Spain (Agren 1994).

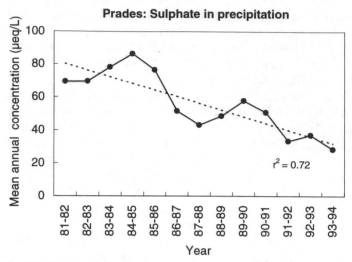

Fig. 14.3. Time trends between 1981 and 1994 for annual volume-weighted mean concentration of SO_4^{2-} in bulk deposition at Prades. The linear regression is significant ($P < 0.01$)

On the other hand, trends for the N compounds, base cations, and alkalinity in precipitation were not significant at Montseny (Àvila 1996). Since SO_4^{2-} decreased while NO_3^- did not, the proportional contribution of NO_3^- to the total load of strong acid anions increased, as found elsewhere (Rodhe and Rood 1986; Dillon et al. 1988). The stability of base cations at Montseny contrasts with the steep declines in base cation deposition over central and northern Europe and North America (Hedin et al. 1994).

14.7 Relevance of the Atmospheric Inputs to Ecosystem Nutrient Cycles

The annual deposition of major elements in bulk precipitation is shown in Table 14.3 for Prades and Montseny. Since ion concentrations are similar at both sites (Table 14.1), the higher precipitation at Montseny (ca. 300 mm year^{-1} more than at Prades) results in higher precipitation inputs for most elements at Montseny (Table 14.3). The variability between years reflects differences in concentrations as well as in the amount of precipitation and is similar at both sites (Table 14.3). Inputs of base cations and Cl$^-$ in bulk precipitation at Prades and Montseny are within the range measured in many other European and North American sites (Table 14.3). On the other hand, inputs of S and inorganic N to our holm oak sites are very low. Organic N was not measured in the precipitation at Prades and Montseny, but it was analyzed at Filabres (SE Spain; Domingo 1991): from a total input of 6.5 kg N

Table 14.3. Mean annual fluxes in bulk precipitation (kg ha^{-1} year^{-1}) for 1983–1994 (Montseny) and 1981–1994 (Prades), and in several forested areas of Europe and North America

Locality	Reference[a]	H$^+$	Na$^+$	K$^+$	Ca^{2+}	Mg^{2+}	NH$_4$-N	NO$_3$-N	SO$_4$-S	Cl$^-$
Montseny	This study Mean	0.13[b]	4.5	1.4	9.9	1.0	2.8	2.5	6.4	8.9
	(SD)	(0.05)	(1.3)	(0.6)	(4.6)	(0.3)	(0.6)	(0.6)	(1.6)	(2.8)
Prades	This study Mean	0.06[b]	3.2	1.8	6.3	0.7	1.8	1.7	4.8	5.7
	(SD)	(0.05)	(1.2)	(0.9)	(1.8)	(0.5)	(1.3)	(0.5)	(1.6)	(1.6)
Hubbard Brook (NE USA)	1	0.96	1.6	0.9	2.2	0.6	2.9	19.7	12.8	6.2
Coweeta (SE USA)	2	0.45	3.3	3.6	4.4	2.2	1.95	2.9	10.7	5.6
Karpenissi (Greece)	3	–	5.9	4.0	33.1	4.3	7.6	3.0	20.7	29.6
Vosgos-Strengbach (N. France)	4	0.5	3.3	1.5	3.4	0.8	3.8	4.8	9.3	6.3
Mont-Lozère (S. France)	5	1.8	13	3.1	13	2.4	6.2	6.2	19.2	20.4
La Robinette (Belgium)	6	1.4	9.3	4.8	6.6	2.2	15.5	10.6	23.0	–
Birkenes (Norway)	7	0.8	17.2	2.3	2.6	2.3	7.6	7.4	15.5	28.7
Plynlimon Hore (Wales)	8	0.53	47.3	1.6	4.1	6.1	7.3	18	53.6	91

[a] References: 1 Likens et al. (1995); 2 Swank and Waide (1987); 3 Nakos and Vouzaras (1988); 4 Probst et al. (1990); 5 Durand et al. (1992); 6 Hambuckers (1987); 7 Abrahamsen et al. (1989); 8 Reynolds et al. (1989).

[b] Computed from pH of each sample, i.e. without taking into account neutralization by alkaline samples. If this is included, resultant H$^+$ fluxes are much lower: 0.0032 kg ha^{-1} year^{-1} at Montseny and 0.0012 at Prades.

ha^{-1} year^{-1} in bulk deposition, 46% corresponded to NH$_4$-N, 28% to NO$_3$-N, and 26% to organic N. If the proportion of organic N was similar in our holm oak forests, the total N input in bulk deposition should be 7.0 kg ha^{-1} year^{-1} at Montseny and 4.6 kg ha^{-1} year^{-1} at Prades. Even these estimates corrected for organic N are rather low in a European context, and indicate that excessive N deposition is not a major threat to these Mediterranean forests.

The importance of atmospheric inputs in the nutrient dynamics of these holm oak forests can be evaluated by comparing the magnitude of the input flux with the intrasystemic nutrient fluxes or pools. Here, we will use the annual nutrient retention in new aboveground woody tissues (Chap. 18) as the relevant flux for comparison. For N, the input in bulk deposition (corrected for organic N as explained above) amounted to 152% (Montseny) and 139% (Prades) of the N yearly incorporated in the aboveground woody biomass. For S, the relationship between the inputs and the biomass net increment calculated for the mid 1980s was 74% at Prades, but in later years it should be lower due to the decreasing trend in S deposition. Bulk precipitation is also an important input for base cations to these holm oak forests, as dissolved inputs amounted to the following percentages of the annual aboveground retention: 28% for K$^+$, 60% for Ca^{2+}, and 71% for Mg^{2+} at Montseny, and 62, 55 and 100%, respectively, at Prades. Red rains inputs in particulate plus dissolved form accounted for around 60% of the total measured bulk precipitation inputs for these cations (Àvila et al. 1997).

In conclusion, bulk precipitation delivers nutrients in amounts that can be relevant in relation to the needs of these holm oak forests. It is convenient to remember that bulk precipitation measures only part of the atmospheric inputs to a forest. Dry deposition of gases and aerosols onto forest canopies is poorly captured by bulk deposition collectors. These additional atmospheric inputs can be, however, reflected in throughfall and stemflow, which are the subjects of the next chapter.

References

Aber JD (1992) Nitrogen cycling and nitrogen saturation in temperate forest ecosystems. Trends Ecol Evol 7:220–224

Abrahamsen G, Seip HM, Semb A (1989) Long term acidic precipitation studies in Norway. In: Adriano DC, Havas D (eds) Acid precipitation, vol 1. Case studies. Springer, Berlin, pp 137–179

Agren C (1994) New figures presented. Acid News 5:14–15

Àvila A (1988) Balanç d'aigua i nutrients en una conca d'alzinar del Montseny. Estudis i Monografies 13, Diputació de Barcelona, Barcelona

Àvila A (1996) Time trends in the precipitation chemistry at a mountain site in north-eastern Spain for the period 1983–1994. Atmos Environ 30:1363–1373

Àvila A, Queralt I, Alarcón M, Martin-Vide J (1996) African dust over north-eastern Spain: mineralogy and source regions. In: Guerzoni S, Chester R (eds) The impact of African dust across the Mediterranean. Kluwer, Dordrecht, pp 201–205

Àvila A, Queralt I, Alarcón M (1997) Mineralogical composition of African dust delivered by red rains over northeastern Spain. J Geophys Res 102, D18:21977–21996

Bellot J (1989) Análisis de los flujos de deposición global, trascolación, escorrentía cortical y deposición seca en el encinar mediterráneo de L´Avic (Sierra de Prades, Tarragona). PhD Thesis, University of Alicante, Alicante

Bellot J, Escarré A (1989) Balances de nutrientes en pequeñas cuencas de encinar II. Quimismo de la precipitación y aportes de origen atmosférico. Mediterr Ser Estud Biol 10:63–85

Berge E (1988) Time-trends of sulphate and nitrate in precipitation in Norway (1972–1982). Atmos Environ 22:333–338

Blumenthal DL, White WH, Smith B (1978) Anatomy of Los Angeles smog episode: pollutant transport in the daytime sea breeze regime. Atmos Environ 12:893–907

Buishand TA, Kempen GT, Frantzen AJ, Reijnders HFR, Van den Eshof AJ (1988) Trend and seasonal variation of precipitation chemistry data in The Netherlands. Atmos Environ 22:339–348

Caboi R, Cidu R, Cristini A, Fanfani L, Zuddas P (1992) Influence of Saharan dust and marine spray on the chemical composition of rain in Sardinia, Italy. In: Kharaka YK, Maest AS (eds) Water–rock interaction. Balkema, Rotterdam, pp 469–472

Camarero L, Catalan J (1993) Chemistry of bulk precipitation in the central and eastern Pyrenees, northeast Spain. Atmos Environ 27A:83–94

Camuffo D, Bernardi A, Bacci P (1991) Transboundary transport of atmospheric pollutants through the eastern Alps. Atmos Environ 25:2863–2871

Carratalá A (1993) Caracterización química de la precipitación en la Comunidad Valenciana: distribución espacial y temporal. PhD Thesis, University of Alicante, Alicante

Carratalá A, Bellot J, Gómez A, Millán MM (1996) African dust influence on rainwater on the eastern coast of Spain. In: Guerzoni S, Chester R (eds) The impact of African dust across the Mediterranean. Kluwer, Dordrecht, pp 323–332

Cogbill CV, Likens GE (1974) Acid precipitation in the northeastern United States. Water Resour Res 10:1133–1137

Dillon PJ, Lusis M, Reid R, Yap D (1988) Ten-year trends in sulphate, nitrate and hydrogen deposition in central Ontario. Atmos Environ 22:901–905

Domingo F (1991) Papel de la cubierta vegetal en los aportes de nutrientes y salidas por avenamiento en una cuenca reforestada con coníferas en la Sierra Filabres (Almeria). PhD Thesis, University of Alicante, Alicante

Drablos D, Tollan A (eds) (1980) Ecological impact of acid precipitation. SNSF Project, Oslo-As

Durand P, Neal C, Lelong F (1992) Anthropogenic and natural contributions to the rainfall chemistry of a mountainous area in the Cévennes National Park (Mont Lozère, southern France). J Hydrol 130:71–85

Finlayson-Pitts BJ, Pitts JN Jr (1986) Atmospheric chemistry. Wiley, Chichester

Gimeno B, Velissariou D, Barnes JD, Inclán R, Peña JM, Davison A (1992) Daños visibles por ozono en acículas de Pinus halepensis Mill. en Grecia y España. Ecología 6:131–134

Glavas S (1988) A wet-only precipitation study in a Mediterranean site, Patras, Greece. Atmos Environ 22:1175–1178

Gorham E (1955) On the acidity and salinity of rain. Geochim Cosmochim Acta 7:231–239

Gorham E, Martin FB, Litzau JT (1984) Acid rain: ionic correlations in the eastern United States, 1980–1981. Science 225:407–409

Hambuckers A (ed) (1987) Evolution du pool des éléments nutritifs dans les forêts soumises aux pluies acides. Rapport final (1ère partie). Université de Liège, Liège

Hedin LO, Likens GE, Borman FH (1987) Decrease in precipitation acidity resulting from decreased SO_4^{2-} concentration. Nature 325:244–246

Hedin LO, Granat L, Likens GE, Buishand TA, Galloway JN, Rodhe H (1994) Steep declines in atmospheric base cations in regions of Europe and North America. Nature 367:351–354

Hutchinson TC, Havas M (eds) (1980) Effects of acid precipitation on terrestrial ecosystems. Plenum Press, New York

Johnson DW, Lindberg SE (eds) (1992) Atmospheric deposition and forest nutrient cycling. Springer, New York

Lalas DP, Asimakopoulos DN, Deligiorgi DG, Helmis CG (1983) Sea breeze circulation and chemical pollution in Athens, Greece. Atmos Environ 17:1621-1632

Likens GE, Bormann FH (1995) Biogeochemistry of a forested ecosystem, 2nd edn. Springer, New York

Liljestrand HM (1985) Average rainwater pH, concepts of atmospheric acidity, and buffering in open systems. Atmos Environ 19:487-499

Löye-Pilot MD, Morelli J (1988) Fluctuations of ionic composition of precipitations collected in Corsica related to changes in the origins of incoming aerosols. J Aerosol Sci 19:577-585

Löye-Pilot MD, Martin JM, Morelli J (1986) Influence of Saharan dust on the rain acidity and atmospheric input to the Mediterranean. Nature 321:427-428

Millán MM, Sanz MJ (1993) La contaminación atmosférica en la Comunidad Valenciana: Estado de los conocimientos sobre los problemas en el Maestrazgo y Els Ports de Castellón. Informes CEAM 93-1. CEAM, Valencia

Millán MM, Artiñano B, Alonso L, Castro M, Fernández-Patier, Goberna J (1992) Mesometeorological cycles of air pollution in the Iberian Peninsula. Air Pollution Research Report 44, European Commission, Brussels

Millán MM, Salvador R, Mantilla E, Artiñano B (1996) Meteorology and photochemical air pollution in southern Europe: experimental results from EC research projects. Atmos Environ 30:1909-1924

Millán MM, Salvador S, Mantilla E, Kallos K (1997) Photooxidants dynamics in the Mediterranean Basin in summer: results from European research projects. J Geophys Res (Atmos) 102: 8811-8823

Nakos G,Vouzaras A (1988) Budgets of selected cations and anions in two forested experimental watersheds in central Greece. For Ecol Manage 24:85-95

Pedersen U, Schaug J, Skjelmoen JE (1992) EMEP data report 1990. Part 1: annual summaries. EMEP/CPP-Report 2/92. NILU

Piñol J (1990) Hidrología i biogeoquímica de conques forestades de les muntanyes de Prades. PhD Thesis, University of Barcelona, Barcelona

Probst A , Dambrine E, Viville D, Fruth B (1990) Influence of acid atmospheric inputs on surface water chemistry and mineral fluxes in declining spruce stand within a small granitic catchment (Vosges Massif), France. J Hydrol 116:101-124

Reinert RA, Gimeno BS, Salleras JM, Bermejo V, Ochoa MJ, Tarruel A (1991) Ozone effects on watermelon plants at the Ebro Delta (Spain): symptomatology. Agric Ecosyst Environ 38:41-49

Reynolds B, Cape JN, Paterson IS (1989) A comparison of element fluxes in throughfall beneath large and small Sitka spruce at contrasting sites in the United Kingdom. Forestry 62:29-39

Rodà F (1983) Biogeoquímica de les aigües de pluja i drenatge en alguns ecosistemes forestals del Montseny. PhD Thesis, Autonomous University of Barcelona, Bellaterra

Rodà F, Bellot J, Àvila A, Escarré A, Piñol J, Terradas J (1993) Saharan dust and the atmospheric inputs of elements and alkalinity to Mediterranean ecosystems. Water Air Soil Pollut 66:277-288

Rodhe H, Rood MJ (1986) Temporal evolution of nitrogen compounds in Swedish precipitation since 1955. Nature 321:762-764

Sanz MJ, Calatayud V, Calvo E (1995) Diferencias morfológicas en las ceras epistomáticas de varias poblaciones de Pinus sylvestris L. de Castellón y Teruel. Ecología 9:201-211

Schaug J, Pedersen U, Skjelmoen JE, Kvalvognes I (1993) EMEP data report 1991. Part 1: annual summaries. EMEP/CCC-Report 4/93, NILU

Sequeira R (1993) On the large-scale impact of arid dust on precipitation chemistry on the continental northern hemisphere. Atmos Environ 27A:1553-1565

Swank WT, Waide JB (1987) Characterization of baseline precipitation and stream chemistry and nutrient budgets for control watersheds. In: Swank WT, Crossley DA Jr (eds) Forest hydrology and ecology at Coweeta. Springer, New York, pp 57- 80

Ulrich B, Pankrath J (eds) (1983) Effect of accumulation of air pollutants in forest ecosystems. Reidel, Dordrecht

Young JR, Ellis EC, Hidy GM (1988) Deposition of air-borne acidifiers in the Western environment. J Environ Qual 17:1-26

15 Throughfall and Stemflow

Juan Bellot, Anna Àvila and Anselm Rodrigo

15.1 Introduction

Forest canopies affect the amounts of water and nutrients reaching forest soils. Water and nutrient inputs under forest canopies are thus different from those in nearby open areas. Canopies intercept and retain part of the incident precipitation (intercepted water), which is eventually evaporated from the canopy and lost to the atmosphere. Water passing through the canopy can reach the forest floor after dripping from leaves and branches (throughfall) or flowing down the stems of trees (stemflow). The sum of throughfall and stemflow is called net or effective precipitation (Parker 1983). The part of incident precipitation that does not appear on the forest floor by either of these routes is called interception loss. Interception has received special attention in a number of forest hydrology studies using experimental or modelling approaches (e.g. Rutter and Morton 1977; Gash 1979; Massman 1983; Bouten et al. 1996).

The distribution of incident rainwater can be summarized, after Helvey and Patric (1965), by the expression: rainfall = interception loss + throughfall + stemflow (units in mm). The relative importance of each component depends on climatological characteristics (amount of rain, rain intensity, wind velocity during rain events, rain duration, temperature) and stand structure characteristics (canopy closure, crown homogeneity and density, age, tree species). In the Mediterranean climate, high temperatures and high potential evapotranspiration favour high interception losses. On the other hand, the heavy storms that characterize the Mediterranean rainfall regime produce less interception.

The alteration of the chemistry of rainfall after contact with the canopy has long been recognized (Eaton et al. 1973; Parker 1983; Lindberg et al. 1986; Lovett 1994). The chemistry of throughfall and stemflow is influenced by the incident rain chemistry, canopy evaporation, washing of dry deposition from the surfaces of leaves and barks, and canopy leaching (Lovett and Lindberg 1984). Usually, ion concentrations in throughfall and stemflow are higher

Ecological Studies, Vol. 137
Ferran Rodà et al. (eds) Ecology of Mediterranean
Evergreen Oak Forests
© Springer-Verlag, Berlin Heidelberg 1999

than in rainwater. The total flux of nutrients to the forest floor in net precipitation is the sum of nutrient fluxes in throughfall and stemflow. When deposition of any nutrient in incident rainwater is subtracted from that in throughfall or stemflow, the remainder is called net throughfall or net stemflow. The relative effect of the canopy on these depositions can be estimated by obtaining the enrichment factor for each element. The main processes of enrichment are washing of dry deposition and leaching of intercellular solutes from leaves. However, some nutrients do not show such enrichments because they are absorbed by foliar surfaces. This can be important for nutrients in low supply and has been widely reported for nitrogen (Schlesinger et al. 1982; Parker 1983; Lindberg et al. 1986; Bynterowicz and Fenn 1996). Chemical changes as rainwater crosses the canopy by throughfall can be expressed in terms of element fluxes (units in kg ha^{-1} year^{-1}), and summarized in the following expression (Bache 1977):

net throughfall = washout of dry deposition + leaching – canopy uptake.

Much effort has been devoted in the last 20 years to discriminating the contributions to net throughfall nutrients of atmospheric sources (dry deposition) versus that of plant leaching. The distinction is important because dry deposition represents (mostly) an external input of nutrients to the ecosystem while leaching represents (mostly) a within-system recirculation of nutrients previously taken up by trees from the soil (Lovett and Lindberg 1984; Lovett 1994).

In this chapter, we present the results of studies on throughfall, stemflow and interception in the holm oak forests of Prades and Montseny in NE Spain with the aim of comparing the hydrological and chemical fluxes at two sites of different stand structure and climate characteristics. We first deal with the hydrological pathways through the forest canopy in relation to the contrasted forest structure of the sites (Chap. 3). We then discuss the biogeochemical behaviour of the main elements, with emphasis on major nutrients (N and P) and atmospheric pollutants (S).

15.2 The Database

Four throughfall studies have been conducted in the experimental holm oak forests of Prades and Montseny (Chap. 2), two at each location. The first study at each site was intended to assess the role of throughfall in the water and nutrient cycles of holm oak forests. The second Prades study addressed the effects of N fertilization (Chap. 13) in throughfall chemistry and element fluxes. As no effect of fertilization was found on throughfall amounts or chemistry, we have pooled data from treated and untreated plots. The second Montseny study compared the atmospheric deposition onto holm oak canopies (using throughfall as a proxy) in two sites differently exposed to air

Table 15.1. Site and sampling characteristics for the throughfall and stemflow studies conducted in holm oak forests of Prades and Montseny

Data set[a]	Site	Aspect	Topographic position	Sampling period	Sampling schedule	Tree density (stems ha^{-1})[b]	Number of plots	Plot size (m^2)	Number of throughfall collectors		Number of stemflow collectors
									For volume	For chemistry	
Prades-1	Avic	N	Mid slope	Aug 1981–Nov 1983	Event	8460	1	950	50	8	20[c]
Prades-2	Torners	SE	Mid slope	Nov 1991–Nov 1992	Weekly	18200	6	64	24	24[d]	–[e]
Montseny-1	Permanent plot[f]	NNW	Lower slope	Dec 1978–Dec 1980	Biweekly	2008	1	2300	4–8[g]	8	–[e]
Montseny-2	Torrent de la Mina	N	Lower slope	Jun 1995–Jun 1996	Weekly	2645	4	154	32	32[d]	10

[a] Sources: Prades-1: Bellot (1989); Prades-2: A. Àvila (unpubl. data); Montseny-1: Rodà (1983), Rodà et al. (1990); Montseny-2: A. Àvila and A. Rodrigo (unpubl. data).

[b] D$_{50}$ ≥ 2 cm at Prades; dbh ≥ 5 cm at Montseny.

[c] Ten holm oaks, 5 *Arbutus unedo*, and 5 *Phillyrea latifolia*.

[d] The same collectors were used for throughfall volume and chemistry.

[e] Stemflow was not measured in these studies.

[f] Permanent plot at La Castanya, in the Torrent de la Mina catchment (Chap. 2).

[g] Four collectors until January 1980, 8 collectors thereafter.

pollution. From this study, and for consistency with the other three studies, we will only use here data from the less polluted site (La Castanya). Table 15.1 summarizes the site characteristics and sampling layouts. Analytical methods are described in Bellot (1989) and Rodà et al. (1990). Stemflow was measured in the first Prades study and in the second Montseny study. Sample trees were chosen to cover the range of tree diameters at each site. At Montseny, throughfall and stemflow were studied only under holm oak, since it was practically the only tree species. In the more diverse forest of the Avic catchment, the two major companion tree species (*Arbutus unedo* and *Phillyrea latifolia*) were also included in the first Prades study.

15.3 Throughfall and Stemflow Hydrology

Mean water fluxes in throughfall, stemflow, net precipitation and interception loss for the two sites and periods are shown in Table 15.2. It must be noted that at both sites precipitation was much higher in the second period than in the first one. Throughfall estimates range between 66 and 75% of annual rainfall at each site, while stemflow percentages were 2.6% of annual precipitation for Montseny and 12% for Prades.

A similarity in throughfall percentages in Prades and Montseny is expected in view of the fact that holm oak canopies have a similar LAI in both sites (Chap. 9). However, structural differences in tree size and density between sites are reflected in the much higher amount of water channelled by stemflow in Prades. Interception loss is lower at Prades than at Montseny (13 and 22% of annual precipitation, respectively). Since annual throughfall percentages were similar in both studies, differences in interception should be linked to differences in stem density, producing a greater channelling effect

Table 15.2. Average water fluxes (mm year^{-1}) for the Prades and Montseny holm oak forests. The four data sets are defined in Table 15.1. The last three rows give water fluxes as percentage of the mean annual precipitation of each data set

Water flux	Data set			
	Prades-1	Prades-2	Montseny-1	Montseny-2
Precipitation	518	939	857	1275
Throughfall	389	609	565	963
Stemflow	62	–	–	34
Interception loss	67	–	–	279
Net precipitation	451	–	–	997
Precipitation (%)				
Throughfall	75.0	64.9	65.9	75.5
Stemflow	12.1	–	–	2.6
Interception loss	12.9	–	–	21.9

at Prades than at Montseny. These patterns, together with the threshold upon which both flows start (see below) reflect that the canopy and trunk water-storage capacity is higher at Montseny, and, consequently, more water can be evaporated from the wet canopy. The observed stemflow percentages at Prades are similar to those reported for shrub communities, which usually produce substantial stemflow (Návar and Bryan 1990; Puigdefábregas et al. 1996; González and Bellot 1997). On the other hand, Montseny resembles the holm oak forest at Le Rouquet (southern France; Rapp and Romane 1968) in producing little stemflow.

Linear regressions of throughfall and stemflow on precipitation for both forests are shown in Table 15.3. Throughfall equations show similar inter-cepts and regression coefficients at both forests, while stemflow regressions present a higher regression coefficient at Prades, as expected from the higher water fluxes diverted to stemflow at this site. These regressions predict that a slightly lower amount of precipitation is needed to start throughfall in the shorter Prades forest (1.6 mm) than at Montseny (2.1 mm). The difference is much higher for stemflow: 2.2 mm of precipitation is needed to start it at Prades but 6.8 mm at Montseny. Longer and broader crowns and longer trunks at Montseny probably explain that, for a given rainfall amount, less stemflow per tree is produced here than by similar-sized trees at Prades. At the stand scale, the much higher stem density at Prades (Table 15.1) results in enhanced stemflow. At both sites, the relation between interception and pre-cipitation follows a power expression levelling off at high precipitation, but showing a small and continuous increase that reflects the continued evapo-ration during large rainfall events.

Net precipitation on the forest floor is distributed very heterogeneously, because stemflow is usually restricted to a small circle around tree bases (Herwitz 1986; Návar 1993) and because throughfall usually follows prefer-ential routes through the canopy (Parker 1983). Spatial variation of through-fall has been related to distance from the tree trunk and to canopy charac-teristics (Ovington 1954; Aussenac 1970; Ford and Deans 1978; Prebble and Stirk 1980; Herwitz 1986; Johnson 1990). The high sampling intensity (50 collectors in a 950-m^2 plot) used in the first Prades study allows a map of

Table 15.3. Regression analysis of throughfall (T), stemflow (S) and interception (I) versus precipitation (P) for Prades and Montseny in data sets Prades-1 and Montseny-2 (defined in Table 15.1). Units in mm event^{-1} for Prades and mm week^{-1} for Montseny

Data set	Regression	n	r^2
Prades-1	T = − 1.30 + 0.82 P	60	0.995
	S = − 0.29 + 0.13 P	60	0.995
	I = 0.86 × P$^{0.41}$	60	0.561
Montseny-2	T = − 1.74 + 0.82 P	49	0.996
	S = − 0.27 + 0.036 P	49	0.931
	I = 0.79 × P$^{0.64}$	49	0.902

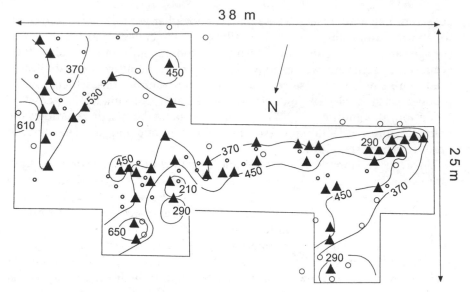

Fig. 15.1. Spatial distribution (in mm year^{-1}) of annual throughfall in a 950-m^2 plot in holm oak forest at Prades. *Solid triangles* are throughfall collectors. *Open circles* are tree stems. Large circles indicate stemflow collectors

annual throughfall amounts to be drawn for this plot (Fig. 15.1). Throughfall varied threefold, ranging from 210 to 650 mm year^{-1}, with a mean of 389 mm year^{-1}. Therefore, very large differences in the amount of infiltrating water probably exist even within a small and relatively homogeneous forest plot. Though some lateral redistribution of infiltrated throughfall likely occurs, the above differences may affect aspects of ecosystem function such as the initiation of subsurface flow or the rates of biogeochemical transformations in the soil. As Loustau et al. (1992) suggested for a pine forest, the 50 throughfall collectors behaved rather independently of each another: each received more or less water than the average depending on their position and the daily amount of precipitation, and, consequently, the shape of the distribution frequency of throughfall volumes changed with each event (Bellot 1989). Small rainfall events presented a reverse-J distribution (strongly skewed to the right) because only a few collectors in favourable positions under dripping branches received large throughfall amounts. In large rainfall events, collectors received more even throughfall amounts, giving a more normal distribution.

15.4 Nutrients in Throughfall and Stemflow

The concentrations of solutes delivered to the forest floor by throughfall and stemflow under holm oak are usually greater than in incident precipitation (Table 15.4). For most ions, concentrations are similar between throughfall

Table 15.4. Volume-weighted means (ion concentrations and alkalinity in μeq L^{-1}, conductivity in μS cm^{-1} at 25 °C) for bulk precipitation, throughfall (T) and stemflow (S) at Prades and Montseny. The four data sets are defined in Table 15.1. For Prades-l, throughfall and stemflow concentrations under different species (*Quercus ilex*, *Arbutus unedo* and *Phillyrea latifolia*) are given

Data set	Conductivity	Na$^+$	K$^+$	Ca^{2+}	Mg^{2+}	NH$_4^+$	NO$_3^-$	SO$_4^{2-}$	Cl$^-$	Alkalinity
Prades-1										
Bulk precipitation	22.1	28.7	8.2	56.4	11.5	27.7	17.9	72.2	35.5	32.6
T – *Q. ilex*	42.6	77.0	76.7	117	37.0	54.9	17.4	135	63.7	35.9
A. unedo	48.7	43.5	114	161	47.7	44.4	19.4	132	67.1	65.1
P. latifolia	37.8	37.4	61.4	114	25.5	45.5	17.6	128	55.0	44.7
S – *Q. ilex*	43.5	49.6	115	107	39.5	–	12.4	165	66.8	22.0
A. unedo	39.8	44.4	95.4	141	33.7	–	12.1	126	59.8	34.4
P. latifolia	36.5	44.8	99.5	136	27.1	–	6.8	137	59.0	29.2
Prades-2										
Bulk precipitation	11.7	15.8	6.9	33.4	5.8	14.9	16.0	34.4	21.4	15.6
T – *Q. ilex*	32.9	29.7	68.0	129	32.7	27.2	40.7	105	44.9	72.5
Montseny-1										
Bulk precipitation	23.2	23.4	3.6	61.1	12.5	–	18.8	–	27.7	–
T – *Q. ilex*	45.4	38.5	86.1	129	45.2	–	33.7	–	61.7	–
Montseny-2										
Bulk precipitation	13.8	18.5	2.9	34.8	7.0	29.5	25.5	36.5	22.8	18.5
T – *Q. ilex*	25.4	26.0	42.3	66.4	22.9	27.1	22.4	54.8	45.0	76.1
S – *Q. ilex*	20.2	18.1	41.3	47.5	16.2	17.5	10.7	41.2	36.1	77.6

and stemflow, and between Prades and Montseny, particularly if the large sampling variabilities and the differences in rainfall, precipitation chemistry, and sampling layout of each study are borne in mind. Also, the three major tree species at Prades show an overall similar chemistry of throughfall and stemflow (Bellot and Escarré 1991; Table 15.4). In contrast with the other ions, NO_3^- concentrations in stemflow are lower than in throughfall for the three species, probably because of biological uptake by cryptogamic epiphytes (mostly crustose lichens) and microbes growing on bark surfaces.

On an equivalent basis, the mean throughfall chemistry at Prades is dominated by Ca^{2+} and SO_4^{2-} (Table 15.4). Calcium was also the dominant cation at Montseny, but the major anion contribution corresponded to alkalinity, at least in the second study. Throughfall and stemflow present average positive alkalinities at both sites, indicating the non-acidic nature of these solutions as opposed to the usual acidic character of throughfall in northern and central Europe and eastern North America (Bredemeier 1988; Lovett 1994; Dambrine et al. 1995). Foliar and bark surfaces of holm oak buffer the incoming precipitation as indicated by the general alkalinity increase in throughfall. Enhancement of the alkalinity by foliar contact can be due to the washout of alkaline dust deposited on the surfaces or to exchange reactions with the intercellular solutes. The relative contribution of each process is unknown. At both sites, SO_4-S concentrations in throughfall are higher than in bulk deposition, but most of this enrichment is balanced by base cations.

The annual nutrient inputs to the forest floor in throughfall and stemflow are shown in Table 15.5. Throughfall fluxes are larger than those in stemflow mostly because throughfall conveys much higher water amounts. On a mass basis, K^+, Ca^{2+}, SO_4-S and Cl^- show the highest gross throughfall fluxes at both sites. Stemflow inputs represent at most 17% of the total inputs (throughfall + stemflow) of each nutrient at Prades. Despite its low proportional contribution, stemflow can play an important role in the forest as it infiltrates near the tree trunks, presumably providing a localized supply of water and nutrients to the trees (Young et al. 1984). At Montseny, stemflow inputs are much smaller due to the lower amounts of water delivered by this pathway. In this case, stemflow nutrient inputs represent only around 2–3% of total dissolved inputs to the soil for all elements.

In Table 15.5, the ratio between gross throughfall and bulk deposition fluxes (T/BD) is also given. Prades seems to receive a higher marine influence as the Na^+ T/BD ratios are higher than at Montseny. The base cation ratios at Montseny are similar for the two periods of study, even though the throughfall quantities were very different. At Prades, net throughfall fluxes of Ca^{2+} and Mg^{2+} and their T/BD ratios were higher in the second period than in the first one (Table 15.5). Since these ions are present in African dusts and the incidence of red-rain events delivered by African air masses (Chap. 14) apparently increased in the early 1990s (second period) compared to the early 1980s (first period), it is tempting to speculate whether this increased removal of Ca^{2+} and Mg^{2+} from holm oak canopies was due to dry-deposited

Table 15.5. Annual fluxes in bulk deposition (BD), throughfall (T) and stemflow (S) at Prades and Montseny in kg ha^{-1} year^{-1}, except for alkalinity (meq m^{-2} year^{-1}). The four data sets are defined in Table 15.1. Prades-1 data of throughfall and stemflow represent the joint contribution of the three major tree species

Data set	Alkalinity	Na$^+$	K$^+$	Ca^{2+}	Mg^{2+}	NH$_4$-N	NO$_3$-N	SO$_4$-S	Cl$^-$	PO$_4$-P
Prades-1										
BD	10.3	3.4	1.6	5.8	0.7	1.9	1.3	6.0	6.5	0.033
T	11.9	5.9	12.3	9.7	1.7	2.8	0.9	8.2	8.7	0.064
S	1.0	0.6	2.6	1.4	0.3	–	0.1	1.4	1.3	0.004
T/BD	1.1	1.7	7.7	1.7	2.4	1.5	0.7	1.4	1.4	1.9
Prades-2										
BD	14.6	3.4	2.5	6.3	0.7	2.0	2.1	5.2	7.1	–
T	44.2	4.2	16.2	15.8	2.4	2.3	3.5	10.2	9.7	0.44
T/BD	3.0	1.2	6.5	2.5	3.4	1.2	1.7	2.0	1.4	–
Montseny-1										
BD	–	4.6	1.2	10.5	1.3	–	2.3	–	8.4	0.054
T	–	5.0	19.0	14.6	3.1	–	2.7	–	12.4	0.71
T/BD	–	1.1	15.8	1.4	2.4	–	1.2	–	1.5	13.1
Montseny-2										
BD[a]	23.6	5.4	1.5	8.9	1.2	5.4	4.7	7.6	10.8	–
T	73.3	5.8	16.0	12.8	2.7	3.7	3.0	8.5	15.4	0.63
S	2.6	0.14	0.54	0.32	0.06	0.08	0.05	0.22	0.4	0.014
T/BD	3.1	1.1	10.7	1.3	2.3	0.7	0.6	1.1	1.4	–

[a] Includes a minor input through gravitational sedimentation of particles onto funnel collectors during rainless sampling periods.

African dust. African dust is transported to Europe typically at altitudes of ca. 2000–5000 m above the Earth's surface. Hence, its removal is generally considered to occur in wet rather than in dry form, and more data are available for wet than for dry deposition fluxes. We have checked our data from individual red-rain events (or sampling periods) to test whether red-rains produce increased net throughfall fluxes and T/BD ratios of Ca^{2+} and Mg^{2+} through dry deposition of African dust onto the forest canopy. Unfortunately, only 5 events were useful for this purpose (2 at Prades and 3 at Montseny). Only one of these showed a markedly enhanced removal of Ca^{2+} and Mg^{2+} from the canopy, while another showed a moderate increase. This is a weak base to test the hypothesis. Dry deposition of African dust onto Mediterranean forests clearly merits further research.

15.5 Net Inputs and Sources of Nutrients in Throughfall and Stemflow

Net annual inputs for both forests are presented in Table 15.6. Potassium, Ca^{2+}, SO_4-S and Cl^- have the highest net fluxes, followed by Mg^{2+} and Na^+. Some of the net fluxes for NO_3-N and NH_4-N were positive and others were negative, though in all cases they were small. Negative fluxes mean that the amount of inorganic N in throughfall or stemflow is lower than in the incoming precipitation. In these cases, inorganic N is being taken up in the canopy, as frequently cited in the literature (Miller et al. 1976; Schlesinger and Hasey 1980; Bynterowicz and Fenn 1996). Canopy N uptake cannot be completely ruled out even when the flux of inorganic N in net throughfall (or

Table 15.6. Annual net fluxes in throughfall (T) and stemflow (S) for Prades and Montseny in kg ha^{-1} $year^{-1}$, except for alkalinity (meq m^{-2} $year^{-1}$) and water flux (mm $year^{-1}$). The four data sets are defined in Table 15.1. Data of Prades-1 represent the sum of fluxes by the three major tree species

Data set	Water	Na^+	K^+	Ca^{2+}	Mg^{2+}	NH_4-N	NO_3-N	SO_4-S	Cl^-	PO_4-P	Alkalinity
Prades-1 T+S	451	3.1	13.3	5.3	1.3	0.9	−0.3	3.6	3.5	0.03	2.6
Prades-2 T	609	0.8	13.7	9.5	1.8	0.4	1.4	5.0	2.6	−0.4[a]	29.6
Montseny-1 T	565	0.4	17.8	4.1	1.8	–	0.4	–	3.9	0.65	–
Montseny-2 T+S	997	0.5	15.0	4.2	1.7	−1.5	−1.5	1.2	5.5	−0.6[a]	52.3

[a] Assuming that PO_4-P input in bulk precipitation was ≤ 0.05 kg ha^{-1} $year^{-1}$.

in net precipitation) is positive, since unmeasured dry deposition of N-containing gases or aerosols could be adding N to throughfall and masking canopy uptake.

The origin of the ions removed from the holm oak canopy by throughfall and stemflow is difficult to determine. In the first Prades study, attempts to differentiate the contribution of dry deposition from canopy leaching were made by Bellot (1989) and López (1989) through three approaches: (1) the use of dry deposition traps, in an experimental setting similar to that described by Miller and Miller (1980), and Lakhani and Miller (1980); (2) the use of statistical relationships between the net nutrient fluxes, the precipitation amount and the number of rainless days before the event (Lovett and Lindberg 1984); and (3) a sequential washing of covered and uncovered branches, as proposed by Johannes et al. (1986) and Kazda (1990). It was concluded that inputs from atmospheric dry deposition accounted for 90% of the Cl^- and Na^+ removed from the canopy by precipitation, for 70–80% of SO_4^{2-} and Ca^{2+}, and for about 50% of Mg^{2+}. The cases of H^+ and NO_3-N were particular, because their atmospheric inputs in wet and dry deposition were neutralized or retained by the holm oak canopy. For K^+, the results indicated that more than 90% of its net flux in throughfall and stemflow originated within the ecosystem through leaching of plant tissues. The alkalinity increase when precipitation washes the holm oak canopy could be produced by leaching of weak anions or by dissolving dry-deposited calcareous dust.

At Montseny, dry deposition of marine and continental aerosols seems to be low, as indicated by the moderate enrichment ratios of Na^+ and Ca^{2+} (Table 15.5) and the moderate net throughfall fluxes of these cations (Table 15.6). For instance, using data from the first Montseny study, even if all Na^+ in net throughfall were of marine origin, sea-salt aerosol impaction would only account for 2.5% of Mg^{2+} and 17% of Cl^- in net throughfall (Rodà et al. 1990). On the other hand, K^+ is mostly derived from leaching at this site since its net throughfall fluxes (1) follow a distinct seasonal pattern, with maxima in late spring and early summer when flowering and leaf senescence of holm oak occur, and (2) are highly correlated with the intensity of organic colour in throughfall. Ions behaving similarly to K^+ could be interpreted as also being derived mainly from leaching (Rodà 1983). This would be the case of Mg^{2+} and PO_4-P whose net throughfall fluxes showed strong positive correlations with those of K^+ at Montseny.

Results from other studies are similar to what is concluded from the Prades and Montseny sites in that K^+ is overwhelmingly and Mg^{2+} partially (around 50%) produced by leaching mechanisms, while dry deposition is thought to be responsible for most Na^+, Cl^- and SO_4^{2-} in net throughfall (Brinson et al. 1980; Miller and Miller 1980; Parker 1983). Lindberg and Garten (1988) studied the recovery in throughfall of ^{35}S injected in trunks and concluded that foliar leaching accounted for only 5–7% of total annual sulphate flux in net throughfall. Since most of the SO_4-S removed from trees by precipitation enters the ecosystem through dry deposition, the SO_4-S flux in net throughfall or net precipitation can be taken as a measure of dry deposition onto the forest canopy, which is notoriously difficult to measure directly. Using this criterion, dry deposition of S is moderate at Prades and very low in the topographically protected site at Montseny (Table 15.6).

In summary, as water crosses the holm oak canopies, fluxes of most elements increase, except for H^+ and in some cases N, in ratios similar to those reported for sites in central Europe or North America affected by acid rains (Dambrine et al. 1995). However, the S enrichment in our holm oak sites is counterbalanced by increases in base cations, so S inputs do not pose a problem of acidification for Mediterranean holm oak forests. Net throughfall fluxes of K^+ can be mostly attributed to leaching, and represent a major contribution to K^+ cycling in the forest (Chap. 18).

References

Aussenac G (1970) Action du couvert forestier sur la distribution au sol des precipitations. Ann Sci For 27:383–399

Àvila A, Alarcón M, Queralt I (1998) The chemical composition of dust transported in red rains – its contribution to the biogeochemical cycle of a holm oak forest in Catalonia (Spain). Atmos Environ 32:179–191

Bache DH (1977) Sulphur dioxide uptake and the leaching of sulfates from a pine forest. J Appl Ecol 14:881–895

Bellot J (1989) Análisis de los flujos de deposición global, trascolación, escorrentía cortical y deposición seca en el encinar mediterráneo de L'Avic (Sierra de Prades, Tarragona). PhD Thesis, University of Alicante, Alicante

Bellot J, Escarré A (1991) Chemical characteristics and temporal variations of nutrients in throughfall and stemflow of three species in Mediterranean holm oak forest. For Ecol Manage 41:125–135

Bouten W, Schaap MG, Aerts J, Vermetten AWM (1996) Monitoring and modelling canopy water amounts in support of atmospheric deposition studies. J Hydrol 181:305–321

Bredemeier M (1988) Forest canopy transformation of atmospheric deposition. Water Air Soil Pollut 40:121–138

Brinson MM, Bradshaw HD, Holmes RN, Elkins JB (1980) Litterfall, stemflow and throughfall nutrient fluxes in an alluvial swamp forest. Ecology 61:827–835

Bynterowicz A, Fenn ME (1996) Nitrogen deposition in California forests: a review. Environ Pollut 92:127–146

Dambrine E, Ulrich E, Cénac, N, Durand P, Gauquelin T, Nfirabel P, Nys C, Probst A, Ranger J, Zéphoris M (1995) Atmospheric deposition in France and possible relation with forest decline. In: Landemann G, Bonneau M (eds) Forest decline and atmospheric deposition effects in the French mountains. Springer, Berlin, pp 177–199

Eaton JS, Likens GE, Bormann FH (1973) Throughfall and stemflow chemistry in a northern hardwood forest. J Ecol 61:495–508

Ford ED, Deans JD (1978) The effects of canopy structure on stemflow, throughfall and interception loss in a young Sitka spruce plantation. J Appl Ecol 15:905–917

Gash JHC (1979) An analytical model of rainfall interception by forest. Q J R Meteorol Soc 105:43–55

González JC, Bellot J (1997) Soil moisture changes under shrub cover (Rosmarinus officinalis) and cleared shrub as response to precipitation in a semiarid environment. Stemflow effects. Arid Soil Res Rehabi 11:187–199

Helvey JD, Patric JH (1965) Canopy and litter interception of rainfall by hardwoods of eastern United States. Water Resour Res 1:193–207

Herwitz SR (1986) Infiltration-excess caused by stemflow in a cyclone-prone tropical rainforest. Earth Surf Proc Landforms 11:401–412

Johannes AH, Chen YL, Dackson K, Suleski T (1986) Modelling of throughfall chemistry and indirect measurements of dry deposition. Water Air Soil Pollut 30:211–216

Johnson RC (1990) The interception, throughfall and stemflow in highland Scotland and the comparison with other upland forests in the UK. J Hydrol 118:281–287

Kazda M (1990) Sequential stemflow sampling for estimation of dry deposition and crown leaching in beech stands. In: Harrison AF, Ineson P, Heal OW (eds) Nutrient cycling in terrestrial ecosystems. Elsevier, London, pp 46–55

Lakhani KH, Miller HG (1980) Assessing the contribution of crown leaching to the element content of rainwater beneath trees. In: Hutchinson TC, Havas M (eds) Effects of acid precipitation on terrestrial ecosystems. Plenum Press, New York, pp 161–172

Lindberg SE, Garten CT Jr (1988) Sources of sulphur in forest canopy throughfall. Nature 336: 148–151

Lindberg SE, Lovett GM, Richter DD, Johnson DW (1986) Atmospheric deposition and canopy interactions of major ions in a forest. Science 231:141–145

López MV (1989) Estimas de las entradas por deposición seca y lixiviación a los bosques de encinas y pinos de la Sierra de Prades, Tarragona. MSc Thesis, CIHEAM, Zaragoza

Loustau D, Berbigier P, Granier A, El Hadj Moussa F (1992) Interception loss, throughfall and stemflow in a maritime pine stand. 1. Variability of throughfall and stemflow beneath the pine canopy. J Hydrol 138:449–467

Lovett GM (1994) Atmospheric deposition of nutrients and pollutants in North America: an ecological perspective. Ecol Appl 4:629–650

Lovett GM, Lindberg SE (1984) Dry deposition and canopy exchange in a mixed oak forest as determined by analysis of throughfall. J Appl Ecol 21:1013–1027

Massman WJ (1983) The derivation and validation of a new model for the interception of rainfall by forests. Agric Meteorol 28:262–286

Miller HG, Miller JD (1980) Collection and retention of atmospheric pollutants by vegetation. In: Drablos D, Tollan A (eds) Ecological impact of acid precipitation. SNSF Project, Oslo-As, pp 33–40

Miller HG, Cooper JM, Miller JD (1976) Effect of nitrogen supply on nutrients in litter fall and crown leaching in a stand of Corsican pine. J Appl Ecol 13:233–248

Návar J (1993) The causes of stemflow variation in three semi-arid growing species of northeastern Mexico. J Hydrol 145:175–190

Návar J, Bryan R (1990) Interception loss and rainfall redistribution by three semiarid shrubs in northeastern Mexico. J Hydrol 115:51–63

Ovington JD (1954) A comparison of rainfall in different woodlands. Forestry 27:41–53

Parker GG (1983) Throughfall and stemflow in the forest nutrient cycle. Adv Ecol Res 13:57–133

Prebble RE, Stirk GB (1980) Throughfall and stemflow on silverleaf ironbark (Eucalyptus melanophloia) trees. Aust J Ecol 5:419–427

Puigdefábregas J, Alonso JM, Delgado L, Domingo F, Cueto M, Gutiérrez L, Lázaro R, Nicolau JM, Sánchez G, Solé A, Torrentó JR, Vidal S, Aguilera C, Brenner AJ, Clark SC, Incoll LD (1996) The Rambla Honda field site: interactions of soil and vegetation along a catena in semiarid Spain. In: Thornes J, Brandt J (eds) Mediterranean desertification and land use. John Wiley, Chichester, pp 137–168

Rapp M, Romane F (1968) Contribution à l'étude du bilan de l'eau dans les écosystèmes méditerranéens. Egouttement des précipitations sous des peuplements de Quercus ilex L. et de Pinus halepensis Mill. Oecol Plant 3:271–283

Rodà F (1983) Biogeoquímica de les aigües de pluja i de drenatge en alguns ecosistemes forestals del Montseny. PhD Thesis, Autonomous University of Barcelona, Bellaterra

Rodà F, Àvila A, Bonilla D (1990) Precipitation, throughfall, soil solution and streamwater chemistry in a holm oak (Quercus ilex) forest. J Hydrol 116:167–183

Rutter AJ, Morton AJ (1977) A predictive model of rainfall interception in forests. III. Sensitivity of the model to stand parameters and meteorological variables. J Appl Ecol 14:567–588

Schlesinger WH, Hasey MM (1980) The nutrient content of precipitation, dry fallout and intercepted aerosols in the chaparral of southern California. Am Midl Nat 103:114–122

Schlesinger WH, Gray JT, Gilliarn FS (1982) Atmospheric deposition processes and their importance as sources of nutrients in a chaparral ecosystem of southern California. Water Resour Res 18:623–629

Young JA, Evans RA, Easi DA (1984) Stemflow on western juniper (*Juniperus occidentalis*) trees. Weed Sci 32:320–327

16 Soil Nitrogen Dynamics

Isabel Serrasolses, Victoria Diego and David Bonilla

16.1 Introduction

Soils are by far the largest nitrogen (N) pool in forest ecosystems, usually exceeding 85% of total ecosystem capital (Cole and Rapp 1981). Most of the soil N is tied to organic matter and only a small percentage of the total N is available to plants through N mineralization. Aboveground, N returns to the soil via litterfall and, after an immobilization period on the forest floor, litter decomposition supplies mineral N to the soil. The forest floor contains large pools of readily mineralizable N and is thus a major supplier of mineral N to trees (Keeney 1980; Casals et al. 1995). The forest floor of Mediterranean forests contains similar or lower amounts of organic matter than temperate forests but higher amounts than tropical forests (Cole and Rapp 1981).

N mineralization in Mediterranean soils is usually highest in periods of favourable temperature and water availability, thus exhibiting a marked seasonality, with most mineralization occurring in spring and autumn (Lemée 1967; Read and Mitchell 1983). Nitrification is more dependent on suitable conditions than ammonification because nitrifier populations are very sensitive to soil moisture, ammonium supply, and drying and rewetting cycles (Schaefer 1973; Davidson et al. 1992).

Nitrate leaching is one of the major pathways for N loss in terrestrial ecosystems because nitrate is relatively mobile in soils and it is easily leached by percolating water. The rate of leaching in a soil profile depends on the nitrate concentration in soil water and on the downward flux of water. Therefore N leaching in Mediterranean ecosystems has also a marked seasonal pattern determined by rainfall, almost coinciding with the N mineralization pattern. In undisturbed holm oak (*Quercus ilex* L.) forests there is a net retention of N on a whole catchment scale (Chap. 20), since the soils retain the N inputs from atmospheric deposition and also retain most of the mineral N produced in the soil. Hence, N leaching should be negligible in these soils.

The aim of this chapter is to evaluate the soil N availability in the holm oak forests of Prades and Montseny. To achieve this purpose the dynamics of

Ecological Studies, Vol. 137
Ferran Rodà et al. (eds) Ecology of Mediterranean
Evergreen Oak Forests
© Springer-Verlag, Berlin Heidelberg 1999

N in the soil profile was studied, particularly the total and mineral N, and the major soil N fluxes such as net mineralization, nitrification and leaching losses.

16.2 Total Nitrogen and C:N Ratio in the Soil Profile

Nitrogen concentration in holm oak leaf litterfall averaged 0.97% in Prades (V. Diego, unpubl. data) and 0.94% in Montseny (Mayor and Rodà 1992), giving a C:N ratio of 53. This high C:N ratio decreases rapidly in the forest floor through N immobilization in the L and F horizons. The F layer is the richest horizon in N: 1.66–1.91% N in Prades and 1.56% in Montseny (Fig. 16.1). In a litterbag decomposition study at Prades, 2.2 kg N ha^{-1} year^{-1} was immobilized during the initial phases of leaf litter decay when the C:N was higher than 35 (Serrasolsas 1994). After this period of N immobilization in the L and F horizons, the litter is no longer N-limited for microbes and net N mineralization begins. Both organic carbon and N decrease at the same

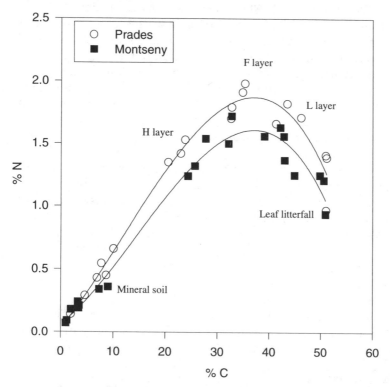

Fig. 16.1. Relationship between organic carbon (% C) and total nitrogen (% N) in leaf litterfall, forest floor and mineral soil at Prades and Montseny. (Data from Hereter 1990, 1992; Serrasolsas 1994)

rate (Fig. 16.1) and, consequently, the C:N ratios of the H horizon and the upper part of the mineral soil remain approximately constant, with values between 20 and 16. These C:N ratios are in the range 15–25 commonly found in surface (0–20 cm) forest soils (Attiwill and Leeper 1987). In deeper mineral soil at our sites, the C:N ratio decreases to 6–8 below 50-cm depth. Most of the organic matter with a low C:N ratio in deep soil is assumed to be humified and it is N recalcitrant (Vallejo and Hereter 1986; Vallejo 1993).

The forest floor of our holm oak forests contains 531–622 kg N ha^{-1} (Table 18.5). Using the litter decomposition model of Jenny et al. (1949), the calculated annual amount of N released from the forest floor is around 35 kg N ha^{-1} in both holm oak forests, with a half-life of N in the forest floor of 11 years. N turnover in Mediterranean forest floors varies widely, from 2.5 years in a calcareous site with a high decomposition rate (Le Rouquet, southern France; Lossaint and Rapp 1978) to 18.2 years in a dry site (Vallejo and Rosich 1990). This range overlaps with the lower range for temperate forests (Cole and Rapp 1981).

16.3 Soil Mineral Nitrogen

Concentrations of soil mineral N and rates of microbial transformations of N were studied at Prades on the upper, SE-facing slope of Torners (Chap. 2) in the context of the fertilization and irrigation experiment (Chap. 13) and the burning and clear-cutting experiment (Chap. 22). At Montseny, they were studied in a midslope position on the N-facing slope adjacent to the TM9 catchment at La Castanya (Chap. 2), with the purpose of characterizing the soil N dynamics in an undisturbed mesic holm oak forest (Bonilla and Rodà 1992) and of assessing the effects of root uptake through a trenching experiment (Bonilla and Rodà 1989, 1990). Unless otherwise stated, all results on soil N in this chapter refer to untreated (control) holm oak plots.

At Prades, mineral N concentrations were high in the H horizon (Fig 16.2), probably due to the high mineralization and nitrification rates of this horizon, with a mean of 119 (SE 16) mg N kg^{-1}. On average, this mineral N was 65% ammonium and 35% nitrate, though these proportions varied seasonally. The amount of mineral N in the H horizon was 2.8 ± 0.5 kg N ha^{-1}, less than 1% of the total N is this layer. Similar amounts of mineral N were found in the litter layer of the chaparral (2–6 kg N ha^{-1}; DeBano et al. 1979), and in the H horizon of a Scots pine forest near Montseny (3–6 kg N ha^{-1}, Casals et al. 1995). The H horizon was not sampled at Montseny because it was practically absent in the studied slope. However, since a high concentration of nitrate was leached out of the forest floor in the permanent plot (see Sect. 16.5), high mineralization and nitrification rates probably also occurred at Montseny in sites where the H horizon was well developed.

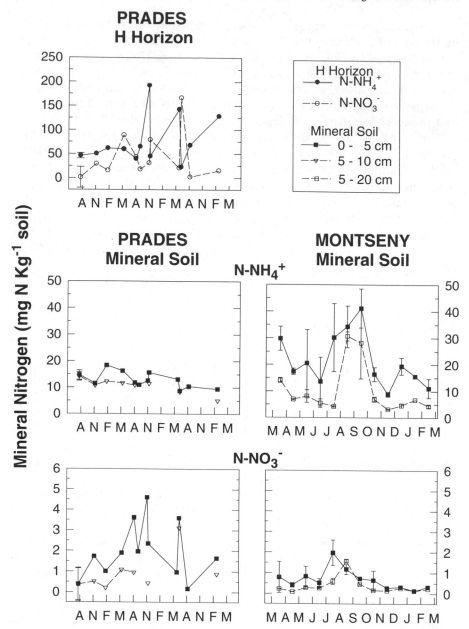

Fig. 16.2. Soil ammonium-N and nitrate-N concentrations throughout the year at Montseny and over 3 years at Prades. Each point is the mean of 3–4 (Prades) and 2 (Montseny, ± SE) composite samples

Mineral N concentrations decreased with soil depth (Fig. 16.2). In the upper mineral soil (0–5 cm), mineral N was higher at Montseny (22.3 ± 2.6 mg N kg^{-1}) than at Prades (14.8 ± 0.8 mg N kg^{-1}), while similar concentrations (around 10 mg N kg^{-1}) were found in deeper layers (Fig. 16.2). Ammonium was always the dominant form of mineral N at both sites, accounting on average for 87–97% of mineral N. Mean nitrate concentrations were very low at both sites, but they were higher in Prades than in Montseny, especially in the upper layer (Fig. 16.2). This probably reflects the lower leaching under the drier Prades climate. The amount of mineral N averaged 3.2 ± 0.2 kg N ha^{-1} at 0–5 cm and 2.3 ± 0.2 at 5–10 cm depth at Prades. At Montseny it was much higher, especially in the lower mineral soil, with 4.4 ± 1.0 kg N ha^{-1} at 0–5 cm and 11.9 ± 3.5 at 5–20 cm depth. Casals et al. (1995) found similar amounts of mineral N in the surface (0–10 cm) soil of a Scots pine forest (3–8 kg N ha^{-1}).

In spite of the high temporal variability in mineral N concentrations, seasonal trends were similar for all soil horizons in each site, but differed slightly between sites (Fig. 16.2). The highest content of mineral N occurred in spring and autumn at Prades, and in summer and early autumn at Montseny. Nitrate content reached a maximum in spring and autumn (and occasionally in summer) at Prades, and in summer at Montseny. Both ammonium and nitrate were low in winter at both sites (Fig. 16.2). This site-specific seasonal variability could be due to the different climatic conditions, the drier site (Prades) being more strongly influenced by the summer drought than the wetter site (Montseny). These temporal patterns in soil mineral N content reflect in part the changing rates of microbial N transformations, which are the subject of the next section.

16.4 Nitrogen Mineralization

Net N mineralization and nitrification rates determined by the in situ plastic bag incubation method (Pastor et al. 1984) were highly variable within and between incubation periods, especially in the H horizon and the 0–5 cm mineral soil (Fig. 16.3). The forest floor is heterogeneous in both amount and composition (Joffre et al. 1996; Fons et al. 1997) and together with the upper mineral soil is strongly influenced by changes in temperature and moisture, which often results in short pulses of mineralization and/or nitrification that are heterogeneously distributed in space and time. In general, mineralization rates per unit weight of soil were higher in the H horizon and 0–5 cm soil than in 5–20 cm soil in both sites, although the H horizon in Prades also showed occasionally the highest N immobilization (i.e. negative net N mineralization; Fig. 16.3).

At both sites N mineralization and nitrification rates had a seasonal pattern, especially in the H horizon and the upper mineral soil. In these near-surface layers, the highest rates occurred when soil temperature and moisture were high. In the lower mineral soil, N mineralization and nitrification rates at Prades were less variable throughout the year than in the upper layers (Fig. 16.3), whereas at Montseny high rates of net N mineralization in July

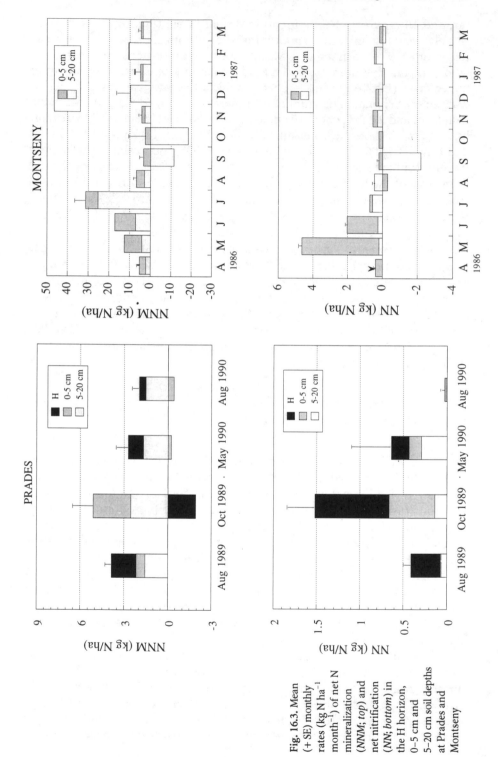

Fig. 16.3. Mean (+ SE) monthly rates (kg N ha^{-1} month^{-1}) of net N mineralization (*NNM; top*) and net nitrification (*NN; bottom*) in the H horizon, 0–5 cm and 5–20 cm soil depths at Prades and Montseny

were followed by high rates of N immobilization in September and October. At Montseny, net N mineralization in the upper mineral soil was weakly but positively correlated with soil temperature (r = 0.47, $P < 0.05$, $n = 23$). Net nitrification showed more marked seasonality than net N mineralization, probably because of the high environmental sensitivity of nitrifier populations (Schaefer 1973; Johnson 1992). Nitrification showed a positive relationship with soil moisture, both at Montseny (r = 0.52, $P < 0.05$, $n = 23$) and at Prades (r = 0.82, $P < 0.001$, $n = 12$) in the surface horizons.

Considering the mineral soil of 0–20 cm depth, the Prades forest had on a surface area basis a much lower mean net N mineralization than Montseny: 2.7 versus 6.7 kg N ha^{-1} month^{-1}, respectively. On an annual basis, about half of the mineralized N at Montseny was produced in each of the two mineral soil layers, whereas at Prades 71% of the net N mineralization in the mineral soil took place in the 5–20 cm layer (Fig. 16.3). These results imply that the lower mineral soil cannot be ignored in ecosystem N dynamics.

Net nitrification was low in both holm oak forests and occurred mostly in the forest floor and in the top mineral soil. At Prades, net nitrification in the H horizon averaged 0.40 kg N ha^{-1} month^{-1}, and represented 52% of the nitrified N in the soil profile (H and 0–20 cm of mineral soil). In the mineral soil (0–20 cm), mean net nitrification rates were lower at Prades than at Montseny (0.37 and 0.63 kg N ha^{-1} month^{-1}, respectively).

16.5 Nitrogen Leaching

At Montseny, the movement of mineral N in solution through the soil profile was studied in a 3-m-long trench in the permanent plot at La Castanya (Chap. 2) using zero-tension throughflow collectors placed below the H horizon and at 160-cm depth at the bedrock contact (Àvila et al. 1995). At Prades, and in the context of the burning and clear-cutting experiment referred to above, percolating water was collected in zero-tension lysimeters under the forest floor, and at 10- and 30-cm depths in the mineral soil (the latter being the local bedrock contact). Leached N was also collected using anionic and cationic exchange resins (Binkley and Matson 1983) in the same Prades site.

At Prades the maximum nitrate concentrations in the water leached from all depths occurred in summer, whilst other peaks occurred in spring and autumn. Concentrations were lowest in winter when the soil nitrate content was also very low. The amount of nitrate collected in resins was strongly related to the volume of rainfall during the incubation period (r = 0.75, $P < 0.0001$), so leaching of nitrate occurred during the major rainfall periods (spring and autumn). At Montseny, nitrate concentrations in throughflow water were highly variable, with little seasonal pattern (Àvila et al. 1995).

Figure 16.4 shows the mean nitrate concentrations as water percolates from rainfall to streamwater through each holm oak ecosystem. Infiltrated

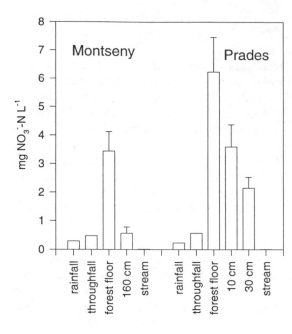

Fig. 16.4. Mean (+ SE) annual nitrate concentrations in rainfall and throughfall (Chap. 18: Montseny-1 and Prades-2), in percolating water collected at different soil depths (Poch 1992; Àvila et al. 1995; Gázquez 1995), and in streamwater (Piñol et al. 1992; Àvila et al. 1995)

Table 16.1. Solution fluxes (in kg N ha^{-1} year^{-1}) of mineral nitrogen in the Prades holm oak forest. Fluxes are given for rainfall, throughfall, percolating water, and streamwater. Nitrogen in percolating water was sampled with zero-tension lysimeters and by resin bags at the indicated soil depths.

Flux		Nitrogen form				
		NO$_3$-N		NH$_4$-N	Mineral N	
Rainfall[a]		1.7		1.8	3.5	
Throughfall[b]		3.6		2.0	5.6	
Percolating water		Lysim.[c]	Resin	Resin	Lysim.[c]	Resin
Below forest floor		21.7	14.2	1.2	21.7	15.5
Mineral soil	5 cm	–	13.5	0.8	–	14.2
	10 cm	8.1	–	–	8.1	–
	30 cm	16.5	–	–	16.5	–
Streamwater[d]		–	–	–	0.01	

[a] Bulk precipitation at the Avic catchment (Chap. 14).
[b] On the SE-facing Torners slope (Prades-2; Chap. 15).
[c] From Poch (1992) and Gázquez (1995). Water collected with lysimeters was not analyzed for ammonium.
[d] At the outlet of the Avic catchment (Chap. 20).

throughfall and stemflow become highly enriched in nitrate at both sites after passing through the forest floor (Fig. 16.4). The highest nitrate concentration of the solution profile was found in the H horizon, which was also where the highest rate of nitrification took place. Table 16.1 shows the fluxes of nitrate in percolating water at Prades. In zero-tension lysimeters, the nitrate flux below the forest floor was 22 kg NO_3-N ha^{-1} $year^{-1}$, a figure of the same order of magnitude as the release of N from the forest floor estimated by the Jenny decomposition model (35 kg N ha^{-1} $year^{-1}$).

Since the nitrate flux in throughfall was minor (Table 16.1), the above result means that more nitrate was leached from the forest floor than was produced in the H horizon through nitrification. Some of this extra nitrate may be produced by nitrification in the F horizon, since, although freshly fallen litter may act as a net sink for N (Berg and Ekbohm 1983), the forest floor layers L and F may also act as a net source of N (Qualls et al. 1991; Casals et al. 1995). However, comparisons between nitrification and nitrate leaching fluxes must bear in mind that our nitrification estimate at Prades is based on only four monthly incubations, and that small zero-tension lysimeters (and throughflow collectors) may overestimate leaching fluxes through drawdown of lateral water. Casals et al. (1995) and Cortina et al. (1995) measured similar solution fluxes of N (14–20 kg N ha^{-1} $year^{-1}$) under the forest floor in pine forests near Montseny.

In the mineral soil, nitrate concentrations in percolating water decreased at both Prades and Montseny (Fig. 16.4), probably due to plant uptake, microbial immobilization, and denitrification. Nitrate concentrations in streamwater were very low at both sites, and were lower than in the water percolating through the deep soil (Fig. 16.4). This was not due to nitrate retention in the streams (through uptake by the aquatic biota or other in-channel processes) since a seep under the permanent plot at Montseny already showed very low nitrate concentrations (Àvila 1988). Therefore, the reduction in nitrate concentrations between the deep percolating water and the streamwater arose either from spatial heterogeneity of soils within the catchments, from streamwaters being a mixture of waters with a mean residence time longer than that of gravitational soil water (Neal et al. 1992), or from terrestrial processes acting at depth such as uptake by holm oak roots growing in rock fissures or deep denitrification.

The Prades lysimeters, placed on an upper slope site, measured a significant water flow at a soil depth of 30 cm. Thus, even though nitrate concentration in percolating water decreased with soil depth, the amount of nitrate leached through the mineral soil at the bedrock contact was still quite high (16.5 ± 3.0 kg N ha^{-1} $year^{-1}$) compared with the output in streamwater (Table 16.1). In a holm oak forest on calcareous soil, Rapp (1990) and Lossaint and Rapp (1978) found nitrate leaching losses of 2.6 and 10 kg N ha^{-1} $year^{-1}$ at 15- and 30-cm depths, respectively. The fact that at Prades more nitrate was leached from the soil than exported from the catchment, if not totally accounted for by the soil heterogeneity and the possible overestimation mentioned above, raises the interesting possibility that nitrate is lost from draining water as it moves downslope. This would translate into a fertility flux from the upper to the lower parts of the catchment.

16.6 Ecosystem Nitrogen Availability

On average, the soil content of mineral N and the rates of net N mineralization and nitrification were higher at Montseny than at Prades, probably due to its wetter climate and shorter summer water stress. At Montseny, net N mineralization in the 0-20 cm soil yielded a flux of 80 kg of plant available N ha^{-1} year^{-1}, similar to the mean N mineralization rate in temperate deciduous forests (Bonilla and Rodà 1992; Reich et al. 1997). This is a relatively high rate, indicating that N availability is quite high in this holm oak forest, a fact that would not be evident from its moderate N flux in litterfall (37 kg N ha^{-1} year^{-1}; Table 18.6).

At Prades, estimates of the annual rates of soil N transformations are uncertain since only four monthly incubations were conducted (Fig. 16.3). However, the low monthly rates of net N mineralization measured at Prades (Fig. 16.3) may indicate that N availability is much lower than at Montseny, and that N could limit primary production. In fact, when the Prades holm oak forest was fertilized with ammonium nitrate, the net N mineralization during the first year after treatment increased four times over that of unfertilized plots (V. Diego, unpubl. data). In this experiment, aboveground net primary production was enhanced by N fertilization (Chap. 13), demonstrating the role of N availability in this dry Mediterranean forest.

Nitrification in the 0–20 cm soil at Montseny amounted to 7.5 kg N ha^{-1} year^{-1}, i.e. only 9% of the N mineralized annually. This result shows that in this mesic holm oak forest, growing on an acid soil, nitrification is of limited importance in soil N dynamics. It also means that the mineral N nutrition of holm oak at this site is based on ammonium rather than nitrate. Low nitrification is a key ecosystem property involved in maintaining a tight N cycle and in keeping low the leaching losses of nitrate (Sect. 16.5).

In the drier Prades forest, growing on a circumneutral soil, nitrification seems to play a somewhat major role since on average 14% of the mineralized N was nitrified in the 0–20 cm mineral soil (Fig. 16.3). When the H layer is included, this percentage raised to 28%. However, nitrification pulses can be very localized in space and time, and thus the role of nitrate in the nutrition of holm oak can be very variable. For example, a further single field incubation at Prades (Serrasolsas 1994) showed that in the H horizon N was taken up by trees as nitrate, whereas in the mineral soil both nitrate (57%) and ammonium (43%) were taken up. In particular, fast nitrification may occur when the soils are moist and warm. High nitrate concentrations are often found in the soil solution after the first rainfall following the summer drought at both Prades and Montseny (Chap. 17). Microbial N is one of the main sources of mineralized N after soil drying and rewetting (Okano 1990; Serrasolsas and Khanna 1995). At Prades, microbial N is a labile reserve of easily mineralizable N, corresponding to 3% of the total N in the H horizon, and between 1 and 2% in the mineral soil (0–5 cm), thus exceeding mineral N (data not shown).

In calcareous soils nitrate can make a larger contribution to the mineral N nutrition of holm oak than in our silicate soils. For instance, in the holm oak forest at Puéchabon near Montpellier (southern France), the net N mineralization was 28–70 kg N ha^{-1} year^{-1}, and 57% of it was nitrified (Merzouki et al. 1990).

In summary, the mesic Montseny site had a substantially higher availability of soil N than its xeric counterpart at Prades. As we will see in Chapter 18, this difference has at most a moderate effect on the respective forest nutrient cycles. Despite the differences in N availability, the losses of nitrate leached out to streams were very low at both sites, revealing that both ecosystems keep the tight N cycle characteristic of undisturbed forests. A root trenching experiment in the Montseny site clearly demonstrated the major role that N uptake by tree roots has in maintaining this tight cycle in holm oak forests (Bonilla and Rodà 1989, 1990). Disturbances such as clear-cutting and burning (Chap. 22) may increase nitrate leaching from the soil in holm oak forests (Sect. 22.6), though nutrient uptake by regrowing vegetation contributes to limit nitrogen losses.

References

Attiwill PM, Leeper GW (1987) Forest soils and nutrient cycles. Melbourne University Press. Carlton, Victoria

Àvila A (1988) Balanç d'aigua i nutrients en una conca d'alzinar del Montseny. Estudis i Monografies 13, Diputació de Barcelona, Barcelona

Àvila A, Bonilla D, Rodà F, Piñol J, Neal C (1995) Soilwater chemistry in a holm-oak (*Quercus ilex*) forest: inferences on biogeochemical processes for a montane Mediterranean area. J Hydrol 166:15–35

Berg B, Ekbohm G (1983) Nitrogen immobilization in decomposing needle litter at variable carbon:nitrogen ratios. Ecology 64:63–67

Binkley D, Matson P (1983) Ion exchange resin bag method for assessing forest soil nitrogen availability. Soil Sci Soc Am J 47:1050–1052

Bonilla D, Rodà F (1989) Dinámica del nitrógeno en el suelo de un encinar montano: respuesta a una perturbación experimental. Options Medit Sér Sémin 3:187.190

Bonilla D, Rodà F (1990) Nitrogen cycling responses to disturbance: trenching experiments in an evergreen oak forest. In: Harrison AF, Ineson P, Heal OW (eds) Nutrient cycling in terrestrial ecosystems. Elsevier, London, pp 179–189

Bonilla D, Rodà F (1992) Soil nitrogen dynamics in a holm oak forest. Vegetatio 99/100:247–257

Casals P, Romanyà J, Cortina J, Fons J, Bode M, Vallejo VR (1995) Nitrogen supply rate in Scots pine (*Pinus sylvestris* L.) forests of contrasting slope aspect. Plant Soil 169:67–73

Cole DW, Rapp M (1981) Elemental cycling in forest ecosystems. In: Reichle DE (ed) Dynamic properties of forest ecosystems. Cambridge University Press, Cambridge, pp 341–409

Cortina J, Romanyà J, Vallejo V R (1995) Nitrogen and phosphorus leaching from the forest floor of a mature *Pinus radiata* stand. Geoderma 66:321–330

Davidson EA, Hart SC, Firestone MK (1992) Internal cycling of nitrate in soils of a mature coniferous forest. Ecology 73:1148–1156

DeBano LF, Eberlein GE, Dunn PH (1979) Effects of burning on chaparral soils: I. Soil nitrogen. Soil Sci Soc Am J 43:504–509

Fons J, Sauras T, Romanyà J, Vallejo VR (1997) Sampling strategies in forest soils. Ann Sci For 54:493–499

Gázquez R (1995) Seguimiento lisimétrico del efecto del fuego y de la tala en suelos forestales de la Serra de Prades, Tarragona. MSc Thesis, Escola Superior d'Agricultura de Barcelona, Barcelona

Hereter A (1990) Els sòls forestals del Montseny. Avaluació de la seva fertilitat. PhD Thesis, University of Barcelona, Barcelona

Hereter A (1992) Organic horizons in Mediterranean forest soils: characteristics and influence of the parent material. In: Teller A, Mathy P, Jeffers JNR (eds) Responses of forest ecosystems to environmental changes. Elsevier Applied Science, London, pp 776–777

Jenny H, Gessel SP, Birgham FT (1949) Comparative study of decomposition rate of organic matter in temperate and tropical regions. Soil Sci 68:419–432

Joffre R, Rambal S, Romane F (1996) Local variations of ecosystem functions in Mediterranean evergreen oak woodland. Ann Sci For 53:561–570

Johnson DW (1992) Nitrogen retention in forest soils. J Environ Qual 21:1–12

Keeney DR (1980) Prediction of soil nitrogen availability in forest ecosystems: a literature review. For Sci 26:159–171

Lemée G (1967) Investigations sur la minéralisation de l'azote et son évolution annuelle dans les humus forestiers in situ. Oecol Plant 2:285–324

Lossaint P, Rapp M (1978) La forêt méditerranéenne de chênes verts (Quercus ilex L.). In: Lamotte M, Bourlière F (eds) Problèmes d'écologie: structure et fonctionnement des écosystèmes terrestres. Masson, Paris, pp 129–185

Mayor X, Rodà F (1992) Is primary production in holm oak forest nutrient limited? A correlational approach. Vegetatio 99/100:209–217

Merzouki A, Lossaint P, Billès G, Rapp M (1990) The impact of deforestation on the mineral nitrogen available during restoration of the holm oak (Quercus ilex L.) coppice. Ann Sci For 21: 633–641

Neal C, Neal M, Warrington A, Àvila A, Piñol J, Rodà F (1992) Stable hydrogen and oxygen isotope studies of rainfall and streamwaters for two contrasting holm oak areas of Catalonia, northeastern Spain. J Hydrol 140:163–178

Okano S (1990) Availability of mineralized N from microbial biomass and organic matter after drying and heating of grassland soils. Plant Soil 129:219–225

Pastor J, Aber JD, McClaugherty CA, Melillo JM (1984) Aboveground production and N and P cycling along a nitrogen mineralization gradient on Blackhawk Island, Wisconsin. Ecology 65:256–268

Piñol J, Àvila A, Rodà F (1992) The seasonal variation of streamwater chemistry in three forested Mediterranean catchments. J Hydrol 140:119–141

Poch R (1992) Seguiment lisimètric en sòls d'alzinars de Prades afectats per foc i tala. MSc Thesis, Escola Superior d'Agricultura de Barcelona, Barcelona

Qualls G, Haines BL, Swank WT (1991) Fluxes of dissolved organic nutrients and humic substances in a deciduous forest. Ecology 72:254–266

Rapp M (1990) Nitrogen status and mineralization in natural and disturbed Mediterranean forests and coppices. Plant Soil 128:21–30

Read DJ, Mitchell DT (1983) Decomposition and mineralization processes in Mediterranean-type ecosystems and in heathlands of similar structure. In: Kruger FJ, Mitchell PT, Jarvis JUM (eds) Mediterranean type ecosystems. The role of nutrients. Springer, Berlin, pp 208–232

Reich PB, Grigal DF, Aber JD, Gower ST (1997) Nitrogen mineralization and productivity in 50 hardwood and conifer stands on diverse soils. Ecology 78:335–347

Rodà F, Àvila A, Bonilla D (1990) Precipitation, throughfall, soil solution and streamwater chemistry in a holm oak (Quercus ilex) forest. J Hydrol 116:167–183

Schaefer R (1973) Microbial activity under seasonal conditions of drought in Mediterranean climates. In: Di Castri F, Mooney HA (eds) Mediterranean type ecosystems. Origin and structure. Springer, Berlin, pp 191–198

Serrasolsas I (1994) Fertilitat de sòls forestals afectats pel foc. Dinàmica del nitrogen i del fòsfor. PhD Thesis, University of Barcelona, Barcelona

Serrasolsas I, Khanna PK (1995) Changes in heated and autoclaved forest soils of SE Australia. I. Carbon and nitrogen. Biogeochemistry 29:3–24

Vallejo VR (1993) Evaluation of C:N ratio as a parameter of N mineralization. Mitt Österr Bodenk Ges 47:71–78

Vallejo VR, Hereter A (1986) Analysis of the C/N ratio distribution in Mediterranean forest soils. Trans XIII Congr of the ISSS, Hambourg, vol. II, pp 650–651

Vallejo VR, Rosich D (1990) Nitrogen evolution in Mediterranean forest soils. Trans XIV Congr of the ISSS, Kyoto, pp 438–439

17 Soil Water Chemistry

Núria Melià, Juan Bellot and V. Ramon Vallejo

17.1 Introduction

Soil solution is the medium in which most terrestrial biogeochemical processes occur. Its chemical composition results from the dynamic equilibria between the solid, liquid and gas phases of the soil (Wolt 1994). Soil solution chemistry provides valuable information on the dynamics of nutrient cycling and availability to plants, mineral stability, weathering trends, and pollutant mobility within the soil profile. Moreover, the effects of environmental changes induced by anthropogenic disturbances on the ecosystem may be promptly reflected in the soil solution composition. However, the range of variability found in soil solution may mask the extent of these effects.

Many authors have reported spatial and temporal patterns of variation in soil solution chemistry (Zabowski and Ugolini 1990; Giesler and Lundström 1993; Manderscheid and Matzner 1995). Various soil properties, e.g. CEC, pH, organic matter and clay content, may lead to differences in processes at each soil horizon and, therefore, to spatial (both vertical and lateral) differences in the chemical composition of the soil solution (Tokuchi et al. 1993). Thus, water that moves down through the soil profile carries solutes from the forest floor and the topsoil, where nutrient concentrations are high and organic matter decomposition is the major process, to the deeper horizons, which have lower nutrient concentrations and where plant uptake and exchange processes predominate. Temporal changes in soil solution may be attributed to the effect of seasonal patterns on soil temperature, moisture and biological activity in these soil horizons.

This chapter summarizes the main characteristics of soil water chemistry in the holm oak forests of Montseny and Prades. Chapter 16 dealt with nitrate leaching in gravitational soil water. In the present chapter vertical changes within the soil profile and their seasonal pattern are described for all major ions both in free-flowing soil water (gravitational water) and in soil water held at higher tensions (soil solution). Comparison of these soil water fractions between both sites and the relationship with the main processes occurring in these soils are emphasized.

Ecological Studies, Vol. 137
Ferran Rodà et al. (eds) Ecology of Mediterranean
Evergreen Oak Forests
© Springer-Verlag, Berlin Heidelberg 1999

17.2 Gravitational Water

Gravitational water percolating through the soil was collected at the sites and by the methods described in Section 16.5. Briefly, zero-tension lysimeters were used at Prades and throughflow collectors at Montseny. Detailed descriptions of these studies can be found in Gázquez (1995) and Poch (1992) for Prades and in Àvila et al. (1995) for Montseny.

At both sites much higher water volumes were collected at the deepest sampled depth than under the forest floor, revealing the importance of deep subsurface flow in these highly permeable soils. At Prades, the mean water volume per sample was 25 L m^{-2} below the forest floor, 16 L m^{-2} at 10-cm depth, and 50 L m^{-2} at 30-cm depth, although the coefficient of variation was higher than 100% at all depths. Similarly, at Montseny, the mean water volume per sample was 0.8 L below the forest floor and 30 L at 160 cm (Àvila et al. 1995); an intermediate sampler at 50 cm never collected water. This suggests that the throughflow trench collected basically the horizontal water flow at the soil–bedrock contact, as opposed to the vertical flow collected with the zero-tension lysimeters at Prades.

17.2.1 General Characteristics and Variations Within the Soil Profile

Gravitational water in both holm oak forests showed chemical changes along the soil profile, as also found in other studies (Ranger et al. 1993; Tokuchi et al. 1993; Menéndez et al. 1995). In general, water collected below the forest floor presented higher volume-weighted ion concentrations than at lower depths, although the extent of the changes with depth varied according to the site and the ion being considered (Fig. 17.1). At Prades, the soil water pH was nearly neutral below the forest floor, with a volume-weighted mean of 6.63, which differed significantly from the lower, slightly acidic values found in the mineral layers (5.96 at 10 cm and 6.04 at 30 cm). Lower values were observed at Montseny, where the mildly acidic mean pH of 6.28 below the forest floor decreased in the deep subsurface flow to a pH of 5.53 (Àvila et al. 1995).

Calcium was the dominant cation at all depths at both sites, representing about 50% of the total cationic charge in soil water collected below the forest floor (Fig. 17.1). Among the strong anions, dominance alternated between Cl$^-$ and SO$_4^{2-}$ at Prades, whereas SO$_4^{2-}$ dominated at Montseny. However, weak anions determined as alkalinity made the largest contribution to anion charge in throughflow below the forest floor at Montseny (Fig. 17.1b), which had a volume-weighted mean alkalinity of 481 μeq L^{-1} (29% of the total cation charge). This high alkalinity was provided mainly by dissolved organic anions. There was a substantial charge imbalance (Fig. 17.1b), resulting from an anion deficit of 20% of the total cation charge. This deficit arose probably (1) from organic anions that were not weak enough to protonate during the alkalinity titration, and (2) from part of the cations that could be

(a) Prades

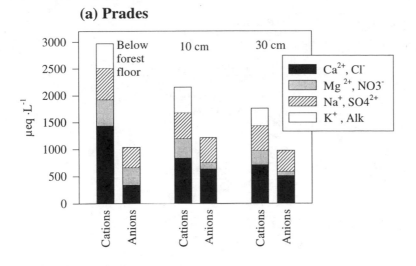

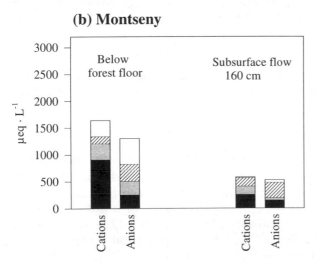

(b) Montseny

Fig. 17.1. Volume-weighted mean concentration of main cations and anions by depth in the soil water collected from zero-tension lysimeters at a Prades and from throughflow collectors at b Montseny. Samples are from June 1990 to July 1991 at Prades after Poch (1992), and from May 1985 to May 1988 at Montseny after Àvila et al. (1995). Alkalinity was not analyzed at Prades

complexed in soil water but that were included in the atomic absorption determinations. In deep subsurface flow at Montseny, both alkalinity and the anion deficit were drastically reduced (Fig. 17.1b), probably because the involved organic compounds had been retained or metabolized in the soil. At Prades, where alkalinity was not analyzed in gravitational water, the anion charge deficit accounted for more than 50% of the total cation charge at all depths, and also decreased with depth (Fig. 17.1a). Studies in temperate for-

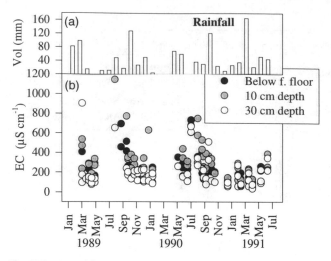

Fig. 17.2. a Monthly precipitation; b temporal variation of electrical conductivity (*EC*) from soil water samples collected with zero-tension lysimeter replicates below the forest floor, at 10- and 30-cm depths at Prades

ests have also observed a sizeable anion deficit in soil water, and it was related to dissolved organic anions (Rustad et al. 1993; Fernandez et al. 1995).

17.2.2 Temporal Variability

The occurrence of gravitational water flow in these Mediterranean ecosystems is highly dependent on the precipitation regime. At Prades, despite the rainfall heterogeneity during the study period, a seasonal concentration-dilution pattern was identified in the gravitational water collected at each depth: just after the summer drought soil water was highly concentrated, and ion concentrations gradually fell through late autumn and winter (Fig. 17.2). This same seasonal pattern occurs in streamwaters at Prades (Piñol et al. 1992). In contrast to these results, the chemistry of soil water collected at both depths at Montseny did not follow any clear seasonal pattern (Àvila et al. 1995), with only K^+ concentration and alkalinity below the forest floor peaking sometimes in summer.

17.2.3 Ion Relationships

Considering that Cl^- is scarcely affected by exchange processes or biological uptake in the soil, ion ratios with respect to Cl^- in the soil water can be used to discriminate the seasonal concentration pattern due to evapotranspiration, from other processes affecting the chemistry of soil water. Two ion ratios are shown in Fig. 17.3 to illustrate these changes occurring in the soil

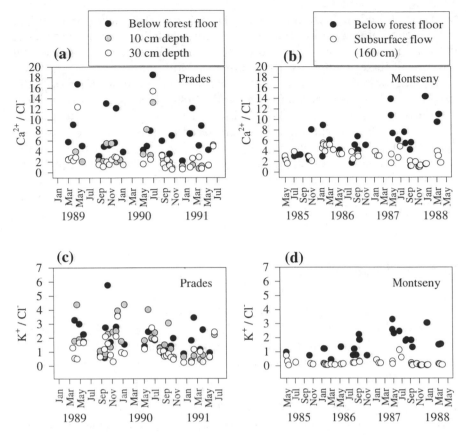

Fig. 17.3. Temporal variation by depth of ion ratios relative to Cl⁻ in gravitational water collected at Prades and Montseny. Ion ratios computed on an equivalent basis. Ca^{2+}/Cl^- (**a**) and K^+/Cl^- (**c**) at Prades. Ca^{2+}/Cl^- (**b**) and K^+/Cl^- (**d**) at Montseny. Montseny data from Bonilla (1990)

profile. Both Montseny and Prades soil waters showed higher Ca^{2+}/Cl^- ratios below the forest floor than at the deeper layers during most of the sampling period (Fig. 17.3a,b). Further, a seasonal pattern could be identified below the forest floor: high ratios were present after the summer drought and decreased with time. The K^+/Cl^- ratio also showed a similar seasonal structure at both sites, although differences with depth were not clearly apparent in the Prades soil water, and no seasonal trend was observed at the lower depth at Montseny (Fig. 17.3c, d).

This temporal variability in soil water chemistry could be related to seasonal patterns of (1) rainfall and throughfall, (2) leaching of ions from fresh litter, (3) dissolution of forest floor decomposition products and (4) cation exchange and uptake. Unlike results from other areas, where trends in soil water chemistry could be associated with patterns of atmospheric frontal systems (Hughes et al. 1994), little seasonality in the rainfall chemistry could be distinguished in these holm-oak forests (Chap. 14). Bellot and Escarré (1991) found higher ion concentrations in throughfall during the summer at Prades and Rodà et al. (1990) also reported a distinct maximum in net throughfall fluxes of K^+ in early summer at Montseny (Chap. 15). Furthermore, maximum litterfall was found in late spring or early summer at Prades (Bellot et al. 1992) and at Montseny (Verdú 1984). The highest carbon dioxide efflux from soil at both areas was observed in late spring and late summer (Piñol et al. 1995), suggesting that high decomposition rate occurs during these periods. These results suggest that the ion enrichment of soil water from throughfall and leaching from fresh litter may occur especially during late spring and late summer rainfall events. Added to these processes, the high decomposition rate occurring in late summer may provide an important source of soluble ions for the first autumn rainfalls.

17.3 Soil Solution

The chemistry of water held in the soil at low hydric potentials was studied at Montseny with low-tension lysimetry and at Prades with drainage centrifugation. At Montseny, soil solution was sampled weekly with ceramic-cup lysimeters evacuated manually to –65 kPa. Replicate lysimeters were installed at 20- and 40-cm depths in or near the permanent plot at La Castanya (Chap. 2). Detailed descriptions of this study were provided by Rodà et al. (1990) and Àvila et al. (1995). At Prades, research into soil solution composition focused on the spatial variability within the Avic catchment (Chap. 2). Sixteen pits at each altitudinal extreme of the catchment were sampled at 0- to 2.5-cm, 2.5- to 15-cm and 15- to 30-cm depths. Field moist samples were centrifuged using double-bottomed cups (Gillman 1976) at an equivalent pressure of 3 MPa and the extracted solutions were analyzed after filtration.

17.3.1 Low-Tension Lysimeter Solution at Montseny

The mean solution chemistry was similar at 20- and 40-cm depths (Table 17.1). The soil solution in the mineral soil was generally less concentrated, and had more acidic pH values, than the gravitational water collected below the forest floor (Fig. 17.1, Table 17.1). As in gravitational water, soil solution was dominated by Ca^{2+} and SO_4^{2-}, which accounted for more than 50% of the cation and anion charge respectively. Potassium and NO_3^- in soil solution were the ions with the lowest mean concentrations (Table 17.1). The order of abundance of cations followed the same sequence as that in the cation exchange complex of the soil (Àvila et al. 1995): $Ca^{2+} > Mg^{2+} > Na^+ > K^+$. The anion charge deficit computed from the analyzed ions was much lower than that found in the soil water flow below the forest floor, but it still represented

Table 17.1. Chemical composition of soil solution at Montseny from low-tension lysimeters[a] and at Prades from centrifugation extracts[b]. Arithmetic mean and standard deviation (in parentheses) for all variables except pH. Geometric mean is shown for the latter[c]

	Cond.[d]	pH	Alkal.[e]	DOC[c]	Ca^{2+}	Mg^{2+}	Na^+	K^+	NH_4^+	Cl^-	NO_3^-	SO_4^{2-}
Montseny												
20 cm	75 (23)	5.66	65 (15)	–	417 (123)	205 (67)	142 (39)	35 (27)	–	175 (132)	1.7 (4.4)	394 (105)
40 cm	94 (55)	5.62	60 (11)	–	491 (291)	248 (147)	176 (89)	35 (39)	–	205 (172)	6.0 (19.4)	539 (389)
Prades												
0–2.5 cm	433 (203)	6.98	963 (910)	20.2 (14.6)	3733 (2286)	913 (598)	227 (118)	634 (271)	121 (104)	1134 (1031)	319 (657)	1341 (1055)
2.5–15 cm	339 (216)	6.05	194 (364)	16.3 (17.2)	2325 (1752)	633 (424)	264 (158)	297 (161)	63 (80)	1333 (889)	122 (390)	1131 (742)
15–30 cm	212 (117)	6.19	169 (132)	7.2 (5.0)	1323 (1175)	342 (272)	244 (175)	174 (124)	48 (91)	845 (580)	54 (115)	722 (510)

[a] Data from Àvila et al. (1995), from December 1984 to June 1986.
[b] Data from Prades from 32 pits sampled on four occasions (November and December 1993, April and September 1994).
[c] Alkalinity and ion concentrations in µeq L⁻¹. Dissolved organic carbon, in mmol L⁻¹. Dashes indicate not analyzed.
[d] Conductivity in µS cm⁻¹ at 20 °C.
[e] Total alkalinity at Montseny, bicarbonate alkalinity at Prades.

a high percentage of the total charge if compared with that from the deep subsurface flow (Fig. 17.1, Table 17.1).

17.3.1.1 Temporal Variability
Ion concentrations in the soil solution varied seasonally (Àvila et al. 1995): they were highest at both depths in the first samples collected after the summer drought, tended to decrease gradually through the winter, and remained low or recovered somewhat through the spring until the lysimeters dried out in summer. This pattern applied to all ions at both depths, except by K^+ and NO_3^- whose variations were more erratic.

17.3.2 Centrifuge Solution at Prades

The soil solutions obtained by centrifugation at Prades were highly concentrated (Table 17.1). The solution chemistry was dominated by Ca^{2+} which, averaged across soil horizons, made up 62% of the total cation charge, and by SO_4^{2-} and Cl^-, with 38% and 39% of the total analyzed anion concentrations, respectively. The cation sequence, in order of abundance, was $Ca^{2+} > Mg^{2+} > K^+ > Na^+ > NH_4^+$, and, similar to Montseny paralelled the abundances observed in the cation exchange complex of the soil matrix (Chap. 2). The positive charge imbalance found in soil solution modified the anion abundance sequence with respect to the total anionic charge. This anion charge deficit contributed to 27% of the total charge, as opposed to 28% of Cl^- and 28% of SO_4^{2-}, and showed a strong linear relationship with the analyzed dissolved organic carbon ($r = 0.81, P < 0.0001$).

17.3.2.1 Changes Along the Profile
Consistent changes in ion concentration and ion dominance occurred with depth. All ion concentrations except Na^+ decreased significantly along the profile ($P < 0.0001$; Table 17.1). pH was also significantly higher in the upper horizon. This decreasing pattern with depth modified the distribution of the anion dominance in the horizons studied. The anion charge deficit, which we assume is mostly due to DOC charge, made the major contribution (33%) to the total anion charge in the upper horizon. This 'DOC buffer capacity', added to the 17% bicarbonate alkalinity, made up nearly 50% of the total anion charge in this horizon. The anion charge deficit declined in the deeper horizons, where the Cl^- and SO_4^{2-} codominance arose.

17.3.2.2 Spatial Variability
Few differences were observed between the two altitudinal extremes of the catchment. Only Mg^{2+} concentrations were significantly lower on the upper slopes, with mean values of 698, 549 and 267 $\mu eq\ L^{-1}$ at 0–2.5 cm, 2.5–15 cm and 15–30 cm depths, respectively, than in the valley bottom (1125, 717 and 410 $\mu eq\ L^{-1}$ at the same three depths). This finding agrees with the higher Mg^{2+} concentration in holm oak leaves found in the valley bottom of the catchment (Sabaté 1993).

17.3.2.3 TEMPORAL VARIABILITY

As at Montseny, the soil solution chemistry at Prades varied temporally. The variations in ion concentrations were not simply related to the inverse of soil moisture (Fig. 17.4). The April and September samplings tended to yield the

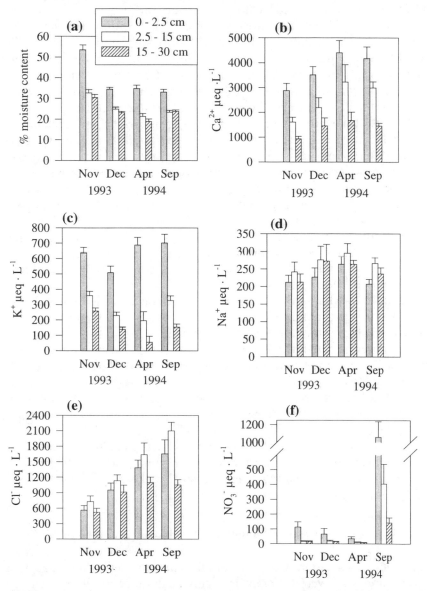

Fig. 17.4. a Mean soil moisture content (dry mass basis), and **b-f** soil solution ion concentrations by sampling occasion and horizon at Avic (Prades), averaged for upper and lower catchment positions. Soil solution extracted by centrifugation. Data for Mg^{2+} and SO_4^{2+} are not shown since they have similar patterns to Ca^{2+} (**b**) and Cl^- (**e**), respectively. Each *bar* is the mean (+ SE) of 32 soil pits

largest differences between the two upper horizons and the deepest layer for Ca^{2+}, Mg^{2+}, SO_4^{2-} and Cl^-. The long and severe drought of 1994 probably resulted in an accumulation of salts in the upper horizons through evapotranspiration. This concentration effect could mainly be seen for Cl^- and SO_4^{2-} (Fig. 17.4e), which showed an increasing pattern from the first to the last sampling. In contrast, Na^+ failed to present any differences between samplings or depths. The effects of the drought on soil solution chemistry must have been higher than evidenced by Fig. 17.4 since soil solution could not be obtained by the centrifugation method during the driest months until the September sampling, which took place 3–4 days after a 100-mm storm.

Nitrate concentrations in soil solution were extremely high in September, at the end of the drought (Fig. 17.4f). Other authors also found high peaks of NO_3^- concentrations after the summer drought in forest floor leachates in Scots pine forests (Casals et al. 1995) and in *Pinus radiata* plantations (Cortina et al. 1995). Microbial processes occurring after the rewetting of dried soils increase nitrogen mineralization (van Gestel et al. 1993), especially at high temperatures (Chap. 16). In addition, the drought of 1994 produced a high mortality of trees and fine roots, and nitrogen uptake was probably reduced. The role of nutrient demand by trees in keeping low levels of NO_3^- in the soil solution of holm oak forests was clearly demonstrated by a trenching experiment at Montseny: after severing all tree roots entering otherwise undisturbed 2×2-m plots, NO_3^- concentrations in soil solution skyrocketed (Bonilla and Rodà 1990).

Potassium showed a particular behaviour. The concentration effect due to the summer drought, found for most other ions, was not apparent for K^+. On the contrary, K^+ concentrations in November, when the soil was wettest, were similar to those in September (Fig. 17.4c). These results seem to indicate that leaching of K^+ from the forest floor to deeper soil water during wet periods could be as important as the concentration effect observed in other cations in dry periods. Furthermore, K^+ concentrations were lowest in December or April, while those of Ca^{2+} or Cl^- were lowest in the wettest month (November).

Those ions most affected by the biological activity in soils (K^+, NH_4^+ and NO_3^-) also differed from other ions in having minimal concentrations in the deeper horizons in April (Fig. 17.4c,f). Biological demand for these ions was probably involved in this depletion.

17.3.3 Differences in Soil Solution Between Prades and Montseny

The previous soil solution data from Prades and Montseny show that both holm oak forests are very different in terms of the magnitude of their ion concentrations. Soil solution at Prades was highly mineralized, with higher ion concentrations, alkalinity and pH than at Montseny (Table 17.1). Moreover, ion concentrations decreased with depth at Prades, whilst lower differences between depths were found at Montseny, partly because here the soil solution in the upper 15 cm of soil was not sampled. Both sites followed a cation abundance sequence in soil solution according to their respective cation exchange complex, indicating a strong interaction with the soil matrix.

The higher concentrations observed at Prades cannot be completely attributed to its drier climate. Other factors may contribute to the differences in soil solution chemistry between Prades and Montseny. First, the cation exchange capacity of the soil influences the soil solution chemistry: the base saturation at Montseny is much lower than at Prades (Chap. 2; Àvila et al. 1995), thus providing less solutes to the soil solution at equilibrium.

Second, the different methods of collection used at each area sampled different phases of the soil water. Several studies obtained lower ion concentrations in soil water flow collected with zero-tension lysimeters than in soil solution collected with low-tension lysimeters (Swistock et al. 1990; Ranger et al. 1993; Fernandez et al. 1995). Hence, differences between the low-tension lysimetry used at Montseny and the centrifugation method used at Prades could be expected, since the latter extracted soil water held in micropores at much higher tensions. Zabowski and Ugolini (1990) reported larger differences between both methods in the seasonal pattern than in ion concentrations.

Third, the mean concentration of carbon dioxide in the entire soil profile was much higher at Prades (1.7% on the upper slopes and 1.1% on the lower slopes of the Avic catchment) than at Montseny (0.4%; Piñol et al. 1995). These results suggest a major influence of pCO_2 on soil water chemistry at Prades through the effect of carbonic acid on weathering processes, which is reflected in the higher alkalinity of soil waters and streamwaters at this site. The lower pCO_2 observed at Montseny suggests a higher diffusivity of the soils. Piñol et al. (1995) found that the soil of the permanent plot at Montseny was richer in sand and poorer in silt and clay than the soils of Avic at Prades. These features, added to the wetter climate at Montseny, suggest that water at this site has lower residence time in soil and weaker interaction with the soil matrix, and consequently ion concentrations in soil water are lower than at Prades.

17.4 Soil Water Flow and Soil Solution: Soil processes and General Remarks

Changes occurring along the profile from all collected water types indicate a decreasing trend with depth in ion concentration at both holm oak forests. Gravitational water collected below the forest floor was highly enriched in K^+, Ca^{2+} and NO_3^- compared to the concentrations found in the lower layers (Fig. 17.1; Table 17.1). Further, although soil water at Prades had higher ion concentrations, both sites showed similar ion relationships below the forest floor (Table 17.2), and followed similar temporal patterns related to enrichment by throughfall, leaching and dissolution of decomposition products.

Differences between sites become apparent at lower depths, as can be seen in the cation abundances and ion ratios of the soil solution (Tables 17.1 and 17.2). Montseny soil solution was proportionally poorer in K^+ compared to Prades, and each site followed a cation abundance sequence according to that found in their respective cation exchange complexes.

Water that has entered the soil can either bypass the soil matrix through preferential flow pathways or infiltrate the soil micropore system, increasing its residence time and, consequently, the interaction with the soil matrix. Àvila et al. (1995) showed that the chemical composition of the deep subsurface flow at Montseny was intermediate between that from the water collected below the forest floor and that coming from the soil solution sampled with low-tension lysimeters. From this observation, they suggested that free-flowing water collected in the lower levels of the soil profile is a mixture of the macropore water flow coming from the forest floor and pre-event displaced soil solution from the deeper horizons. Ion ratios in these three water types broadly support this interpretation (Table 17.2).

Table 17.2. Mean ion ratios, on equivalent basis, in gravitational soil water, soil solution, and streamwater at Montseny and Prades. Gravitational waters are underlined

		Ca^{2+}/Mg^{2+}	Ca^{2+}/K^+	Ca^{2+}/Cl^-	K^+/Cl^-
Montseny					
Below forest floor[a]		3.99	4.77	6.04	1.60
Soil solution	20 cm[b]	2.05	19.7	3.18	0.37
	40 cm[b]	2.02	96.9	2.72	0.25
160 cm[a]		2.40	29.3	2.98	0.19
Baseflow[c]		1.53	31.4	3.97	0.13
Prades					
Below forest floor		4.19	4.30	7.04	1.62
Soil solution	0–2.5 cm[d]	4.22	6.26	4.88	1.02
	2.5–15 cm[d]	3.78	10.5	2.21	0.36
30 cm		3.63	2.70	2.79	1.00
Streamwater[e]		4.74	78.4	8.09	0.10

[a] Arithmetic mean ratios from throughflow collectors (Bonilla 1990), May 1985 to May 1988.
[b] Arithmetic mean ion ratios from low-tension lysimeters (Bonilla 1990), December 1984 to June 1986.
[c] Ion ratios of volume-weighted means in streamwater of the TM9 catchment (Àvila et al. 1995).
[d] Arithmetic mean ion ratios from centrifuge soil solution.
[e] Ion ratios of volume-weighted means in streamwater of the Avic catchment (Piñol and Àvila 1992).

17.5   Relationship Between Soil Water
and Streamwater Chemistries

The chemistry of the soil water at both sites was quite different from that of the streamwater (Fig. 17.5). During baseflow, groundwater is the major source feeding the stream. This water is enriched with rock weathering products, with high alkalinity, and base cation concentrations. Soil water chemistry is expected to contribute to the stormflow composition to some extent. The streamwater of the Avic catchment at Prades is more mineralized than that from the TM9 catchment in Montseny (Piñol et al. 1992), coinciding with the results found in the soil water concentrations from both sites. Calcium was the major cation in Prades streamwaters, whereas at Montseny Na^+ and Ca^{2+} codominate in streamwater.

At Montseny, all analyzed ions except Na^+ were higher in soil water than in both the baseflow and the stormflow (Fig. 17.5). In addition, the mean ion

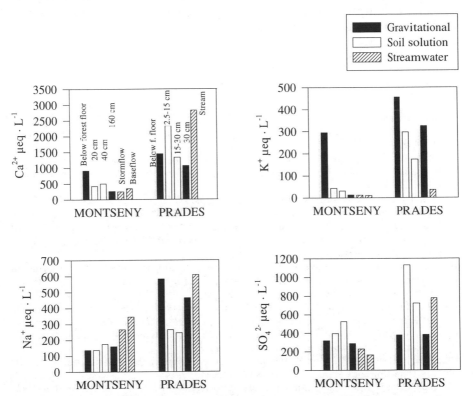

Fig. 17.5. Water chemistry profiles from gravitational soil water below forest floor to streamwater at Prades and Montseny. Volume-weighted means are given for gravitational water and streamwater, and arithmetic means for soil solution. Baseflow and stormflow at Montseny are for the TM9 catchment (from Àvila et al. 1995). Streamwater at Prades is for the Avic catchment (from Piñol and Àvila 1992)

concentrations in the deep subsurface flow were similar to those in the stormflow in Ca^{2+}, Mg^{2+} and K^+, but lower in Na^+ and slightly higher in SO_4^{2-} and Cl^-. According to the two-component mixing model from Àvila et al. (1995), an approximately equal proportion of groundwater and deep subsurface flow was found to contribute to the stormflow.

At Prades, a different pattern of concentration was observed between the soil water types and the streamwater depending on the behaviour of the ion considered (Fig. 17.5). Thus, for Ca^{2+} and Mg^{2+}, which were more abundant in the soil matrix, and less mobile, the concentrations in the streamwater were closer to the soil solution composition than to the gravitational water, especially in the topsoil (0–15 cm). For SO_4^{2-}, the situation of streamwater concentration was intermediate between that from the gravitational water and from the soil solution. The more mobile Na^+ and Cl^-, which have a low biological demand, presented similar concentrations in the streamwater and the gravitational water percolating through the soil. Finally, for the most biophilic ions, K^+ and especially NO_3^-, both soil solution and gravitational water showed much higher concentrations than streamwater, as happened also at Montseny, indicating that these nutrients were retained by the soil–plant system before reaching the stream.

References

Àvila A, Bonilla D, Rodà F, Piñol J, Neal C (1995) Soilwater chemistry in a holm oak (*Quercus ilex*) forest: inferences on biogeochemical processes for a montane-Mediterranean area. J Hydrol 166:15–35

Bellot J, Escarré A (1991) Chemical characteristics and temporal variations of nutrients in throughfall and stemflow of three species in Mediterranean holm oak forest. For Ecol Manage 41:125–135

Bellot J, Sánchez JR, Lledó MJ, Martínez P, Escarré A (1992) Litterfall as a measure of primary production in Mediterranean holm-oak forest. Vegetatio 99/100:69–76

Bonilla D (1990) Biogeoquímica del suelo en un encinar y landas del Montseny: la solución del suelo, la dinámica del nitrógeno y la respuesta a una perturbación experimental. PhD Thesis, Autonomous University of Barcelona, Bellaterra

Bonilla D, Rodà F (1990) Nitrogen cycling responses to disturbance: trenching experiments in an evergreen oak forest. In: Harrison AF, Ineson P, Heal OW (eds) Nutrient cycling in terrestrial ecosystems. Elsevier, London, pp 179–189

Casals P, Romanyà J, Cortina J, Fons J, Bode M, Vallejo VR (1995) Nitrogen supply rate in Scots pine (*Pinus sylvestris* L.) forests of contrasting slope aspect. Plant Soil 168–169:67–73

Cortina J, Romanyà J, Vallejo VR (1995) Nitrogen and phosphorus leaching from the forest floor of a mature *Pinus radiata* stand. Geoderma 66:321–330

Fernandez IJ, Lawrence GB, Son Y (1995) Soil solution chemistry in a low-elevation spruce-fir ecosystem, Howland, Maine. Water Air Soil Pollut 84:129–145

Gázquez R (1995) Seguimiento lisimétrico del efecto del fuego y la tala en suelos forestales de la Serra de Prades, Tarragona. MSc Thesis, Escola Superior Agricultura de Barcelona, Barcelona

Giesler R, Lundström U (1993) Soil solution chemistry: effects of bulking soil samples. Soil Sci Soc Am J 57:1283–1288

Gillman GP (1976) A centrifuge method for obtaining soil solutions. Div of Soils Rep 16. CSIRO, Melbourne, Australia

Hughes S DA, Norris DA, Stevens PA, Reynolds B, Williams TG, Woods C (1994) Effects of forest age on surface drainage water and soil solution aluminium chemistry in stagnopodzols in Wales. Water Air Soil Pollut 77:115–139

Manderscheid B, Matzner E (1995) Spatial and temporal variation of soil solution chemistry and ion fluxes through the soil in a mature Norway spruce (*Picea abies* (L.) Karst.) stand. Biogeochemistry 30:99–114

Menéndez I, Moreno G, Gallardo JF, Saavedra J (1995) Soil solution composition in forest soils of Sierra de Gata mountains, central-western Spain: relationship with soil water content. Arid Soil Res Rehabil 9:495–502

Piñol J, Àvila A (1992) Streamwater pH, alkalinity, pCO_2 and discharge relationships in some forested Mediterranean catchments. J Hydrol 131:205–225

Piñol J, Àvila A, Rodà F (1992) The seasonal variation of streamwater chemistry in three forested Mediterranean catchments. J Hydrol 140:119–141

Piñol J, Alcañiz JM, Rodà F (1995) Carbon dioxide efflux and pCO_2 in soils of three *Quercus ilex* montane forests. Biogeochemistry 30:191–215

Poch R (1992) Seguiment lisimètric en sòls d'alzinars de Prades afectats per foc i tala. MSc Thesis, Escola Superior Agricultura de Barcelona, Bellaterra

Ranger J, Discours D, Mohamed Ahamed D, Moares C, Dambrine É, Merlet D, Rouiller J (1993) Comparaison des eaux liées et des eaux libres des sols de 3 peuplements d'épicéa (*Picea abies* Karst.) des Vosges. Application à l'étude du fonctionnement actuel des sols et conséquences pour l'état sanitaire des peuplements. Ann Sci For 50:425–444

Rodà F, Àvila A, Bonilla D (1990) Precipitation, throughfall, soil solution and streamwater chemistry in a holm oak (*Quercus ilex*) forest. J Hydrol 116:167–183

Rustad LE, Fernandez IJ, Fuller RD, David MB, Nodvin SC, Halteman WA (1993) Soil solution response to acidic deposition in a northern hardwood forest. Agric Ecosyst Environ 47:117–134

Sabaté S (1993) Canopy structure and nutrient content in a *Quercus ilex* L. forest of Prades mountains: effect of natural and experimental manipulation of growth conditions. PhD Thesis, University of Barcelona, Barcelona

Swistock BR, Yamona JJ, Dewalle DR, Sharpe WE (1990) Comparison of soil water chemistry and sample requirements for pan vs tension lysimeters. Water Air Soil Pollut 50:387–396

Tokuchi N, Takeda H, Iwatsubo G (1993) Vertical changes in soil solution chemistry in soil profiles under coniferous forest. Geoderma 59:57–73

Van Gestel M, Merckx R, Vlassak K (1993) Microbial biomass responses to soil drying and rewetting: the fate of fast- and slow-growing microorganisms in soils from different climates. Soil Biol Biochem 25:109–123

Verdú AMC (1984) Circulació de nutrients en ecosistemes naturals del Montseny: Caiguda de virosta i descomposició de les fulles. PhD Thesis, Autonomous University of Barcelona, Bellaterra

Wolt JD (1994) Soil solution chemistry: applications to environmental science and agriculture. John Wiley, New York

Zabowski D, Ugolini FC (1990) Lysimeter and centrifuge soil solutions: seasonal differences between methods. Soil Sci Soc Am J 54:1130–1135

18 Nutrient Distribution and Cycling

Antoni Escarré, Ferran Rodà, Jaume Terradas and Xavier Mayor

18.1 Introduction

Nutrient cycling is one of the major functional properties of ecosystems. The amounts of nutrients contained within the ecosystem, their distribution among ecosystem compartments, and their rates of flow between compartments reflect the interactions between organisms and their environment, and are an expression of the extent to which ecosystems affect and are affected by the availability and fate of nutrients (Attiwill and Adams 1993).

Two characteristics of Mediterranean holm oak forests are likely to affect the distribution and cycling of nutrients: these forests have moderate biomass and low production (Chap. 3) and they have been exploited since ancient times, being repeatedly coppiced or burnt. Low production leads to reduced nutrient demands from the soil, while long-term exploitation may have resulted in nutrient depletion. The associated nutrient syndrome for dominant tree species, such as holm oak, should correspond to slow-growing, nutrient-frugal species that respond parsimoniously to increases in nutrient availability, and with moderate intensity of nutrient cycling between the vegetation and the soil.

In this chapter we analyze the nutrient cycling information available for holm oak stands in the experimental areas of Prades and Montseny in NE Spain (Chap. 2). We first summarize nutrient concentrations in holm oak trees. We then focus on a stand-level view of the intrasystemic nutrient cycle, i.e. the major nutrient pools and fluxes within the forest ecosystem. Throughout, we will highlight the similarities and differences in nutrient cycling patterns between holm oak forests and other forest ecosystems.

18.2 The Database

Nutrient concentrations for holm oak at Montseny are selected from Ferrés (1984, unpubl. data), Mayor (1990) and Canadell (1988) and at Prades from Clemente (1983), Sabaté (1993), Djema (1995) and C.A. Gracia and S. Sabaté

Ecological Studies, Vol. 137
Ferran Rodà et al. (eds) Ecology of Mediterranean
Evergreen Oak Forests
© Springer-Verlag, Berlin Heidelberg 1999

(unpubl. data). Nutrient concentrations in litterfall are taken from Verdú (1984) and Mayor (1990) for Montseny and from Clemente (1983) and Bellot et al. (1992) for Prades. Nutrient concentrations in companion species at Prades (*Phillyrea latifolia*, *Arbutus unedo* and *Viburnum tinus*) are from Clemente (1983).

Nutrient pools and fluxes considered in this chapter are average estimates for holm oak stands in the Avic catchment at Prades and in closed-canopy plots in the Torrent de la Mina catchment at Montseny (Chap. 2). At both sites, aboveground stand biomass and production were obtained by repeated forest surveys and dimensional analysis of sample plots (Chap. 3). At Prades 69 25-m² plots were used, while at Montseny 18 closed-canopy 154-m² plots were considered. Data for belowground biomass and production correspond to Chapter 4, and the estimates for canopy leaching are taken from Chapter 15 (data sets Prades-1 and Montseny-2).

Published comprehensive data on nutrient concentrations in holm oak are scarce (cf. the review by Rundel 1988). In our studies on nutrient cycling in the holm oak forests of Prades and Montseny large data sets have been obtained on nutrient concentrations in all major tree components, and on how they vary with age, crown position, individual trees, season, year, topographic position, and so on. For this chapter we have selected or computed those average nutrient concentrations for each tree component that we judged to better represent the mean conditions in closed-canopy holm oak forests at Prades and Montseny. Here we will deal mainly with those tree components that have been used for computing nutrient pools and fluxes associated with stand biomass and production.

18.3 Nutrient Concentrations in Holm Oak Trees

18.3.1 Differences Among Tree Parts

Nutrient concentrations in the main components of holm oak trees at Montseny and Prades are shown in Table 18.1 for aboveground parts, and in Table 18.2 for belowground parts. Differences between both sites are not great, though Ca levels tend to be higher in Prades, where soil pH and base saturation are much higher. Potassium concentrations in inflorescences and fruits are also higher at Prades, but these can be affected by the degree of ripening at the time of falling and by leaching. At both sites, mature holm oak leaves have moderate nutrient concentrations, e.g. a mean foliar N concentration between 12 and 15 mg g^{-1} (all nutrient concentrations are on a dry weight basis). Foliar nutrient concentrations decrease rapidly during the first weeks after budbreak, but then they remain relatively stable in mature leaves for about 2 years. Foliar Ca increases throughout the leaf lifespan while N and P tend to decrease, particularly in leaves older than 2 years. Twigs are as rich or

Table 18.1. Mean nutrient concentrations (mg g^{-1}) in holm oak in Prades[a] (P) and Montseny[b] (M) experimental areas

Tree component		N	P	K	Ca	Mg
Leaves	P	13.3	1.1	6.2	9.8	1.0
	M	12.8 ± 0.3	1.0 ± 0.2	5.4 ± 1.3	5.8 ± 1.4	1.1 ± 0.3
Current-year twigs	P	9.2	1.2	7.9	13.8	1.4
	M[c]	9.6	1.3	5.1	4.3	0.9
Branch wood (Ø <5 cm)	M	2.1	0.8	2.4	3.3	0.5
Branch bark (Ø <5 cm)	M	4.9	0.4	3.0	27.0	1.1
Bole wood (Ø >5 cm)	P[d]	1.4 ± 0.2	0.7 ± 0.2	2.1 ± 0.6	3.3 ± 1.0	0.5 ± 0.2
	M	1.2	0.3	1.8	2.5	0.4
Bole bark (Ø >5 cm)	P[d]	4.3 ± 0.8	0.2 ± 0.07	2.3 ± 0.5	42.4 ± 11	0.8 ± 0.2
	M	4.5	0.3	3.5	30.9	1.7
Male inflorescences	P[e]	15.0	0.4	13.5	5.0	1.4
	M[f]	14.7	1.2	4.7	5.1	0.8
Fruits	P[e,g]	7.0	0.6	8.3	3.0	1.0
	M[f,g]	7.5	0.7	4.8	2.3	0.7

[a] Unless otherwise stated, data for Prades are mean figures for the Avic catchment.
[b] Unless otherwise stated, data for Montseny are mean figures for the permanent plot near outlet of the Torrent de la Mina catchment (La Castanya).
[c] Mean of 11 trees sampled on 29 April 1981.
[d] Values from the north-facing slope of Torners (C. Gracia and S. Sabaté, unpubl. data).
[e] From Bellot et al. (1992) and Clemente (1983).
[f] N, P, K and Mg measured in 18 closed-canopy plots in the Torrent de la Mina catchment, after Mayor (1990). Ca measured in the permanent plot, after Verdú (1984).
[g] Acorns and cups of non-aborted fruits.

slightly richer in P than leaves of their same age. Bark is specially rich in Ca, with concentrations up to 4.2%. These high Ca levels in bark seem to be a characteristic of many *Quercus* species, even in non-calcareous soils such as those of Montseny and Prades. Small branches thus have a high Ca concentration since they have a high proportion of bark. Nutrient concentrations in wood tend to decrease with increasing diameter of roots, boles and branches. As for reproductive organs, inflorescences are richer than leaves in N, in P (only at Montseny) and in K (only at Prades). Fruits (acorns+cups) are richer than leaves only in K at Prades.

The above pattern for holm oak broadly agrees with the generalization that nutrients more actively involved in metabolism (N, P, K and Mg) are found in higher concentrations in the most active organs – leaves, current-year twigs, and inflorescences – while Ca is more abundant in structural tissues. Fine roots, not analyzed here, can also have moderately high nutrient concentrations in oaks (Yin et al. 1991).

Table 18.2. Mean nutrient concentrations (mg g^{-1}) in belowground components of holm oak in Prades[a] (P) and Montseny[b] (M) experimental areas

Component		N	P	K	Ca	Mg
Root crown wood	P	1.94	0.62	2.17	4.69	0.55
	M	1.67	0.62	3.43	3.97	0.87
Root crown bark	P	4.31	0.19	1.69	45.7	0.71
	M	–	–	–	–	–
Large-roots[c] wood	P	1.90	1.08	1.80	3.49	0.96
	M	4.14	0.44	3.96	4.14	1.08
Large-roots[c] bark	P	3.30	0.21	1.62	25.7	0.53
	M	4.47	0.12	4.30	11.1	1.10
Medium-roots[d] wood	P	2.19	1.38	1.91	3.60	1.12
	M	2.85	0.77	3.58	3.93	1.05
Medium-roots[d] bark	P	3.39	0.25	1.76	20.9	0.62
	M	4.00	0.24	4.00	10.4	0.93
Thin-roots[e] wood	P	2.52	1.84	2.80	5.43	1.72
	M	–	–	–	–	–
Thin-roots[e] bark	P	3.77	0.34	2.16	19.7	0.90
	M	–	–	–	–	–

[a] Data from the north-facing slope of Torners, after Djema (1995).
[b] Data from the Torrent de la Mina catchment, after Canadell (1988).
[c] Roots of Ø > 5 cm.
[d] Roots of Ø 1–5 cm.
[e] Roots of Ø < 1 cm.

Table 18.3. Mean foliar nutrient concentrations (mg g^{-1}) in holm oak and other evergreen and deciduous woody species in the Avic catchment at Prades[a]

Tree species	N	P	K	Mg	Ca
Evergreen					
Quercus ilex	14.1	0.98	10.3	1.6	9.0[b]
Phillyrea latifolia[c]	11.1	0.51	9.0	1.1	10.5
Arbutus unedo[c]	11.6	0.75	9.7	2.2	9.9
Deciduous					
Quercus faginea	17.4	1.2	8.1	2.1	8.0
Acer monpessulanum	16.0	1.7	10.9	2.7	16.6

[a] From Clemente (1983). Leaves sampled on 5–6 August 1981 from 15–28 individuals of each species in each of five major areas within the Avic catchment. Three other samplings in early January, April and October of the same year have not been considered here (but see below) since these are not adequate for the deciduous trees.
[b] Ca concentration reported by Clemente (1983) for holm oak on this sampling date was anomalously low (3.2 mg g^{-1}). Mean value obtained in the three other samplings is instead given here.
[c] Mainly understory large shrubs.

Holm oak leaves have similar or somewhat higher nutrient concentrations than those of the major evergreen companion species at Prades (Table 18.3). A coexisting deciduous oak (*Quercus faginea*) has higher foliar N concentrations than holm oak, reflecting the usual difference in photosynthetic capacity between evergreen and deciduous species.

18.3.2 Spatial Variations

Variations in tree nutrient concentrations among sites are usually related to soil characteristics, which in turn vary with topographic position and other factors. At Prades, Sabaté et al. (1995) reported differences in mineral concentrations in holm oak leaves in two endpoints along a topographic gradient within the Avic catchment: they found higher foliar concentrations of N and P, and lower concentrations of K, in a valley-bottom site than in a ridge-top site. Differences appeared both when nutrient concentrations were expressed on a dry weight basis and on a leaf area basis. Higher foliar N and P contents in the valley bottom can be, in part, explained by the higher nutrient supply provided by a deeper soil: soil depth averaged 85 cm in the valley bottom and only 45 cm in the ridge top. In contrast, the higher foliar K in the ridge top site could derive from the stronger need to regulate the stomatal aperture in this more xeric site, or from local differences in bedrock mineralogy.

Bedrock characteristics may contribute significantly to variations in soil chemistry that can affect nutrient concentrations in tree tissues. In 15 different Catalan sites with contrasting bedrock (limestone and marl, volcanic rock, and metamorphic schist) N, P and K concentrations in current-year leaves of holm oak varied very little among sites, less so than in older leaves, current-year twigs, or older twigs (Canadell and Vilà 1992). This suggests a tight regulation of the concentrations of these most important nutrients in the youngest leaf cohort, which is the best illuminated. The higher variability of nutrient concentrations in older leaves and in twigs found by Canadell and Vilà (1992) and Sabaté (1993) can in part result from the higher heterogeneity of the light climate they experience, and from nutrient redistribution within tree crowns. Mobile nutrients could be redistributed depending on growth demands, with the different sites being sampled at different points of the resultant depletion and replenishment cycles.

18.3.3 Foliar Retranslocation

When compared to mature leaves, senescent holm oak leaves show lower concentrations of N, P, and K, and sometimes also of Mg, and higher concentrations of Ca. These are usual changes in nutrient contents during leaf senescence in most tree species. In our holm oak sites, from 30 to 40% of N, P, K, and Mg in mature leaves is retranslocated during leaf senescence (Mayor and Rodà 1992; Sabaté 1993). These are low retranslocation rates (retranslocation of up to 60–80% of N and P in mature leaves is not uncommon, particularly in deciduous trees). Nutrient retranslocation (or resorption) is an active process that removes nutrients from senescing leaves before abscission. Resorbed nutrients are transported to growing organs (sinks) or to storage in perennial tissues (Chapin 1980). Retranslocation means for the plant a higher control of nutrient resources, and allows it to reutilize them.

Though retranslocation from senescing leaves has received most attention, retranslocation is not restricted to senescing organs, but it can occur at other times of the year, particularly during growth pulses in evergreens. At Prades, N and P concentrations in holm oak leaves were lower during summer and larger in autumn and spring, while the contrary was found for stems (Sabaté 1993; Chap. 9) suggesting that intra-annual nutrient concentration changes are linked to nutrient retranslocation process occurring in spring and autumn in association with growth periods. If retranslocation is computed by substracting nutrient concentrations (on a leaf area basis) in leaf litterfall from the maximum seasonal concentrations attained by mature leaves, holm oak at Prades retranslocates 34–52% of the maximum foliar N contents, and 47–65% of those of P (Sabaté 1993), i.e. a substantially higher resorption than when only retranslocation during leaf senescence is considered.

Independently of when it occurs, retranslocation could be ecologically significant for Mediterranean evergreen species. A plant is expected to increase its leaf longevity when nutrient status is low, and the proportion of supporting tissue increases with leaf longevity (Chabot and Hicks 1982). Retranslocation is linked to nutrient mobility, which depends on tissue composition. Therefore, in situations where growth is limited by resource availability the retranslocation capacity is expected to decrease (Del Arco et al. 1991; Escudero et al. 1992). As noted above, evergreen species like holm oak tend to have smaller retranslocation rates than deciduous trees. This limits the adaptive capacity of evergreen species for increasing their nutrient-use efficiency through retranslocation in nutrient-poor soils. In fact, increasing leaf longevity seems to be much more effective for increasing nutrient-use efficiency in these sites (Escudero et al. 1992). In this context, Pugnaire and Chapin (1993) concluded that retranslocation is not an adaptation to low nutrient availability but a phenotypic response to variations in nutrient status. Killingbeck (1996) distinguished two components of such response: potential resorption (the maximum *amount* of nutrients than can be withdrawn from senescing leaves) and resorption proficiency (the levels to which nutrients have been reduced in senesced leaves). According to Killingbeck (1996), plants reducing N and P concentrations in their senesced leaves to levels below 0.7% for N and 0.05% for P can be considered highly proficient. Mean nutrient concentrations in holm oak leaf litterfall were 0.87–0.97% for N and 0.055–0.010% for P, depending on site and period (Verdú 1984; Mayor 1990; Bellot et al. 1992; V. Diego, unpubl. data). This means that holm oak is not very proficient at resorbing N and P.

After the extreme drought period in the summer of 1994, withered holm oak crowns were abundant in Prades due to cavitation, and this blocked the normal process of leaf abscission. As a consequence, dead leaves remained on the crowns for more than 2 years. Interestingly, their N and P concentrations were similar to those in living leaves, demonstrating the importance of retranslocation in reducing N and P levels in naturally senesced leaves. On the other hand, K concentrations in the dead, standing leaves were, as expected, lower than in living leaves due to leaching by rain.

18.3.4 Comparison with Other Forest Ecosystems

Whole-tree nutrient concentrations can be used as indicators of patterns of nutrient use and demand in different biomes. In Table 18.4 average aboveground concentrations in whole trees are listed for three holm oak forests (Prades, Montseny, and Le Rouquet in southern France) and for temperate coniferous and temperate deciduous forests (Cole and Rapp 1981). For N, holm oak stands have intermediate whole-tree concentrations between those of coniferous and deciduous temperate forests, but P, K, Ca and Mg concentrations are higher in holm oak than in either type of temperate forest. Calcium concentrations are 3–7 times higher in holm oak, largely because it has a thick and Ca-rich bark. For P, K, and Mg (and partly also for Ca) the higher concentrations in holm oak result mainly from it having a wood that is not as poor in nutrients as most temperate trees.

Table 18.4. Mean whole-tree nutrient concentrations (mg g^{-1}) in different biomes. Figures refer only to aboveground biomass

Forest biome	Mean aboveground biomass (Mg ha^{-1})	N	P	K	Ca	Mg
Holm oak forests[a] ($n = 3$)	162	2.1	0.4	1.7	12.0	0.6
Temperate coniferous[b] ($n = 13$)	307	1.6	0.2	1.1	1.6	0.2
Temperate deciduous[b] ($n = 14$)	152	2.9	0.2	1.5	3.6	0.4

[a] La Castanya (Montseny), Avic (Prades) and Le Rouquet (southern France; after Cole and Rapp 1981).
[b] Computed from data in Cole and Rapp (1981).

18.4 Nutrient Pools

The vegetation of our holm oak forests contains substantial amounts of nutrients (Table 18.5), since high nutrient concentrations compensate for the moderate biomass. Nutrient contents in aboveground biomass are about similar at both sites, but belowground nutrient contents are much higher at Prades, mainly as a result of the estimated belowground biomass being

Table 18.5. Nutrient contents (kg ha^{-1}) in different compartments of holm oak forest ecosystems at Prades (Avic catchment) and Montseny (Torrent de la Mina catchment)

Nutrients in	N	P	K	Ca	Mg
Aboveground biomass					
Prades	335	77	286	1070	72
Montseny	275	56	266	814	71
Belowground biomass					
Prades	276	81	222	1318	83
Montseny	129	22	156	232	41
Total biomass					
Prades	611	158	508	2388	155
Montseny	404	77	422	1046	113
Forest floor					
Prades[a]	531	41	144	705	154
Montseny[b]	622	–	–	–	–
Mineral soil					
Prades[c]	5224[d]	–	300[e]	3710[e]	306[e]
Montseny[f]	9164[d]	–	227[e]	1426[e]	785[e]

[a] From Serrasolsas (1994), one control plot on the SE-facing slope of Torners, sampled in July 1988.
[b] Computed from Hereter (1990), plots No. 11, 12, 13 and 14, on phyllites and schists.
[c] 0–30 cm depth; data from I. Serrasolsas, N. Melià and J. Piñol.
[d] Total nitrogen.
[e] Exchangeable cations.
[f] Down to bedrock; from D. Bonilla and F. Rodà (unpubl. data).

2.8 times higher at Prades than at Montseny (Chap. 4). Our estimate of belowground biomass at Montseny assumes all trees to be derived from seed, i.e. without the large stools of resprouted tress. Taking into account the fraction of resprouted trees in the Torrent de la Mina catchment at Montseny (63% of all stems with dbh $\geq$ 5 cm) would increase the estimate of belowground biomass there, but it would still fall short of that at Prades since only 60% of the resprout-derived holm oak genets at Montseny have stools that are appreciably larger than those of seed-derived trees.

Nutrient contents in the forest floor are also substantial, contradicting the widespread notion that the forest floor of Mediterranean forests is poorly developed. As much nitrogen and magnesium are stored in the forest floor as in the total plant biomass (Table 18.5). The mineral soil, despite being shallow and very stony, harbours large nutrient pools in both holm oak sites. For example, at Montseny the mineral soil holds 90% of the total ecosystem nitrogen, the forest floor 6%, and the vegetation the remaining 4%. This pattern of nutrient compartmentation is akin to that of many cool-temperate forests, and tells holm oak forests apart from tropical rainforests.

18.5 Nutrient Fluxes

Most nutrient fluxes are surprisingly similar at both holm oak sites, considering the differences in climate, methods, and study periods (Table 18.6). Nitrogen fluxes in litterfall are moderate, similar to those in temperate conifers and lower than in temperate deciduous forests. Litterfall N is lower and K higher at Prades than at Montseny. Canopy leaching has been computed assuming that leaching accounts for 90% of K, 50% of Mg, and 25% of Ca contents in net throughfall plus stemflow (Chap. 15). At Prades, a small amount of inorganic N is annually removed from the canopy by rainfall, while at Montseny there is a net canopy uptake of 3 kg N ha^{-1} year^{-1} (Table 18.6), though this estimate is based on a single year of measurement.

As primary production is low in these holm oak forests (Chap. 3), the amounts of nutrients annually incorporated into the production of new tissues are correspondingly low, particularly in woody tissues (Table 18.6). More Ca is needed to construct the annual cohort of leaves at Prades than at

Table 18.6. Mean annual nutrient fluxes (kg ha^{-1} year^{-1}) in the holm oak forest ecosystems at Prades (Avic) and Montseny (Torrent de la Mina). Within each flux, upper figure refers to Prades and lower one to Montseny

Nutrient flux	N	P	K	Ca	Mg
(1) Retention in boles	2.0	0.6	2.0	7.7	0.5
	2.6	0.5	3.2	9.6	0.9
(2) Retention in branches	1.3	0.3	0.9	3.7	0.2
	2.0	0.5	1.8	7.0	0.5
(1+2) Aboveground retention	3.3	0.9	2.9	11.4	0.7
	4.6	1.0	5.0	16.6	1.4
(3) Retention in coarse roots	4.1	1.1	3.3	19.6	1.1
	2.9	0.5	3.5	5.2	0.9
(4) Incorporated into new leaves	31.5	2.6	14.9	23.5	2.5
	32.7	2.6	13.1	15.4	2.8
(5) Incorporated into flowers and fruits	2.5	0.1	2.3	1.0	0.3
	8.6	0.7	2.9	3.1	0.6
(6) Litterfall	25.5	2.1	20.9	31.8	3.1
	37.0	2.5	11.7	33.0	3.5
(7) Canopy leaching	1.2	0.08	15.2	3.1	1.1
	−3.0	0.65	13.6	1.1	0.8
(6+7) Aboveground return	26.7	2.2	36.1	34.9	4.2
	37.0	3.2	25.3	34.1	4.3
(1+2)+(6+7) Uptake for sustaining current aboveground production	30.0	3.1	39.0	46.3	4.9
	41.6	4.2	30.3	50.7	5.7
(1+2)+3+(6+7) Total uptake[a]	34.1	4.2	42.3	65.9	6.0
	44.5	4.7	33.8	55.9	6.6

[a] Except for fine roots.

Montseny and, even more spectacularly, about four times as much Ca goes annually into new coarse roots at Prades than at Montseny. This difference is a result of the higher base saturation of the soil, the higher Ca concentration in coarse roots, and the higher estimated root production at Prades.

The above-mentioned nutrient fluxes can be integrated into three major derived fluxes: return, retention, and uptake. We have computed them for the aboveground components, for which data are reasonably complete. Aboveground nutrient return is defined as the sum of nutrient fluxes in litterfall and canopy leaching. Nutrient return measures the intensity of nutrient cycling between the trees and the soil. Nutrient return fluxes are moderate at both sites. Aboveground retention is defined as the amount of nutrient annually incorporated into the production of new woody tissues in boles and branches. Nutrient retention is rather low at both sites, but except for P it is notably higher at Montseny than at Prades, reflecting the higher current production of boles plus branches at the first site (2.4 vs. 1.6 Mg ha^{-1} year^{-1}). Annual nutrient uptake needed to sustain the functioning of aboveground biomass is defined as the sum of aboveground retention and return. The involved nutrient fluxes are again broadly similar at both sites, and they reach (kg ha^{-1} year^{-1}) 46–51 for Ca, 30–39 for K, and 30–42 for N. If the nutrients needed to construct coarse roots are added to these figures, total net nutrient uptake (except for fine roots) reaches (kg ha^{-1} year^{-1}) 56–66 kg for Ca, 34–42 for K, and 34–45 for N.

The overall picture that emerges from these data is that these holm oak forests exhibit a parsimonious nutrient cycle, i.e. that they exert a low to moderate demand on the site nutrient capital. However, given the long-lived nature of holm oak and particularly of its stools, the amounts of nutrients accumulated in the biomass can be quite high, as we have seen in the previous section. Figure 18.1 shows for four major nutrients how uptake is split between retention and return. Return makes up a higher proportion of annual uptake of N and K than of P and Ca, reflecting the higher relative allocation of N and K to leaves and reproductive structures, of Ca to bark, and of P (surprisingly) to wood. As a result, N and K cycle faster between trees and the soil than P and Ca.

The intrasystemic nutrient fluxes discussed above can be compared to ecosystem input and output nutrient fluxes (Chap. 20). In common with other undisturbed forest ecosystems (e.g. Monk and Day 1985; Johnson and Henderson 1989; Likens et al. 1995), our holm oak forest sites show much larger within-ecosystem nutrient fluxes than between-ecosystem fluxes. Nutrient demand by trees is determinant in providing this degree of closure of the forest nutrient cycle. The trenching experiment mentioned in Chapters 16 and 17 revealed indeed that holm oak forests have a large potential to lose nitrogen through nitrate leaching after disturbance (Bonilla and Rodà 1989, 1990).

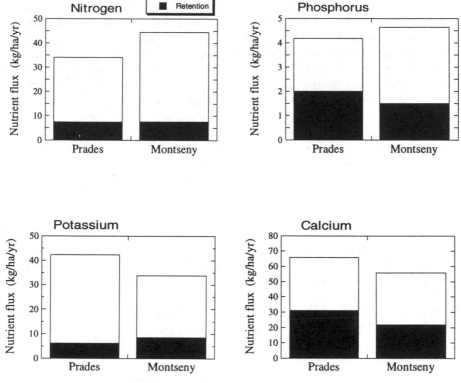

Fig. 18.1. Partition of total annual nutrient uptake into retention and return. Retention includes nutrients incorporated into current production of woody tissues (branches+boles+coarse roots). Return refers to aboveground return (litterfall+canopy leaching)

18.6 Holm Oak Forests and Other Forest Biomes

We have compared the aboveground nutrient retention and uptake of our holm oak sites with temperate coniferous and temperate deciduous stands compiled by Cole and Rapp (1981). For N, P, K, Ca and Mg, mean retention and uptake are almost always higher in deciduous than in coniferous temperate forests. Our holm oak sites usually are closer to the average of coniferous stands, or particularly for retention lie even below the latter. Two examples will illustrate this pattern. Calcium uptake for aboveground components (kg ha^{-1} year^{-1}) is 46–51 at our sites, averages 45 in coniferous temperate stands and 85 in deciduous ones. Aboveground N retention (kg ha^{-1} year^{-1}) is 3.3–4.6 at our holm oak sites, and averages 9.7 and 14.0 for coniferous and deciduous forests, respectively.

As a further exercise to compare the Mediterranean, evergreen, sclerophyllous holm oak forests to boreal and temperate forests, and to search for biogeochemical trends within and among biomes, we conducted a principal component analysis of forest stands described by their nutrient fluxes. We used 31 forest stands: our two holm oak sites plus 29 stands from the appendix of Cole and Rapp (1981), one of which was also a holm oak forest (Le Rouquet, southern France). The 16 variables used were the dry weight of litterfall and nutrient (N, P, K, Ca, Mg) fluxes (kg ha^{-1} year^{-1}) in the annual production of leaves, in the current annual production of branches+boles, and in litterfall. These variables were chosen to reflect the intensity of nutrient cycling in a forest ecosystem. Variables related to tree nutrient content were not included because they depend heavily on stand biomass. Of the 32 stands listed by Cole and Rapp (1981), we retained 29 having complete data for all 16 variables.

Principal component 1 accounted for 52% of the variance and it was a "size" factor, all variables having positive loadings on it. Deciduous stands showed a higher mean score for the first component than coniferous stands, with the three holm oak stands in between (Fig. 18.2). The first component can be interpreted to be a "trophic axis", with eutrophic stands with high rates of nutrient demand for new production and intense nutrient cycling in litterfall plotting at the positive end, and oligotrophic forests at the negative end. The highest score was for a nitrogen-fixing *Alnus rubra* stand and the three lowest for boreal spruce forests. The French holm oak stand, which grows on calcareous soil, appears more eutrophic than our two holm oak sites on silicate soils (Fig. 18.2).

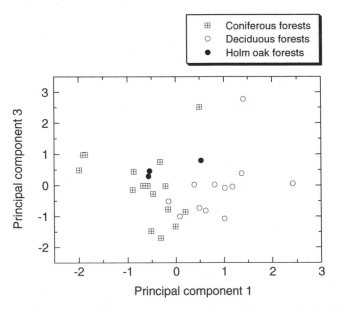

Fig. 18.2. Scatterplot of principal components 1 and 3 in the principal component analysis of 31 forest stands described by their nutrient fluxes. Each symbol is a forest stand. The three holm oak stands are from *left* to *right*: Montseny, Prades, and Le Rouquet (France)

Principal component 2 (17% of the variance) was positively related to Ca and Mg increments in foliage, and negatively related to P increment in branches+boles and litterfall P. The three holm oak stands appeared centered (i.e. scores near zero) on component 2. Principal component 3 (9% of the variance) was positively related to Ca, Mg and K increment in branches + boles, and negatively related to P increment in foliage and litterfall P. The three holm oak stands score moderately high in component 3.

Principal component 2 seems to be related to base cation demand for foliage production, while component 3 seems related to base cation demand for woody production. Both components suggest an antagonism between P nutrition and base-cation richness, at least when the effects of overall nutrient richness (component 1) are removed.

18.7 Nutrient Use Efficiency

Nutrient cycling data can give insight into the nutrient use efficiency (NUE) at the stand level. Among the many definitions of NUE, we have chosen that of Berendse and Aerts (1987) because they split the concept into two biologically meaningful components which can vary independently of each other: A, the nutrient productivity or amount of annual production per unit of nutrient contained in the vegetation (g g^{-1} year^{-1}); and $1/L$, the mean residence time of the nutrient in the vegetation (year). The overall NUE is the product of both components ($A \times 1/L$; g g^{-1}), and gives the amount of production per unit of nutrient in the vegetation, integrated over the lifetime of this nutrient unit. Again, we have restricted ourselves to the aboveground NUE since the available fine root data were not enough for this purpose. For these aggrading holm oak forests we computed the mean residence time of nutrients in the aboveground biomass as the nutrient pool in this biomass divided by the aboveground nutrient return. Nutrient productivity is higher at Montseny than at Prades for all five major nutrients (Fig. 18.3a), the highest relative difference being for P. This higher nutrient productivity is mostly a reflection of the higher aboveground net primary production at Montseny. Probably, the higher rainfall there allows for better returns in terms of carbon gain of each gram of nutrient invested in biomass. Mean residence time, which is positively related to NUE, is much higher at Prades than at Montseny for N and P, and somewhat higher for Ca (Fig. 18.3b). Two facts account for the higher mean residence time of N and P in the aboveground biomass at Prades: (1) the higher mean concentrations of N and P in the biomass at Prades, themselves reflecting either a more concentrated soil solution under a drier climate or a lesser dilution by growth; and (2) a higher return of N and P in litterfall (and for P also in canopy leaching) at Montseny, resulting largely from the higher production of flowers and fruits at Montseny, at least during the respective study periods.

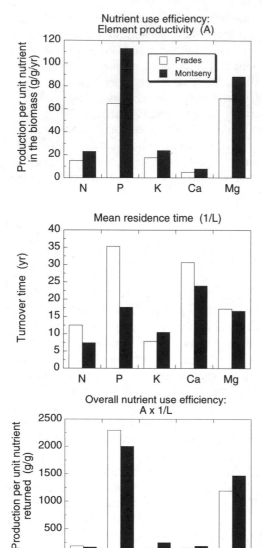

Fig. 18.3. Aboveground nutrient use efficiency (ANUE) for five major elements. **a** Element productivity; **b** mean residence time of the element in aboveground biomass; **c** the product of both components, i.e. overall ANUE

When the overall aboveground NUE is considered, the opposite differences in *A* and in *1/L* cancel each other, and both holm oak sites show similar NUE (Fig. 18.3c), with Montseny having somewhat higher NUE than Prades for K and perhaps Ca and Mg, and slightly lower NUE for P.

The overall NUE of our holm oak sites falls in the range observed for temperate coniferous and deciduous forests (Cole and Rapp 1981), though the NUE for K at Prades is rather low. Interestingly, these holm oak forests do not show particularly high NUE for any of the five considered elements.

18.8 Nutrient Cycling and Stand Structure

The relationships between nutrient cycling and stand structure were given scant attention in the pioneer works on forest nutrient dynamics (Ovington 1959; Rodin and Bazilevich 1967; Duvigneaud and Denaeyer-De Smet 1970), and even in most of the International Biological Programme studies (Cole and Rapp 1981). Our own earlier work on nutrient cycling in holm oak forests at Montseny was concentrated on a single, 0.23-ha permanent plot (Rodà 1983; Ferrés 1984; Verdú 1984); a summary of its nutrient pools and fluxes was given by Escarré et al. (1987). Since this plot was a high-quality site, with an aboveground tree biomass of 160 Mg ha^{-1} (compared to an average of 104 Mg ha^{-1} for the closed-canopy Torrent de la Mina plots), its nutrient cycle differed in some aspects from those reported here for these plots. For instance, N contents in the aboveground tree biomass were 411 kg ha^{-1} in the permanent plot compared with an average of 275 kg ha^{-1} in closed-canopy plots within the Torrent de la Mina catchment. Within a given tree species, nutrient pools in the vegetation are closely related to biomass amounts, while nutrient fluxes linked to production of new tissues are related to the rate of primary production of the stand. Other nutrient fluxes, such as those in litterfall, are usually less dependent on stand structure, at least for closed-canopy stands. For example, litterfall in the high-biomass permanent plot was very close to the average of the 18 closed-canopy plots during 1 year of simultaneous sampling (Mayor 1990).

18.9 Spatial Variation in Nutrient Cycling

Within a given ecosystem type, spatial variations in nutrient pools and fluxes are expected, mainly as a result of spatial differences in either site conditions or stage of stand development. Spatial patterns in nutrient cycling can reveal the underlying controlling factors at the landscape scale. Also, knowledge of such patterns is important for predicting the functioning of forest ecosystems above the plot scale, since given the non-linear nature of many ecological phenomena, the behaviour of a set of cells (or patches) may differ from the average of individual behaviour.

As an example of the kind and degree of spatial variation in nutrient cycling observed in holm oak forests, we will consider here nutrient fluxes in litterfall because they integrate the intensity of nutrient cycling within a forest ecosystem. Litterfall nutrient fluxes were measured for 1 year in the 18 closed-canopy plots at Montseny systematically distributed over the 100-ha holm oak stand of the Torrent de la Mina catchment.

N and P fluxes in litterfall varied over the 18 plots by a factor of 2, while those of K and Mg varied by a factor of 3. The relative variabilities of N and P fluxes, as measured by the coefficient of variation, were 19%, the same as for the amount of litterfall; for K and Mg the coefficient of variation was 30–33%, i.e. the spatial variation of litterfall fluxes for these cations was much higher than for the amount of litterfall.

N and P fluxes in litterfall did not vary significantly either among the range of altitudes considered (800–1000 m a.s.l.) or between the two contrasting slope aspects within the catchment (north-facing and east-facing). K and Mg fluxes varied significantly with altitude, the highest mean fluxes being obtained at 1000 m a.s.l., where the plots having the highest basal area were located.

Litterfall nutrient fluxes were positively correlated to aboveground net primary production (ANPP) across the 18 plots (Mayor and Rodà 1992), but this correlation was probably forced by the participation of litterfall amount in estimating both ANPP and litterfall nutrient fluxes. Moreover, spatial variations in ANPP over the Montseny catchment were uncorrelated with nutrient concentrations in mature leaves and in leaf litterfall, with absolute and relative nutrient retranslocation from senescing leaves, and with the concentration and content (kg ha^{-1}) of nutrients in the upper mineral soil. Thus, no correlational evidence was found that primary production was nutrient-limited in this forest. This is not to say that nutrients are of little importance in holm oak forests, as the strong response of canopy dynamics to N fertilization clearly demonstrates (Chap. 13).

References

Attiwill PM, Adams MA (1993) Tansley review no. 50. Nutrient cycling in forests. New Phytol 124:561–582

Bellot J, Sánchez JR, Lledó MJ, Martínez P, Escarré A (1992) Litterfall as a measure of primary production in Mediterranean holm oak forest. Vegetatio 99/100:69–76

Berendse F, Aerts R (1987) Nitrogen-use-efficiency: a biologically meaningful definition? Funct Ecol 1:293–296

Bonilla D, Rodà F (1989) Dinámica del nitrógeno en el suelo de un encinar montano: respuesta a una perturbación experimental. Options Medit Sér Sémin 3:187.190

Bonilla D, Rodà F (1990) Nitrogen cycling responses to disturbance: trenching experiments in an evergreen oak forest. In: Harrison AF, Ineson P, Heal OW (eds) Nutrient cycling in terrestrial ecosystems. Elsevier, London, pp 179–189

Canadell J (1988) Biomassa aèria i subterrània de l'alzinar de La Castanya, Montseny. MSc Thesis, Autonomous University of Barcelona, Bellaterra

Canadell J, Vilà M (1992) Variation in tissue element concentrations in Quercus ilex L. over a range of different soils. Vegetatio: 99/100:273–282

Chabot BF, Hicks DJ (1982) The ecology of leaf life spans. Annu Rev Ecol Syst 13:229–259

Chapin III FS (1980) The mineral nutrition of wild plants. Annu Rev Ecol Syst 11:233–259

Clemente A (1983) Componentes específico y estacional en la variación de contenidos en elementos químicos de las especies y formas biológicas del encinar mediterráneo. MSc Thesis, University of Alicante, Alicante

Cole DW, Rapp M (1981) Elemental cycling in forest ecosystems. In: Reichle D (ed) Dynamic properties of forest ecosystems. Cambridge University Press, Cambridge, pp 341–409

Del Arco JM, Escudero A, Vega Garrido M (1991) Effects of site characteristics on nitrogen retranslocation from senescing leaves. Ecology 72:701–708

Djema A (1995) Cuantificación de la biomasa y mineralomasa subterránea de un bosque de *Quercus ilex* L. MSc Thesis, Instituto Agronómico Mediterráneo de Zaragoza, Zaragoza

Duvigneaud P, Denaeyer-De Smet S (1970) Biological cycling of minerals in temperate deciduous forests. In: Reichle DE (ed) Analysis of temperate forest ecosystems. Springer, New York, pp 199–225

Escarré A, Ferrés L, López R, Martín J, Rodà F, Terradas J (1987) Nutrient use strategy by evergreen-oak (*Quercus ilex* ssp. *ilex*) in NE Spain. In: Tenhunen JD, Catarino FM, Lange OL, Oechel WC (eds) Plant response to stress. Springer, Berlin, pp 429–435

Escudero A, del Arco JM, Sanz IC, Ayala J (1992) Effects of leaf longevity and retranslocation efficiency on the retention time of nutrients in the leaf biomass of different woody species. Oecologia 90:80–87

Ferrés L (1984) Biomasa, producción y mineralomasas del encinar montano de La Castanya (Montseny). PhD Thesis, Autonomous University of Barcelona, Bellaterra

Hereter A (1990) Els sòls forestals del Montseny. Avaluació de la seva fertilitat. PhD Thesis, University of Barcelona, Barcelona

Johnson DW, Henderson GS (1989) Terrestrial nutrient cycling. In: Johnson DW, Van Hook RI (eds) Analysis of biogeochemical cycling processes in Walker Branch Watershed. Springer, New York, pp 233–300

Killingbeck KT (1996) Nutrients in senesced leaves: keys to the search for potential resorption and resorption proficiency. Ecology 77:1716–1727

Likens GE, Bormann FH (1995) Biogeochemistry of a forested ecosystem, 2nd edn. Springer, New York

Mayor X (1990) El paper dels nutrients com a factors limitants de la producció primària de l'alzinar de la conca del Torrent de la Mina (Montseny). MSc Thesis, Autonomous University of Barcelona, Bellaterra

Mayor X, Rodà F (1992) Is primary production in holm oak forests nutrient limited?: a correlational approach. Vegetatio 99/100:209–217

Monk CD, Day FP (1985) Vegetation analysis, primary production and selected nutrient budgets for a southern Appalachian oak forest: a synthesis of IBP studies at Coweeta. For Ecol Manage 10:87 113

Ovington JD (1959) The circulation of minerals in plantation of *Pinus silvestris* L. Ann Bot 23: 229–239

Pugnaire FI, Chapin FS III (1993) Controls over nutrient resorption from leaves of evergreen Mediterranean species. Ecology 74:124–129

Rodà F (1983) Biogeoquímica de les aigües de pluja i de drenatge en alguns ecosistemes forestals del Montseny. PhD Thesis, Autonomous University of Barcelona, Bellaterra

Rodin LE, Bazilevich NI (1967) Production and mineral cyling in terrestrial vegetation. Oliver and Boyd, London

Rundel PW (1988) Vegetation, nutrition and climate – data tables: (2) foliar analysis. In: Specht RL (ed) Mediterranean-type ecosystems: a data source book. Kluwer, Dordrecht, pp 63–80

Sabaté S (1993) Canopy structure and nutrient content in a *Quercus ilex* L. forest of Prades mountains: effect of natural and experimental manipulation of growth conditions. PhD Thesis, University of Barcelona, Barcelona

Sabaté S, Sala A, Gracia C (1995) Nutrient content in *Quercus ilex* canopies: seasonal and spatial variation within a catchment. Plant Soil 168/169:297–304

Serrasolsas I (1994) Fertilitat de sòls forestals afectats pel foc. Dinàmica del nitrogen i del fòsfor. PhD Thesis, University of Barcelona, Barcelona

Verdú AMC (1984) Circulació de nutrients en ecosistemes forestals del Montseny: caiguda de virosta i descomposició de la fullaraca. PhD Thesis, Autonomous University of Barcelona, Bellaterra

Yin X, Perry JA, Dixon RK (1991) Temporal changes in nutrient concentrations and contents of fine roots in a *Quercus* forest. For Ecol Manage 44:175–184

Part 5 Hydrology and
Biogeochemistry of Catchments

19 Water Balance in Catchments

Josep Piñol, Anna Àvila and Antoni Escarré

19.1 Introduction

Water balances in small catchments are widely used in forest science. Since the watershed research at Wagon Wheel Gap (Colorado, USA) in 1910 (Cameron 1928), many other catchments have been instrumented to measure precipitation inputs and runoff outputs. The main goal of this approach is to evaluate the influence of vegetation cover on water yield, both in autocalibrated and in paired catchments. Reviews by Bosch and Hewlett (1982) and Trimble and Weirich (1987) showed the negative relationship between forest cover and water yield in catchments. The same result is obtained when forest manipulations like clear-cutting or revegetation affect the entire catchment (Hibbert 1971; Likens et al. 1977). When deep water losses are negligible because of the underlying lithology, the study of water balances in catchments is the best technique to get accurate estimates of actual evapotranspiration. Water precipitation input and streamflow output also provide the basic data for the estimation of catchment element budgets (Chap. 20).

The objectives of this chapter are: (1) to compare the basic hydrology of the holm oak (*Quercus ilex* L.) forest catchments at Prades and Montseny with that from more humid areas; (2) to search for regularities in the interannual variation of streamflow and actual evapotranspiration; and (3) to briefly discuss the validity of water balances obtained in small catchments for larger areas, and the possible hydrological changes under a warmer climate.

The chapter summarizes the studies on the water balance conducted at Prades and Montseny over the last 15 years. Part of the results have been published (Àvila and Rodà 1990; Piñol et al. 1991, 1992, 1995), and the present work is mainly based on these previous studies. Nevertheless, data have been updated by the inclusion of a new catchment (TM0 at Montseny) and more hydrological years both at Prades and Montseny. See Chapter 2 for a description of the experimental catchments.

Ecological Studies, Vol. 137
Ferran Rodà et al. (eds) Ecology of Mediterranean
Evergreen Oak Forests
© Springer-Verlag, Berlin Heidelberg 1999

Table 19.1. Annual precipitation (P), runoff (Q), and actual evapotranspiration (E_A, computed as $E_A = P - Q$) (mm year^{-1}) at Prades and Montseny experimental catchments. The water year was taken to start on 1 August at Montseny, and on 1 October at Prades

Location and Catchment	Period	Number of years	P			Q			E_A
			Mean	Min	Max	Mean	Min	Max	Mean
Montseny									
TM9	1984–1996[a]	11	891	627	1236	311	94	556	579
TM9	1990–1996[a]	5	1004	627	1236	382	158	556	622
TM0	1990–1996[a]	5	1004	627	1236	382	174	583	622
Prades									
Avic	1981–1994	13	547	339	929	41	8	95	506
Avic	1986–1994	8	566	339	929	39	8	73	527
Teula	1986–1994	8	566	339	929	58	11	107	508

[a] Hydrological year 1994–1995 not available at Montseny.

19.2 Water Balance at Prades and Montseny Compared to Humid Catchments

The distribution of annual precipitation (P) between actual evapotranspiration (E_A) and streamflow (Q) was very different at Prades and Montseny (Table 19.1): the water yield (Q) of Prades and Montseny catchments was approximately 10 and 35% of annual precipitation, respectively. Figures for water partition reported in other sites of Mediterranean climate fall between those of Prades and Montseny, but are closer to those found at Prades. In California, USA, Lewis (1968) measured streamflow percentages of 8, 19, and 25% for three catchments. Burch et al. (1987) in Victoria, Australia, estimated a streamwater loss of 14% of P. In colder, wetter countries, the absolute values of P and Q are higher, and the distribution of P between Q and E_A is biased towards Q: at Hubbard Brook (New Hampshire, USA) around 65% of P is lost as streamflow in five control catchments (Federer et al. 1990). In the even wetter climate of Plynlimon, mid Wales, UK, Hudson (1988) reported a streamwater loss of 77 and 85% for the Severn and Wye headwaters.

The main difference between the water balance in Mediterranean catchments compared to temperate catchments is the dominance of E_A above Q. The relative importance of the two main fluxes that compose E_A, forest interception and transpiration, is also remarkably different in Mediterranean and temperate climates. At Prades, forest transpiration is the larger water flux of the hydrological cycle (81% of P at Prades estimated at the Avic catchment using hydrological methodology by Piñol, 1990, or 86% of P estimated using ecophysiological methodology by Sala and Tenhunen, 1996) and interception only represents 13% of E_A. At Montseny, the importance of interception increases to 30% of E_A (Àvila and Rodrigo 1996; Chap. 15), whereas in wetter climates it can be an even higher proportion of the total E_A (Gash and Stewart 1977). Thus, in Mediterranean climates, the vegetation controls to a high degree the hydrological cycle (see also Chaps. 10–12 and 15).

To correctly understand the relationship between the main components of the water cycle at the catchment level, the potential evapotranspiration (E_T) also has to be considered. In fact, the key variable to explain the water partition under different climates is the relationship between the available water and the energy available to evaporate that water. This relationship can be roughly estimated by the ratio P/E_T. This ratio is approximately 0.5 at Prades, 1 at Montseny, and greater than 2 in cool-temperate humid areas like the UK or northeastern USA.

19.3 Interannual Variability of Streamflow and Evapotranspiration

The main factor that introduces variability in the annual water balance at our holm oak sites is the variability in annual precipitation. In comparison with it, annual variations of E_T are much lower (Piñol et al. 1998). In rainy years more water is available both to be evaporated and to be lost by streamflow, so we could expect a positive relationship between annual P and annual Q, and between annual P and annual E_A. However, as we will see, the interannual variation of Q and E_A depends strongly on climatic characteristics, summarized in the ratio P/E_T, in addition to their evident dependence on P.

At Prades, the relationship E_A-P is highly significant, whereas the relationship Q-P is only slightly significant for the Avic and Teula catchments (Table 19.2). At Montseny, both relationships are statistically significant, but the first one (E_A-P) is less significant and the second one (Q-P) more significant than at Prades (Table 19.2). In wetter catchments, like those of Hubbard Brook in northeastern USA, Q is linearly dependent on P, but E_A and P are uncorrelated (Likens et al. 1977). If the same relationships are derived for different catchments around the world (Table 19.2), there is a continuum of situations between those observed at Prades and Hubbard Brook. The slopes for linear regressions of Q on P range between 0 (Q is not dependent on P) and 1 (all the Q variation can be attributed to P). Table 19.2 shows the regressions for a selected group of catchments arranged according to values for regression slopes of Q on P, and ordered from Mediterranean to humid temperate catchments. The Montseny catchments occupy an intermediate position between the two groups.

Table 19.2. Relationship between annual Q and P, and between annual E_A and P for a group of catchments selected from all over the world

Location	Catchment	Vegetation	Years	Q = a + b P			E_A = a + b P			P (mm)	P/E_T	Reference
				a	b	r^2	a	b	r^2			
Queensland, Australia		Open woodland	5	57	-0.01	0.00ns	-57	1.01	0.96**	702	0.4[a]	1
Western Australia	4L	Jarrah forest	7	-16	0.02	0.34ns	16	0.98	1.00***	880	-	8
Prades, Spain	Avic	Holm oak forest	13	-32	0.13	0.42*	32	0.87	0.97***	547	0.6[b]	This study
Prades, Spain	Teula	Holm oak forest	8	-47	0.19	0.78**	47	0.81	0.99***	566	0.6[b]	This study
California, USA	B	Oak woodland	10	-202	0.41	0.90**	202	0.59	0.96***	623	0.5[c]	2
Montseny, Spain	TM0	70% Holm oak forest	5	-186	0.56	0.86*	186	0.44	0.77**	1004	1.1	This study
France		Fagus sylvatica	7	15	0.67	0.94***	-15	0.33	0.79**	1849	-	3
Montseny, Spain	TM9	Holm oak forest	11	-326	0.72	0.92***	326	0.28	0.62**	891	1.0	This study
France		Picea excelsa	7	-75	0.74	0.92***	75	0.26	0.58*	1944	-	3
California, USA	A	Oak woodland	10	-359	0.77	0.92***	359	0.23	0.55ns	623	0.5[c]	2
UK	Severn	68% Forest	8	-381	0.91	0.96***	381	0.09	0.15ns	2510	-	4
California, USA	Tomales	Variate	100	-284	0.92	0.95***	-284			862	1.1[d]	7
Greece	7	Abies cephalonica	4	-646	0.92	0.92*	646	0.08	0.07ns	1615	-	5
Hubbard Brook, New Hampshire, USA	6	Hardwood forest	24	-400	0.91	0.96***	400	0.09	0.17ns	1403	2.5[d]	6
Greece	5	Abies cephalonica	4	-745	1.06	0.94*	745	-0.06	0.05ns	1615	-	5

Reference: 1, Prebble and Stirk (1988); 2, Lewis (1968); 3, Durand (1989); 4, Hudson (1988); 5, Nakos and Vouzaras (1988); 6, Likens et al. (1977), Federer et al. (1990); 7, Fischer et al. (1996), Stoneman (1993).
ns, not significant; * $P < 0.05$; ** $P < 0.01$; *** $P < 0.001$.
[a] E_T measured with a class A pan-evaporimeter.
[b] E_T estimated by the Samani–Hargreaves method (Samani and Hargreaves 1985, quoted in McKenney and Rosenberg (1993).
[c] E_T estimated with the Penmann equation.
[d] E_T estimated with the Thornwaite equation.

19.4 The Quotient P/E_T: Key Variable for Determining Water Balance

The above-mentioned relationships between the annual values of the hydrological variables can be understood when the quotient P/E_T is considered. When P is much greater than E_T, as is the case at Hubbard Brook, the evaporative demand can be totally satisfied and the remaining water becomes streamflow. In consequence, there is a positive relationship between P and Q, and annual E_A can be considered constant (Likens et al. 1977). In contrast, when P is lower than E_T, as in Prades, the water supply in the wetter years is still not enough to satisfy the transpiration and evaporation possible in this climate. Thus, at Prades, the water surplus of rainy years is evaporated (E_A increases) rather than lost by streamflow; streamflow is more dependent on the rainfall distribution in time than on the annual volume. For instance, the Avic catchment had its maximum annual Q (95 mm) in the hydrological year 1982–1983 but not the highest precipitation (only 576 mm) because a major part of the streamflow (75 mm) occurred during 2 wet months (185 mm rain in October and 151 mm in November). In the Avic catchment, linear correlation between annual Q and the rainfall of the 3 months with the highest precipitation is much higher ($r^2 = 0.71$) than between annual Q and annual P ($r^2 = 0.42$; Table 19.2). At Montseny, P is only slightly greater than E_T: these catchments present an intermediate behaviour between those of Prades and those of more humid regions.

In dry climates like that of Prades, annual E_A can be very different in different years, and neither annual E_A nor annual transpiration can be considered as conservative hydrological variables, as Roberts (1983) considered them for temperate climates. The higher transpiration rate that can be maintained in humid years allows a higher CO_2 fixation rate in dry sites. At Prades this effect was clearly shown by the much higher rate of stem growth of holm oak trees in the rainy summer of 1992 than in the average of the 3 preceding years, which had normal dry summers (Mayor et al. 1994). Thus, the study of the hydrologic cycle in dry areas is important not only by itself, but also because of its influence on many other ecological and biogeochemical processes occurring in the catchment. The present book contains plenty of examples showing that water availability is the single most important factor in the ecology and nutrient cycling in Mediterranean forests.

The above relationship between the components of the water balance under different climates can be described more formally. It is a usual practice in hydrology to take into account E_T in order to estimate E_A. The classical expressions of Turc (1954) and Budyko (1974) both include E_T in their formulation. Piñol et al. (1991) developed a new expression, based on a previous work of Hsuen-Chun (1988), to predict E_A as a function of E_T and P. This expression includes an additional parameter (k) that can be considered as a lumped parameter of the non-climatic catchment characteristics relevant for the water balance. The expression was:

$$e_A = (p^k/(1 + p^k))^{(1/k)} \tag{19.1}$$

and, given that $p = q + e_A$,

$$q = p - (p^k/(1 + p^k))^{(1/k)}, \tag{19.2}$$

where $p = P/E_T$, $e_A = E_A/E_T$, and $q = Q/E_T$. The derivation of the equations and their characteristics can be found in Hsuen-Chun (1988) and Piñol et al. (1991, 1995).

By using Eqs. (19.1) and (19.2) with a k value of 2.0, the different relationships between E_A and P and Q and P found at Prades, Montseny, and Hubbard Brook are shown to follow the same conceptual model (Fig. 19.1). The different partition of P between Q and E_A in the different catchments be-

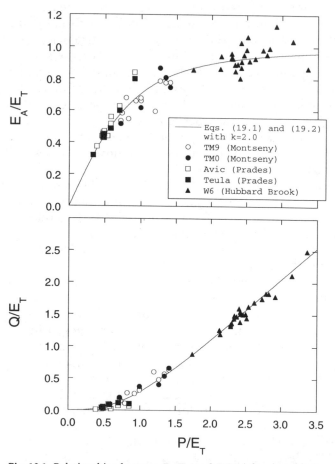

Fig. 19.1. Relationships between E_A/E_T and P/E_T (*above*) and between Q/E_T and P/E_T (*below*) for catchments at Prades, Montseny, and Hubbard Brook (data taken from Federer et al. 1990). Equations (19.1) (*above*) and (19.2) (*below*) are also shown with a k value of 2.0. This is an approximate mean value of those obtained by statistical fitting with actual hydrologic data of different catchments in the world by Piñol et al. (1991, 1995)

comes a simple function of the quotient P/E_T. This simple model also shows graphically why the relationship E_A-P is significant at low values but not at high values of P/E_T, and the opposite for the relationship Q-P. At low values of P/E_T (arid catchments) the slope of Eq. (19.1) (E_A/E_T) is close to the unity and the relationship P-E_A is significant. On the other hand, at low values of P/E_T, the slope of equation 19.2 (Q/E_T) is close to zero, and the natural variability obscures any significant relationship between P and Q. At high values of P/E_T (humid catchments) the slope of Eq. (19.1) is close to zero and that of Eq. (19.2) is close to unity, making non-significant the relationship P-E_A and significant the relationship P-Q.

19.5 Representativeness of the Experimental Catchments for Larger Areas

One of the main concerns about small, experimental catchments is whether the results obtained can be interpreted and extended to larger areas of the same physiographic region. The representativeness of the Prades and Montseny experimental catchments was evaluated in a previous work (Piñol et al. 1992) by comparing them to the streamflow record of two larger, mainly forested, catchments located in both mountainous massifs (Siurana, 60 km^2, at Prades; Llavina, 34 km^2, at Montseny). It was found that the experimental catchments at both Prades and Montseny produced less annual streamflow than larger catchments in the same region. The main reason for this seems to be the larger proportion of forest cover in the experimental catchments (nearly 100%) than in the larger ones (74% at Llavina and 30% at Siurana). Nevertheless, flow duration curves and the annual water balance were more similar between small and large catchments within the same region than between the physiographic regions of Prades and Montseny, thus reinforcing confidence of the results obtained in the small experimental catchments of Prades and Montseny.

19.6 Water Balance in a Warmer Climate

The increase of atmospheric CO_2 and other greenhouse gases during this century is probably changing the Earth's climate. At every new edition of the Intergovernmental Panel on Climatic Change (IPCC) conclusions, more evidence is given of the direct link between the change in the composition of the atmosphere and climate change. It seems clear that the Earth's temperature has already increased in this century and that it will further increase in the near future (Houghton et al. 1996). In the Mediterranean region of Spain this increase of temperature has already been observed (0.10 °C per decade,

Piñol et al. 1998). It seems also clear that the hydrological cycle will be accelerated as a consequence of the higher water holding capacity of warmer air, which will increase evaporation and, consequently, precipitation. Nevertheless, the current predicting capability of general circulation models (GCMs) linked with macroscale and landscape-scale hydrologic models that simulate regional and local hydrologic regimes under global warming scenarios is still poor. In spite of these uncertainties, all the predictions point to potentially worsening conditions for flood control, water storage, and water supply in areas of semiarid mid-latitudinal climates. Little information of this type is currently available for other areas of the world (Loaiciga et al. 1996).

The water balance data of the experimental catchments of Prades and Montseny were used by Piñol et al. (1995) to assess the effect of possible future changes of P and E_T in the Mediterranean region on the main components of the water balance, Q and E_A. The basic hypothesis was that the more humid catchments of Montseny would change their behaviour toward the currently observed behaviour at Prades. Simulations were presented showing changes in the annual E_T, E_A, and Q in scenarios of increased temperature (+2 and +4 °C) and variable annual precipitation (−10, 0, and +10% of current P). Results suggested that in a warmer scenario, the Prades area would hardly be able to maintain a closed-canopy forest such as the present one. Curiously, the summer of 1994 was extremely warm and dry in eastern Spain. During that dry spell, holm oak and pine (mainly *Pinus sylvestris*), the dominant tree species of vast areas of Prades, were severely affected. The long-term effect was much worse for pine than for holm oak, because of the resprouting capacity of the latter that the former does not have. Other secondary species like *Phillyrea latifolia* were less affected. Lloret and Siscart (1995) found similar mortality rates in holm oak and *Phillyrea latifolia* in other areas of the region. Thus, if the climate becomes warmer and drier, it would be possible for other species to take the dominant role of holm oak in some areas with a present climate similar to that of Prades. These changes would probably affect some water fluxes relevant at plot catchment scales, like interception and transpiration. Nevertheless, when the uncertainty in regional estimates of rainfall and temperature increases and the possible shift in species dominance are considered together, the prediction of hydrologic fluxes at local and regional scales is highly inaccurate. At the present stage of knowledge, hydrologic prediction at these spatial scales fits better in the field of prophecy than in science.

References

Àvila A, Rodà F (1990) Water budget of a broadleaved sclerophyllous forested catchment. In : Hooghart JC, Posthumus CWS, Warmerdam PMM (eds) Hydrological research basins and the environment. The Netherlands Organization for Applied Scientific Research, The Hague, pp 29–40

Àvila A, Rodrigo A (1996) Comparison of annual water balance components in Mediterranean catchments (NE Spain). Proc Conf of Experimental and representative basins. 24–26 Sept 1996, Strasbourg, France

Bosch JM, Hewlett JD (1982) A review of catchment experiments to determine the effect of vegetation changes on water yield and evapotranspiration. J Hydrol 55:3–23

Budyko MI (1974) Climate and life. Academic Press, New York

Burch GJ, Bath RK, Moore ID, O'Loughlin EM (1987) Comparative hydrological behaviour of forested and cleared catchments in Southeastern Australia. J Hydrol 90:19–42

Cameron J (1928) The development of governmental forest control in the United States. The Johns Hopkins Press, Baltimore

Durand P (1989) Biogeochimie comparée de trois écosystèmes (pelouse, hetraie, pessière) de moyenne montagne granitique (Mont-Lozère, France). PhD Thesis, University of Orleans, Orleans

Federer CA, Flynn LD, Martin CW, Hornbeck JW, Pierce RS (1990) Thirty years of hydrometeorological data at the Hubbard Brook experimental forest, New Hampshire. USDA For Serv Northeastern For Exp Stn Gen Tech Rep NE-141. Durham, New Hampshire

Fischer DT, Smith SV, Churchill RR (1996) Simulation of a century of runoff across the Tomales watershed, Marin County, California. J Hydrol 186:253–273

Gash JHC, Stewart JB (1977) The evaporation from Thetford forest during 1975. J Hydrol 35: 385–396

Hibbert AR (1971) Increases in streamflow after converting chaparral to grass. Water Resour Res 7:71–80

Houghton JT, Meira Filho LG, Callander BA, Harris N, Kattenberg A, Maskell K (eds) (1996) Climate change 1995. The science of climate change. IPCC. Cambridge University Press, Cambridge

Hsuen-Chun Y (1988) A composite method for estimate annual actual evapotranspiration. Hydrol Sci J 33:345–356

Hudson JA (1988) The contribution of soil moisture storage to the water balances of upland forested and grassland catchments. Hydrol Sci J 33:289–309

Lewis DC (1968) Annual hydrologic response to watershed conversion from oak woodland to annual grassland. Water Resour Res 4:59–72

Likens GE, Bormann FH, Pierce RS, Eaton JS, Johnson NM (1977) Biogeochemistry of a forested ecosystem. Springer, Berlin

Lloret F, Siscart D (1995) Los efectos demográficos de la sequía en poblaciones de encina. Cuad Soc Española Cienc For 2:77–81

Loaiciga HA, Valdes JB, Vogel R, Garvey J, Schwarz H (1996) Global warming and the hydrologic cycle. J Hydrol 174:83–127

McKenney MS, Rosenberg NJ (1993) Sensitivity of some potential evapotranspiration methods to climate change. Agric For Meteorol 64:81–110

Mayor X, Belmonte R, Rodrigo A, Rodà F, Piñol J (1994) Crecimiento diametral de la encina (Quercus ilex L.) en un año de abundante precipitación estival: efecto de la irrigación previa y de la fertilización. Orsis 9:13–23

Nakos G, Vouzaras A (1988) Budgets of selected cations and anions in two forested experimental watersheds in central Greece. For Ecol Manage 24:85–95

Piñol J (1990) Hidrologia i biogeoquímica de conques forestades de les Muntanyes de Prades. PhD Thesis, University of Barcelona, Barcelona

Piñol J, Lledó MJ, Escarré A (1991) Hydrological balance of two Mediterranean forested catchments. Hydrol Sci J 36:95–107

Piñol J, Àvila A, Escarré A, Lledó MJ, Rodà F (1992) Comparison of the hydrological characteristics of three small experimental holm oak forested catchments in NE Spain in relation to larger areas. Vegetatio 99/100:169–176

Piñol J, Terradas J, Àvila A, Rodà F (1995) Using catchments of contrasting hydrological conditions to explore climate change effects on water and nutrient flows in Mediterranean forests. In: Moreno JM, Oechel WC (eds) Global change and Mediterranean-type ecosystems. Springer, Berlin, pp 251–264

Piñol J, Terradas J, Lloret F (1998) Climate warming, wildfire hazard, and wildfire occurrence in coastal eastern Spain. Clim Change 38:345–357

Prebble RE, Stirk GB (1988) Hydrological effects of land use change in small catchments at the Narayen Research Station, Queensland. Aust J Soil Res 26:231–242

Roberts J (1983) Forest transpiration; a conservative hydrological process? J Hydrol 66:133–141

Sala A, Tenhunen JD (1996) Simulations of canopy net photosynthesis and transpiration in *Quercus ilex* L. under the influence of seasonal drought. Agric For Meteorol 78:203–222

Samani ZA, Hargreaves GH (1985) Water requirements, drought and extreme rainfall manual for the United States. International Irrigation Center, Utah State University, Logan, Utah, 105 pp

Stoneman GL (1993) Hydrological response to thinning a small jarrah (*Eucalyptus marginata*) forest catchment. J Hydrol 150: 393–407

Trimble SW, Weirich FH (1987) Reforestation reduces streamflow in the south-eastern United Sates. J Soil Water Conser 42: 274–276

Turc L (1954) Le bilan d'eau des sols: relation entre les précipitations, l'évaporation et l'écoulement. Troisiemes Journées de l'Hydraulique. 12–14 April, Alger

20 Element Budgets in Catchments

Anna Àvila, Juan Bellot and Josep Piñol

20.1 Introduction

The element budget concept for the study of forest ecosystems was first developed by an interdisciplinary group of scientists at the Hubbard Brook Experimental Forest (New Hampshire, USA) in the early 1960s. The reasoning behind the element budget concept was that "in humid regions, the chemical flux and cycling is intimately linked to the hydrologic cycle" as quoted in their seminal book *Biogeochemistry of a Forested Ecosystem* (Likens et al. 1977). In this way, terrestrial ecologists could take advantage of the broad experience in using small catchments for hydrological studies (e.g. the Coweeta watersheds in the USA) in order to test hypotheses regarding the nutrient circulation in small catchments. Input–output budgets were used to provide information on the processes operating in catchments at different time scales and to understand the effects of natural or anthropogenic perturbations on biogeochemical cycles at the ecosystem level.

Two requirements for nutrient balance studies are the same as for hydrological balances: (1) the catchments have to be underlain by water-tight rock so that water losses through deep drainage are negligible, and (2) a hydrological year has to be defined so that interannual changes in the amount of water stored in the catchment (as snow or soil water) are minimized. A third requirement for nutrient balances is that the biological inputs and outputs of nutrients (by animals or plant debris) should be in balance so that there is not a net flux into or out of the system, a condition that is usually met when catchments are part of larger homogeneous landscape units. If these requirements are satisfied, the major element input to the ecosystem is the deposition from the atmosphere and the most important output is usually through the streamwater draining the catchment. Under these conditions, by measuring the element inputs from the atmosphere and the element exports in the streamwater, chemical budgets can be established. These fluxes constitute the external cycle of elements and are usually complemented with detailed studies of the intrasystem circulation within the ecosystem (see

Ecological Studies, Vol. 137
Ferran Rodà et al. (eds) Ecology of Mediterranean
Evergreen Oak Forests
© Springer-Verlag, Berlin Heidelberg 1999

Chap. 18). A representation of the conceptual model for nutrient budgets in catchments is given in Fig. 21.1.

Atmospheric inputs to the ecosystem are in wet and dry form, but quantification of wet inputs does not pose as many problems as the measurement of dry deposition. In fact, the flux of some nutrients with a gaseous phase (e.g. N and S) is quite difficult to evaluate. Approaches to the estimation of dry deposition inputs include micrometeorological techniques (Fowler et al. 1991; Erisman and Baldocchi 1994), measurement of throughfall under natural or surrogate surfaces (Lovett 1994; Butler and Likens 1995) and mass balances in catchments (Eaton et al. 1978; Likens et al. 1990; Hultberg and Grennfelt 1992).

Besides the considerable contribution from atmospheric inputs, a major supply of many mineral nutrients to the soils and the vegetation comes from the weathering of the bedrock. Base cations released to the soil by weathering are either incorporated into the net biomass increment, removed from the soils with the percolating soil water, incorporated into secondary minerals, or retained in the soil exchange complex. By formulating appropriate hypotheses regarding the situation in the soil exchange complex, the net output of base cations from catchments has been used to estimate the rate of weathering of the bedrock under natural conditions (Johnson et al. 1968, 1981; Paces 1985, 1986; Velbel 1985).

Here, we use the nutrient budgets for two Mediterranean holm oak forested sites in the experimental areas of Prades and Montseny (NE Spain; Chap. 2) with similar vegetation and bedrock to understand the role of climate, vegetation and soils in regulating the circulation of nutrients. Also, some inferences on the dry deposition flux for S and on weathering rates are provided.

20.2 Input Fluxes

The atmospheric inputs of nutrients were sampled with bulk deposition collectors. These collectors are permanently open to the atmosphere so they catch a fraction of the dry deposition, the gravitational sedimentation. For base cations, bulk deposition was estimated to increase wet inputs at Montseny by 10% due to gravitational sedimentation (Àvila 1988).

Input fluxes were calculated by multiplying the measured concentrations of each ion in the weekly bulk deposition sample by the weekly volume of precipitation. These weekly amounts were then summed up for the year. Since dry deposition represents an important nutrient input to forested ecosystems, it has been suggested that throughfall gross fluxes are the relevant input fluxes for nutrient cycling studies (Matzner 1986; Lindberg and Lovett 1992). However, nutrient leaching from plants confuses the distinction between the external dry deposition flux and the intrasystem flux (Chap. 15).

For marine-derived elements, net throughfall fluxes mostly represent dry deposition onto canopies because they are scarcely leached from plants. We thus considered all Na$^+$ in net throughfall, and the corresponding sea-salt proportion for Cl$^-$, to represent dry deposition. This amounted (Chap. 15) to 0.4 and 0.6 kg ha^{-1} year^{-1} respectively at Montseny, and 0.7 and 1.2 kg ha^{-1} year^{-1} respectively at Prades (Torners). These fluxes were added to bulk deposition inputs to give the total Na$^+$ and Cl$^-$ inputs to our catchments. Other elements were not corrected for dry deposition. For a wider discussion of the inputs see Chapters 14 and 15.

20.3 Output Fluxes

Output fluxes were measured for two catchments at Montseny (see site description in Chap. 2), named TM9 and TM0 (TM9 for the period 1984–1994 and TM0 for the period 1990–1994) and at the Avic catchment in Prades for the period 1982–1994. Montseny data have been reported in Àvila (1988) and Àvila and Rodà (1988) for the period 1983–1985. The Prades series has been partly analysed by Lledó (1990), Piñol (1990) and Bellot et al. (1992). Some data exist for two other catchments at Prades (Teula and Saucar) but will not be discussed here. Element budgets for the Teula catchment are described for the period 1986–1988 in Piñol (1990).

20.3.1 Streamwater Chemistry

The main chemical features of these streamwaters are summarized here. Prades and Montseny catchments produced highly buffered streamwaters (Table 20.1): pH was rarely below 7 in weekly samples. At the observed pH values, alkalinity corresponds to HCO$_3^-$, and the dominant anion at all sites was bicarbonate which accounted for 76% of the total anion concentration at Avic, and 55 and 52% at TM9 and TM0, respectively. In the Prades catchment, the dominant cation was Ca^{2+}, in similar proportion to HCO$_3^-$. At Montseny, Na$^+$ and Ca^{2+} codominated. The streamwater of the Prades catchments was more mineralized than that of the Montseny streams: the total flow-weighted mean ion content was 10.6 meq L^{-1} at Avic, and only 1.4 and 1.1 meq L^{-1} for TM9 and TM0 (Table 20.1). The higher mineralization of Prades streamwater results from the higher residence time of water within the catchment, and from a concentration effect due to a higher ratio of precipitation to runoff (Piñol et al. 1992; Chap. 19). Weathering rates are probably enhanced at Prades due to the mentioned longer residence time of the water and to higher pCO$_2$ in the soils compared to Montseny (Piñol et al. 1995). The very high concentrations of HCO$_3^-$ and Ca^{2+} at Avic give a calcite-saturated character to these waters: calcite precipitated at some points in the channel where

Table 20.1. Flow-weighted means for solute concentrations in Montseny (TM9 and TM0) and Prades (Avic) streamwaters. Minimum and maximum concentrations in weekly samples are also given. Units are μeq L^{-1}, except for pH (pH units) and SiO$_2$ (mg L^{-1})

	pH	Na$^+$	K$^+$	Ca^{2+}	Mg^{2+}	NO$_3^-$	SO$_4^{2-}$	Cl$^-$	Alkalinity	SiO$_2$
TM9[a]										
Mean	7.53	281	10.1	259	165	1.2	204	111	392	11.0
Min.	6.80	206	6	170	100	0.5	110	60	162	9.0
Max.	8.01	450	50	560	364	118	330	207	940	17
TM0[b]										
Mean	7.44	201	7.1	203	133	1.3	174	93	292	10.4
Min.	7.08	163	4	160	105	0.5	101	66	137	9.5
Max.	7.77	452	22	415	295	14	220	146	870	15
Avic[c]										
Mean	7.4	732	38	3818	779	4.6	908	340	4024	–
Min.	7.2	465	30	1728	385	0.5	603	249	1655	–
Max.	7.7	1112	53	5892	1205	107	2055	620	5628	–

[a] Data period from 1 August 1984 to 31 July 1994. SiO$_2$ was analyzed from August 1991 to July 1995.
[b] Data period from 1 August 1990 to 31 July 1994. SiO$_2$ as for TM9.
[c] Data period from September 1986 to June 1988 (Piñol 1990).

turbulence led to CO$_2$ degassing. This process was not observed at Montseny due to the more dilute character of its streamwater with HCO$_3^-$ and Ca^{2+} concentrations far from the calcite saturation equilibrium (Piñol and Àvila 1992).

Based on their seasonal variation, ions in these streamwaters can be classified into three main groups, reflecting their origin and the major biogeochemical processes affecting them: litophilic ions (HCO$_3^-$, Ca^{2+}, Mg^{2+} and Na$^+$, produced directly or indirectly by weathering), atmophilic ions (SO$_4^{2-}$ and Cl$^-$, deposited from the atmosphere and quantitatively circulating through the catchments) and biophilic ions (NO$_3^-$, NH$_4^+$ and K$^+$, strongly controlled by the living part of the system). The patterns of temporal variation of these ions were strongly linked to discharge variations and were interpreted at Montseny as the outcome of the mixing of varying proportions of waters of different chemistries residing in different compartments within the catchments (Àvila et al. 1992). At Montseny, the streamwater chemistry was described as a mixing of groundwater and soil water (deep subsurface flow) contributing in proportions of 70 and 30% respectively to the streamwater discharge (Àvila et al. 1995; Neal et al. 1995). In the drier Prades catchments, the mixing of different waters could not explain the seasonal variation, which was dominated by the flushing in autumn of soluble salts accumulated in the soils during the summer when water is mostly evapotranspired and the channels do not drain any water (Piñol et al. 1992).

In spite of lying on silicate bedrock, the streamwaters for these Mediterranean sites are highly buffered with high bicarbonate concentrations. Because of their alkalinity, these ecosystems are quite robust to potential acidification processes.

20.3.2 Element Outputs in Streamwater

Element outputs were computed analogously to inputs by multiplying the weekly element concentrations by the amount of runoff corresponding to each sample. While, as seen in the previous section, the chemistry of the Prades and Montseny streamwaters was quite contrasted, the differences were much smaller in terms of output fluxes. The higher element concentrations at Prades were compensated by lower runoff so the gross exports were of the same order of magnitude as those at Montseny (Table 20.2). Because outputs are computed from the product of concentrations and discharge, and weekly concentrations varied in a much smaller range (1 order of magnitude) than discharge (up to 3 orders of magnitude), gross exports were strongly related to water runoff. On an annual scale there were also strong direct relationships between annual element export and water runoff for all elements at Prades and Montseny.

Table 20.2. Average element budgets for Montseny (TM9 and TM0) and Prades (Avic) catchments. Net outputs are (gross) outputs minus inputs. Water fluxes (precipitation and runoff) are in mm year^{-1}; elements and SiO_2 in kg ha^{-1} year^{-1}; alkalinity in keq ha^{-1} year^{-1}

	Water flux	Na$^+$	K$^+$	Ca^{2+}	Mg^{2+}	NH$_4$-N	NO$_3$-N	SO$_4$-S	Cl$^-$	Alkalinity	SiO$_2$
Montseny[a]											
Input											
TM9	858	4.85[b]	1.44	10.2	1.04	2.72	2.52	6.22	9.4[b]	14.2	ind
TM0	1022	5.24[b]	1.77	10.3	1.06	2.72	2.96	6.42	10.8[b]	17.3	ind
Output											
TM9	283	18.3	1.21	14.7	5.69	0	0.05	9.28	11.1	111	34.0
TM0	331	15.3	0.92	13.5	5.37	0	0.06	9.22	10.9	97	31.3
Net output											
TM9	−575	13.5	−0.23	4.5	4.7	−2.7	−2.5	3.1	1.7	97	34.0
TM0	−691	10.1	−0.85	3.2	4.3	−2.7	−2.9	2.8	0.1	80	31.3
Prades (Avic)[c]											
Input	565	3.83[b]	1.78	6.6	0.70	1.63	1.61	4.09	6.56[b]	16.6	–
Output	45	5.04	0.55	23.5	3.15	0	0.01	4.18	4.34	121	–
Net output	−520	1.21	−1.23	16.9	2.45	−1.63	−1.60	0.09	−2.22	104	–

[a] For TM9, budgets are averaged over 10 years (1 August 1984–31 July 1994); for TM0 over 4 years (1 August 1990–31 July 1994). SiO$_2$ data at Montseny correspond to hydrologic years 1991–1994.

[b] Inputs include dry deposition estimated from net throughfall fluxes at our holm oak sites. For Montseny, dry deposition adds 0.4 and 0.6 kg ha^{-1} year^{-1} for Na$^+$ and Cl$^-$ respectively (Rodà 1983). For Prades it adds 0.7 and 1.2 kg ha^{-1} year^{-1} for Na$^+$ and Cl$^-$ respectively (Bellot 1989).

[c] Budgets are averaged over 12 years (1 October 1981–31 September 1994; no data for 1985–1986).

20.4 Net Outputs

Net element outputs were computed as the difference between annual outputs and annual inputs. Inputs for the different elements had a similar behaviour, with significant direct relationships between annual element deposition and annual precipitation, except for Ca^{2+} due to its higher deposition in years with big red rain events, very rich in Ca^{2+} concentrations (Chap. 14).

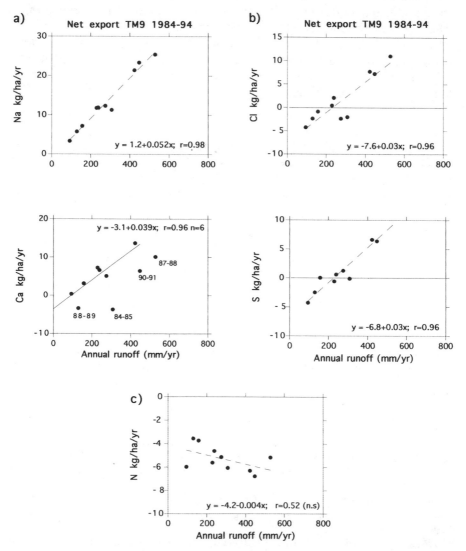

Fig. 20.1. Net exports at Montseny for **a** litophilic ions, **b** atmophilic ions, and **c** biophilic ions. For Na^+, Cl^- and S, the estimated dry deposition contribution to net throughfall (0.4, 0.6 and 1.3 kg ha^{-1} year^{-1}, respectively) has been added to their inputs in bulk deposition

Therefore, net exports, computed from two fluxes (inputs and gross outputs) which are strongly related to water fluxes, presented also a strong relationship with runoff (Figs. 20.1 and 20.2). For Montseny, only TM9 is shown as TM0 behaved similarly for the 4 years of simultaneous record.

The relationship between annual net outputs and annual runoff was significant for all elements ($P < 0.001$ for Montseny, $P < 0.01$ for Prades) except for the more biologically controlled elements, N and K. Plots for Na$^+$, Ca^{2+}, Cl$^-$, S

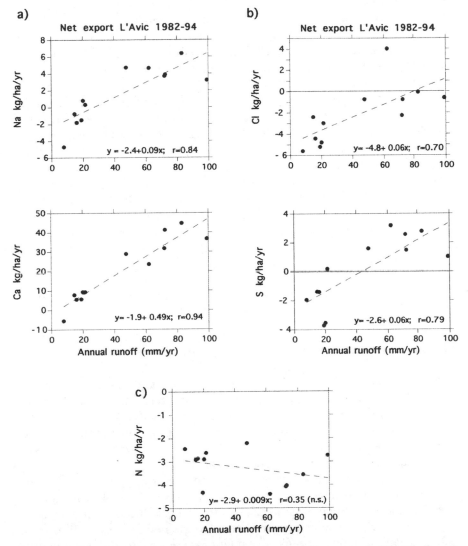

Fig. 20.2. Net exports at Prades for a litophilic ions, b atmophilic ions, and c biophilic ions. For Na$^+$ and Cl$^-$, the estimated dry deposition contribution to net throughfall (0.7 and 1.2 kg ha^{-1} year^{-1}, respectively) has been added to their inputs in bulk deposition. Sulphur has not been corrected for dry deposition

and N against annual runoff are shown in Figs. 20.1 and 20.2, and they illustrate the importance of the amount of water circulating through the catchment as a vector for the export of nutrients from the catchments.

This relationship with water discharge has been widely reported for cool-temperate sites (Likens et al. 1977), but the Mediterranean climate is particularly affected by a high variation of the precipitation regime which will determine a great variability in yearly net exports, as observed. Therefore, measurements taken over a short range of the hydrologic variability would give poor estimates of the average conditions. In consequence, there is a special need for long-term studies in order to obtain representative average balances. We present here 12 and 10 years of nutrient budgets for Prades and Montseny, respectively. The recorded period comprised dry and wet years and we consider the figures provided to be quite representative.

In Figs. 20.1 and 20.2, the patterns of variation of net exports with annual runoff are presented for (a) litophilic, (b) atmophilic and (c) biophilic ions at the Montseny and Prades catchments. The behaviour of some ions was similar at Prades and Montseny: (1) HCO_3^- and Mg^{2+} always had net exports; (2) N always presented net retention; (3) Cl^- and S oscillated around the zero line, with positive exports during wet years and net retention during dry years; and (4) Ca^{2+} was usually exported, but it was retained in years of particularly important inputs in red rains (1984–1985, 1988–1989), a pattern more clearly defined at Montseny than at Prades. On the other hand, K^+ and Na^+ behaved differently at Prades and Montseny. Potassium was always retained at Prades, but had a positive export in 3 years at Montseny. The amounts exported, however, were small (lower than 1 kg ha^{-1} year^{-1}) and on average there was a net retention. Sodium presented a contrasting behaviour at Prades and Montseny: it was always exported from the Montseny catchment, but it was retained during dry years at Prades.

20.4.1 Elements with Net Retention: Nitrogen and Potassium

Nearly all inorganic nitrogen entering the Montseny and Prades catchments in bulk deposition was retained. At both sites, the relationship between net output and runoff was negative (although non-significant; Figs. 20.1 and 20.2), related to the fact that years with more precipitation (and runoff) received higher amounts of inorganic N, but outputs were little affected by increased runoff, resulting in higher retention in the catchment.

It has been proposed that ecosystems in an aggrading phase would efficiently retain those elements necessary for the construction of growing biomass (Vitousek and Reiners 1975). These atmospheric N inputs, in fact, constitute a low-level fertilization for the aggrading holm oak forests (Rodà et al. 1990). Inorganic N inputs in bulk precipitation were similar to the N amounts annually stored in the net growth of aboveground biomass in the holm oak forests of Prades and Montseny (Tables 18.6 and 20.2). The efficiency of N retention in these catchments is demonstrated by the fact that NO_3^- and

NH_4^+ were always below the analytical detection limit (2 µeq L^{-1}) during baseflow.

On average, K^+ was also retained at Prades and Montseny. Inputs of K^+ in bulk deposition were very moderate (Table 20.2), but weathering constituted an important source of K^+ for the system: the K^+ release by weathering at Montseny would amount to 50 kg ha^{-1} $year^{-1}$ if congruent dissolution of the rock occurred. However, muscovite – the major K-bearing mineral at our Montseny site – weathers at a lower rate than sodium plagioclase – the major Na-bearing mineral –, whose content in bedrock was used together with Na^+ net output fluxes for computing congruent dissolution rates (Sect. 20.5). Nonetheless, the actual weathering rate of K^+, jointly with atmospheric deposition (1.4 kg ha^{-1} $year^{-1}$), probably far exceeds the outputs from the soil, which include 5.0 kg ha^{-1} $year^{-1}$ in aboveground biomass retention (Table 18.6) and 1.2 kg ha^{-1} $year^{-1}$ in draining waters. Equilibrium reactions in the soil between the soil solution and the exchange complex must exist to account for the constancy of K^+ concentrations in the streamwaters at low flows (Àvila et al. 1992). The low export in the streamwaters could result from the formation of secondary minerals in the soil, tightly controlling K^+ mobility and circulation. However, the net retention of atmospheric inputs at both sites indicates the tendency to incorporate all available K^+ from bulk deposition into ecosystem compartments, either in the vegetation or the soils.

20.4.2 Elements in Balance: Chloride and Sulphur

Nutrients in balance (exports similar to inputs) are assumed to circulate quantitatively through the system. For Cl^- this is the expected behaviour because there are no important Cl^- sources or sinks in the soils or vegetation of the catchments, and Cl^- budgets have been widely used as a test for the water-tightness of the catchment (Juang and Johnson 1967). At our sites, the average Cl^- budget, which included an estimation of dry deposition onto forest canopies, was close to zero for TM0 and somewhat negative for Avic (Table 20.2; Figs. 20.1 and 20.2). However, the deviation from zero was small, at most 20% of the input or output fluxes. These small deviations could result from an inaccurate estimation of Cl^- input in dry deposition and, for Prades, from Cl^- retention during dry years. The Cl^- balance was slightly positive for TM9, but it must be taken into account that, as explained in Chapter 2, the computed surface area of this catchment was adjusted to give a balanced Cl^- budget.

Sulphur presented a pattern similar to that of Cl^-. At Montseny, there was an S net export of around 3 kg ha^{-1} $year^{-1}$ (averaged for TM9 and TM0; Table 20.2), but at Prades it was close to zero. Because the inputs do not include S dry deposition, and because the phillytes at Montseny are not a significant source of S (Casas 1979), at Montseny this balance could represent the unmeasured dry deposition S flux. Measurements at Montseny indicate a net S flux in throughfall and stemflow of 1.2 kg S ha^{-1} $year^{-1}$ (Chap. 15). Part of this flux could derive from internal foliar leaching but isotopic studies in a temperate deciduous forest revealed that dry deposition accounted for 85% of the net throughfall S flux (Lindberg and Garten 1988). Using evidence from net throughfall fluxes and from the S budgets, it can be concluded that S dry deposition at Montseny lies between 1 and 3 kg ha^{-1} year^{-1}.

As described in Chapter 14, there has been a significant decrease in S inputs in the Mediterranean area over the last 15 years. Because of these diminishing S inputs, one could expect a decreasing trend in the SO_4^{2-} streamwater concentrations or in the S annual exports. However, both at Prades and Montseny, the variation of the annual S exports was more strongly associated with variations in the annual runoff amount ($r = 0.99$, $P < 0.001$ at Montseny; $r = 0.62$, $P < 0.05$ at Prades) than with any temporal trend.

20.4.3 Elements with Net Exports

Magnesium was always exported at Prades and Montseny. Calcium presented net retentions for some years at both sites. For Na$^+$, some differences existed between sites: it was always exported at Montseny, but was retained at Prades in two of the driest years (Fig. 20.1). As discussed above, the chemistry of Prades streamwaters is strongly influenced by the flushing capacity of autumn and winter rainstorms. Although Na$^+$ was retained in some years, there was a net Na$^+$ export from the Avic catchment in Prades over the 12-year period (Table 20.2). The Teula catchment, adjacent to Avic, also presented net Na$^+$ exports during 2 years of record (Piñol 1990).

Calcium net exports were positive, and presented a good correlation with runoff, except for the years with an important contribution of red rains. It is well known that red rains contain high amounts of dissolved calcium (Löye-Pilot et al. 1986; Rodà et al. 1993; Àvila et al. 1998), so that bulk deposition inputs are much higher in years with a number of red rains than in years with negligible red rain. However, years with increased Ca^{2+} inputs due to red rains did not show increased outputs related to this Ca^{2+} availability. In fact, at Montseny it has been clearly observed that net exports were lower in years with red rains compared with other years (Fig. 20.1), implying that the soils effectively retained this extra input, probably increasing their base saturation. At Prades, Ca^{2+} net exports were much higher and the effects of red rains on net outputs were not so evident (Fig. 20.2). Averaging over the whole record, Ca^{2+} presented a clear net export at Prades of 17 kg ha^{-1} year^{-1}, but at Monseny the average net output was only 3–4 kg ha^{-1} year^{-1}. A further consideration is that red rains not only contribute important amounts of dissolved Ca^{2+}, but also transport and deposit a silty–clayish dust rich in calcite (Guerzoni and Chester 1996). The dissolution of this calcite in soils would be another pathway of Ca^{2+} input to the catchments. The amount of dust deposited at Montseny has been estimated as 5 g m^{-2} year^{-1} over an 11-year period (Àvila et al. 1996). However, the calcite content in red rains was highly variable, depending on the source regions providing the dust in North Africa.

By considering the dust amounts from three identified source regions in North Africa (western Sahara, Moroccan Atlas and central Algeria) and their corresponding mean Ca^{2+} concentration, the Ca^{2+} deposited with the African dust was estimated to be 3 kg ha^{-1} year^{-1} (Àvila et al. 1998). If this amount was included as an input in the budgets, then the net Ca^{2+} export would tend to zero at Montseny and would be reduced to around 14 kg ha^{-1} year^{-1} at Prades.

The net export for the main base cations (Ca^{2+}, Mg^{2+} and Na^+) together with HCO_3^- and SiO_2 net outputs indicate active weathering of the bedrock within the catchments (Johnson et al. 1968, 1981). At Prades, Ca^{2+} net export is by far the greatest export, followed by Mg^{2+} and Na^+. At Montseny, the greatest net export is for Na^+, followed by Mg^{2+} and Ca^{2+}. The differences in the release of the different base cations have to be interpreted in view of differences in hydrology (implying the water availability for solute export), bedrock mineralogy and weathering rates at both sites.

20.5 Weathering Rates

Mass balance studies are considered a reliable method for estimating mineral weathering rates in nature (Clayton 1979). The method assumes that the rate of export of base cations from the system indicates the rate at which the bedrock is being weathered. For example, Na^+ has been widely used to obtain weathering estimates from catchment studies (Johnson et al. 1968; Feller 1981; Paces 1985, 1986). This approach assumes that the annual release of Na^+ by weathering equals the net export of Na^+ in the streamwaters plus the small amount of Na^+ annually accumulated in the accreting biomass. Other cations could be used as reference, but Na^+ is not incorporated into secondary minerals or accumulated in the plant biomass to such an extent as K^+, Ca^{2+} or Mg^{2+}. A further assumption is that Na^+ is exported at the same rate as is generated by weathering and that, in consequence, the soil exchange complex is in steady state for Na^+, a reasonable approximation since the pool of exchangeable Na^+ is usually small.

The annual net Na^+ export from our Montseny catchments (11.8 kg ha^{-1} year^{-1}, averaged for TM0 and TM9) plus the amount stored annually in new aboveground biomass (0.06 kg ha^{-1} year^{-1}; Ferrés et al. 1984) produced an annual Na^+ export of 11.9 kg ha^{-1} year^{-1}. The mean bedrock Na_2O content was 0.76% in four samples of phyllites around the experimental catchments (Casas 1979). With these figures, 1500 kg of bedrock ha^{-1} year^{-1} must weather annually to provide for the exported Na^+ in streamwaters and in the growing biomass at the site. For Prades, the weathering rates were estimated at 655 kg ha^{-1} year^{-1} using Na^+ net exports as done at Montseny, and 632 kg ha^{-1} year^{-1} by using K^+ and Mg^{2+} as the relevant elements (Piñol 1990). From these values, weathering seems to proceed at a higher rate at Montseny than at Prades.

However, estimates of rock weathering are subjected to methodological uncertainties concerning the rock composition, bedrock heterogeneity, incongruent dissolution of minerals, element retention in the weathered residuum, and interaction with the soil exchange complex. A more straightforward concept has been recently proposed, the net soil release (Likens and Bormann 1995). Net soil release for cations is the sum of the net outputs of cation equivalents (either in the draining waters or in the growing biomass), and results from the interaction of various processes: atmospheric inputs, mineral weathering, accumulation/depletion of the cation exchange complex, secondary mineral formation, storage in soil organic matter, and plant nutrient uptake (Likens and Bormann 1995). The absolute field rates of some of these processes are difficult to ascertain in catchment studies. From data in Tables 18.6 and 20.2, net soil cation release averaged 2.3 keq ha^{-1} year^{-1} at Montseny (TM9) and 1.8 keq ha^{-1} year^{-1} at Prades (Avic). These fluxes testify to the high cation supply capacity of these silicate soils. Cation accumulation in the aggrading biomass accounted for 47% (Montseny) and 40% (Prades) of the net soil release, the rest being contributed by net outputs in streamwater. Net soil release was not computed for the TM0 catchment, since nutrient retention in biomass is only known for its holm oak forest, which covers 60% of the catchment.

Net cation export in streamwater (also termed cation denudation rate) averaged 1.2 keq ha^{-1} year^{-1} at TM9, 0.93 at TM0 and 1.1 at Avic. Cation denudation rates for 21 temperate and boreal forested catchments ranged from –0.1 to 4.4 keq ha^{-1} year^{-1} with a mean value of 1.03 keq ha^{-1} year^{-1} (Àvila 1988). The figures for Prades and Montseny are close to this mean. Since water is a major vehicle for the export of soil elements, it is surprising that the drier Prades catchment (Avic), whose annual drainage is only 10% of that of both Montseny catchments (Chap. 19), had similar cation denudation rates to the latter. However, at Prades, Ca^{2+} was the main contributor to cation denudation, accounting for 76% of it. This is linked to the higher content of calcium carbonate in the Prades soils (mean carbonate content: 0.10–0.20% in lower and upper slope soils; J. Piñol, unpubl. data) as compared to Montseny (average 0.05% in the permanent plot at the outlet of TM0). The reason for the relatively high carbonate content in the Prades soils, developed on carbonate-free silicate rocks, is intriguing. A possibility is that atmospheric deposition of calcite-containing particles at Prades was more frequent in the past than today (for example, due to an increased transport from North Africa under a drier climate). These deposited carbonates would accumulate in some catchment soils not connected at that time to the drainage contributing area, but which would be presently washed by percolating water as suggested by Piñol (1990).

References

Àvila A (1988) Balanç d'aigua i nutrients en una conca d'alzinar del Montseny. Estudis i Monografies 13, Diputació de Barcelona, Barcelona

Àvila A, Rodà F (1988) Export of dissolved elements in an evergreen-oak forested watershed in the Montseny mountains. Catena Suppl 12:1–11

Àvila A, Piñol J, Rodà F, Neal C (1992) Storm solute behaviour in a montane Mediterranean catchment. J Hydrol 140:143–161

Àvila A, Bonilla D, Rodà F, Piñol J, Neal C (1995) Soilwater chemistry in a holm oak (*Quercus ilex*) forest: inferences on biogeochemical processes for a montane-Mediterranean area. J Hydrol 166:15–35

Àvila A, Queralt I, Gallart F, Martin-Vide J (1996) African dust over northeastern Spain: mineralogy and source regions. In: Guerzoni S, Chester R (eds) The impact of African dust across the Mediterranean. Kluwer, Dordrecht, pp 201–205

Àvila A, Alarcón M, Queralt I (1998) The chemical composition of the dust transported in red rains – its contribution to the biogeochemical cycle of a holm oak forest in Catalonia (Spain). Atmos Environ 32:179–191

Bellot J (1989) Análisis de los flujos de deposición global, trascolación, escorrentía cortical y deposición seca en el encinar mediterráneo de l'Avic (Sierra de Prades, Tarragona). PhD Thesis, University of Alicante, Alicante

Bellot J, Lledó MJ, Piñol J, Escarré A (1992) Hydrochemical budgets, nutrient cycles and hydrological responses in the *Quercus ilex* forest of Prades catchments. In: Teller A, Mathy P, Jeffers JNR (eds) Responses of forest ecosystems to environmental changes. Elsevier, London, pp 389–396

Butler TJ, Likens GE (1995) A direct comparison of throughfall plus stemflow to estimates of dry and total deposition for sulfur and nitrogen. Atmos Environ 29:1253–1265

Casas A (1979) Estudio litogeoquímico del Paleozoico del Montseny. PhD Thesis, University of Barcelona, Barcelona

Clayton JL (1979) Nutrient supply to soil by rock weathering. In: Impact of intensive harvesting on forest nutrient cycling. Environmental science and forestry. State University of New York, Syracuse

Eaton JS, Likens GE, Bormann HF (1978) The input of gaseous and particulate S to a forest ecosystem. Tellus 30:546–551

Erisman JW, Baldocchi D (1994) Modelling dry deposition of SO_2. Tellus 46:159–171

Feller MC (1981) Catchment nutrient budgets and geological weathering in *Eucalyptus regnans* ecosystems in Victoria. Austr J Ecol 6:65–77

Ferrés L, Rodà F, Verdú AMC, Terradas J (1984) Circulación de nutrientes en algunos ecosistemas forestales del Montseny. Mediterr Ser Estud Biol 7:139–166

Fowler D, Duyzer JH, Baldocchi D (1991) Inputs of trace gases, particles and cloud droplets to terrestrial surfaces. Proc R Soc Edinb 97B:35–59

Guerzoni S, Chester R (eds) (1996) Saharan dust across the Mediterranean. Kluwer. Dordrecht

Hultberg H, Grennfelt P (1992) Sulphur and sea salt deposition as reflected by throughfall and runoff chemistry in forested catchments. Environ Pollut 75:215–222

Johnson NM, Likens GE, Bormann FH, Pierce RS (1968) Rate of chemical weathering of silicate minerals in New Hampshire. Geochim Cosmochim Acta 32:531–545

Johnson NM, Driscoll CT, Eaton JS, Likens GE, McDowell WH (1981) Acid rain, dissolved aluminum and chemical weathering at the Hubbard Brook Experimental Forest, New Hampshire. Geochim Cosmochim Acta 45:1421–1437

Juang FHT, Johnson NM (1967) Cycling of chlorine through a forested catchment in New England. J Geophys Res 72:5641–5647

Likens GE, Bormann FH (1995) Biogeochemistry of a forested ecosystem, 2nd edn. Springer, New York

Likens GE, Bormann FH, Pierce RS, Eaton JS, Johnson NM (1977) Biogeochemistry of a forested ecosystem. Springer, New York

Likens GE, Bormann FH, Hedin LO, Driscoll CT, Eaton JS (1990) Dry deposition of sulfur: a 23-year record for the Hubbard Brook Forest Ecosystem. Tellus 42:319–329

Lindberg SE, Garten CT Jr (1988) Sources of sulfur in forest canopy throughfall. Nature 336:148–151

Lindberg SE, Lovett GM (1992) Deposition and canopy interactions of airborne sulfur: results from the Integrated Forest Study. Atmos Environ 26:1477–1492

Lledó MJ (1990) Compartimentos y flujos biogeoquímicos en una cuenca de encinar del Monte Poblet. PhD Thesis, University of Alicante, Alicante

Lovett GM (1994) Atmospheric deposition of nutrients and pollutants in North America: an ecological perpective. Ecol Appl 4:629–650

Löye-Pilot MD, Martin JM, Morelli J (1986) Influence of Saharan dust on the rain acidity and atmospheric input to the Mediterranean. Nature 321:427–428

Matzner E (1986) Deposition/canopy-interactions in two forest ecosystems of northwest Germany. In: Georgii HW (ed) Atmospheric pollutants in forest areas. Reidel, Dordrecht, pp 247–462

Neal C, Àvila A, Rodà F (1995) Long-term modelling of the effects of forest harvest and atmospheric deposition in a sclerophyllous forested catchment in NE Spain. J Hydrol 168:51–71

Paces T (1985) Sources of acidification in central Europe estimated from elemental budgets in small basins. Nature 315:31–36

Paces T (1986) Weathering rates of gneiss and depletion of exchangeable cations in soils under environmental acidification. J Geol Soc Lond 143:673–677

Piñol J (1990) Hidrologia i biogeoquímica de conques forestades de les Muntanyes de Prades. PhD Thesis, University of Barcelona, Barcelona

Piñol J, Àvila A (1992) Streamwater pH, alkalinity and discharge relationships in some Mediterranean catchments. J Hydrol 131:205–225

Piñol J, Àvila A, Rodà F (1992) The seasonal variation of streamwater chemistry in three forested Mediterranean catchments. J Hydrol 140:119–141

Piñol J, Alcañiz JM, Rodà F (1995) Carbon dioxide efflux and pCO_2 in soils of three Quercus ilex montane forests. Biogeochemistry 30:191–215

Rodà F (1983) Biogeoquímica de les aigües de pluja i drenaje en alguns ecosistemes forestals del Montseny. PhD Thesis, Autonomous University of Barcelona, Bellaterra

Rodà F, Àvila A, Bonilla D (1990) Precipitation, throughfall, soil solution and stream chemistry in a holm-oak (Quercus ilex) forest. J Hydrol 116:167–183

Rodà F, Bellot J, Àvila A, Escarré A, Piñol J, Terradas J (1993) Saharan dust and the atmospheric inputs of elements and alkalinity to Mediterranean ecosystems. Water Air Soil Pollut 66:277–288

Velbel MA (1985) Hydrogeochemical constraints on mass balances in forested watersheds of the southern Appalachians. In: Drever JI (ed) The chemistry of weathering. Reidel, Dordrecht, pp 231–247

Vitousek PM, Reiners WA (1975) Ecosystem succession and nutrient retention: a hypothesis. Bioscience 25:376–381

21 Biogeochemical Models

Anna Àvila

21.1 Introduction

Present understanding of terrestrial ecosystems at the catchment scale has enabled models to be developed which reproduce the main hydrological and chemical processes responsible for the dynamics of catchments. Such models have been named biogeochemical models as they seek to link parts of the biosphere and the geosphere through analysis of the chemical circulation of elements.

Besides contributing to the present understanding of the hydrochemical processes operating within terrestrial ecosystems, biogeochemical models have been useful for comparison of a wide range of ecological settings in different catchments and river basins. The specific behaviour of the sites can be interpreted in the model from the relationship of the separate compartments by the established interconnections. Furthermore, biogeochemical models can predict the catchment response under different simulations and they have been instrumental for the prediction of changes in the soil and streamwaters under acidification scenarios related mainly to atmospheric pollution and land use change (Kämäri et al. 1989; Mason 1990).

MAGIC (Model of Acidification of Groundwaters in Catchments) is a biogeochemical model primarily dealing with the effects of acidification. It has been extensively applied in North America and northern Europe (Cosby et al. 1986; Hornberger et al. 1986; Neal et al. 1986; Wright and Cosby 1987; Jenkins et al. 1990). Fewer applications have been undertaken for the Mediterranean area, because of the relatively low acid deposition and high base saturation of most soils. However, an application of MAGIC in a circum-Mediterranean mountain site with catchments covered by beech and spruce predicted risks of acidification resulting from biomass growth and atmospheric scavenging (Durand et al. 1992).

Here, I describe the application of MAGIC to the holm oak (*Quercus ilex* L.) forested catchments of Montseny and Prades. At Montseny this exercise was undertaken with the aim of exploring the effects of forest management

Ecological Studies, Vol. 137
Ferran Rodà et al. (eds) Ecology of Mediterranean
Evergreen Oak Forests
© Springer-Verlag, Berlin Heidelberg 1999

practices and varying loads of atmospheric deposition (Neal et al. 1995) and the influence of climatic change (Àvila et al. 1996) on the soil status and streamwater chemistry. The application of MAGIC in the drier Prades catchments – an environment which is quite distinct from the cold-temperate mid-latitude site environments usually modelled – provided a test of the model's wider applicability (Bellot et al. 1994, 1995).

21.2 Description of MAGIC

MAGIC, developed by J. Cosby and collaborators at the University of Virginia, is one of the more widely used models for long-term prediction of catchment response under different environmental scenarios (Cosby et al. 1985a, b). MAGIC consists of two main sets of equations:

1. Equilibria equations between the soil and the soil solution in which the chemical composition of the soil solution is assumed to be controlled by soil reactions involving sulphate adsorption, cation exchange, dissolution and precipitation of aluminium and dissolution of inorganic carbon.
2. Mass balance equations of the main fluxes to and from the soil and surface waters (Fig. 21.1).

The model's requirements include information regarding key soil characteristics, e.g. the bulk density, porosity, depth, exchange capacity, and the rate of exchange for the basic cations. Mean values for the whole catchment are introduced in the model in spite of the known heterogeneity of soil variables (Neal 1992). The model also requires information on the input fluxes in atmospheric deposition and weathering of the bedrock, and on the output fluxes in streamwater runoff (Fig. 21.1). Aggrading vegetation acts as a nutrient sink: forests in a growing phase present net uptake fluxes. This flux is attenu-

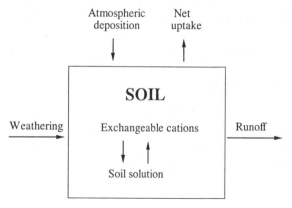

Fig. 21.1. Soil element pools and main element fluxes to and from the soil considered in MAGIC

ated at forest maturity because element retention tends to approach element mineralization. Atmospheric input fluxes and uptake and runoff output fluxes are introduced in the model based on the information of biogeochemical studies at the catchments of interest. The weathering rates are adjusted through an iterative process of approximation until an acceptable fit is attained between model simulated versus observed streamwater chemistry.

Once the data set containing these parameters has been adjusted, the model solves the chemistry of the streamwater from the system of equilibrium equations coupled with the mass balance. The model's time step is 1 year. In the following simulations the runs extended from 140 years ago up to the present day and forecasted a variety of scenarios into the next 140 years. Based on data from the period 1983–1988 for Montseny and 1981–1988 for Prades, the model has been calibrated at both sites for year 1985, which has been taken as the present year.

21.3 Application of MAGIC to Montseny

In Montseny, the TM9 catchment at La Castanya (Chap. 2) was used for calibration of MAGIC. This catchment had been monitored for several years since 1983, with a weekly sampling schedule of bulk deposition, soil solution and streamwater, while information on soil characteristics was readily available (Chaps. 2 and 17). The two-box version of the model was used in an attempt to simulate the water sources contributing to the stream. One box represented the groundwater end-member which generated the baseflow and the other box represented the soilwater end-member. The relative proportions of the groundwater and soilwater end-members were determined by simple mixing relationships between the average composition of three conservative elements (Na^+, SO_4^{2-} and Cl^-) in the streamwater, soilwater and groundwater (Àvila et al. 1995).

The model was calibrated with: (1) fixed parameters: soil characteristics (soil depth, porosity, bulk density, sulphate adsorption), soil chemistry (exchangeable cations, base saturation), hydrological characteristics (mean precipitation, mean runoff), biological uptake, rain chemistry and dry deposition coefficients, and (2) adjustable variables: the weathering rates. Precipitation and throughfall chemistry at Montseny from Rodà et al. (1990) provided estimates of the inputs from the atmosphere (Table 21.1). Precipitation chemistry for the initial year of 1845 was calculated by subtraction from the current precipitation chemistry of the pollutant increment assumed for the industrial period. The holm oak forest at La Castanya was heavily coppiced for charcoal production until the 1950s and currently is accreting biomass. Its net nutrient uptake was assessed from the net tree production at La Castanya (boles and branches > 3–4 cm in diameter) multiplied by the mean nutrient concentrations in these plant fractions, with data from Ferrés

Table 21.1. Atmospheric input concentrations in wet deposition (μeq L^{-1}) and multiplicative coefficients[a] for total deposition applied to Montseny and Prades

	Montseny			Prades		
	Wet deposition		Total deposition coefficient	Wet deposition		Total deposition coefficient
Variable	1845	1985		1845	1985	
Ca^{2+}	30.0	60.0	1.5	47.4	47.4	1.6
Mg^{2+}	8.0	11.0	1.5	8.2	8.2	1.3
Na^+	24.0	24.0	1.5	20.4	20.4	1.3
K^+	0.5	4.0	1.5	5.9	5.9	1.3
NH_4^+	2.0	25.0	1.5	0.0	25.5	1.1
NO_3^-	2.0	20.0	1.5	0.0	15.9	1.1
SO_4^{2-}	10.0	51.0	1.9	21.1	54.2	1.2
Cl^-	27.8	29.0	1.8	16.0	16.0	1.3
Alkalinity	24.9	24.2	–	43.1	19.6	–

[a] Equivalent concentrations in total atmospheric deposition were obtained by multiplying concentrations in wet deposition by the given coefficient.

Table 21.2. Parameters used in the calibration procedure for TM9 (Montseny) and Avic (Prades) catchments

Variable	TM9		Avic
Catchment area (ha)	5.9		51.6
Catchment slope (°)	36		26
Precipitation (mm $year^{-1}$)	884		548
Runoff (mm $year^{-1}$)	305		45
Air temperature (°C)	9		10
Stream CO_2 partial pressure (atm)	0.0003		0.0035
Soils	Layer 1	Layer 2	
Depth (m)	0.4	1.0	0.76
Porosity (frac)	0.55	0.2	0.47
Bulk density (kg m^{-3})	1200	1460	1400
Cation exchange capacity (meq kg^{-1})	100	1	–
Base saturation (%)	56	50	82
Exchangeable Ca (%)	45	30	59
Exchangeable Mg (%)	5	10	5
Exchangeable Na (%)	3	5	8
Exchangeable K (%)	3	5	10
SO_4 ads. max. capacity (meq kg^{-1})	7	2	5
SO_4 ads. half saturation (meq m^{-3})	800	2	800
Weathering rates (meq m^{-2} $year^{-1}$)			
Ca	71	45	121
Mg	30	32	20
Na	13	71	6
K	11	1.5	2.8
Net uptake (meq m^{-2} $year^{-1}$)			
Ca	0	80	46.5
Mg	0	10	5.3
K	0	8	17.9

Table 21.3. Observed and modelled fits (in brackets) for the soilwater end-member, groundwater end-member and streamwater in MAGIC calibration for TM9 (Montseny). Rainfall composition is also shown. Ion concentrations in µeq L^{-1}

Solute	Rainfall	Soilwater		Groundwater		Streamwater	
Ca^{2+}	60.0	258	(256)	467	(463)	256	(279)
Mg^{2+}	11.0	150	(153)	302	(305)	162	(177)
Na$^+$	24.0	159	(160)	490	(493)	275	(262)
K$^+$	4.0	13	(12)	17	(16)	11.5	(10.5)
NH$_4^+$	25.0	0	(0)	0	(0)	0	(0)
SO$_4^{2-}$	51.0	288	(283)	284	(283)	202	(208)
Cl$^-$	29.0	151	(151)	151	(151)	109	(111)
NO$_3^-$	20.0	4	(4)	2	(2)	1	(2)
pH	6.68	5.5	(6.3)	7.7	(7.4)	–	
Alkalinity	24.2	101	(108)	808	(831)	392	(407)

(1984). The net uptake for this forest was (in meq m^{-2} year^{-1}) 80 for Ca^{2+}, 8 for Mg^{2+} and 10 for K$^+$. The soil parameters used in the calibration procedure are shown in Table 21.2.

The initial values in year 1845 for the base saturation and weathering rates were fitted so that, when run forward 140 years into the present, the best fit to the present-day observed soil chemistry and runoff chemistry values was obtained (Table 21.3). The calibrated model was then used to predict soil and water changes over the next 140 years for various scenarios of atmospheric deposition, biomass uptake and climatic variation.

21.3.1 Forest Management and Atmospheric Deposition Scenarios

Forest growth enhances soil and water acidification through the uptake of base cations from the exchange sites in the soil and their allocation into the net biomass increment. Tree harvest is thus a factor of acidification when the rate of replacement of base cations in the soil is lower than the rate of export in harvest. Here, three management scenarios were defined to simulate the current forestry practices in the Montseny holm oak forests: (1) continuous and constant tree harvesting equal to the annual net forest production: the forest biomass was maintained at a steady state; (2) cycles of tree felling and recovery, consisting of a harvest of 33% of tree biomass every 41 years; and (3) the previous harvest cycles discontinued from 1985 into the future. These forestry practices were modelled as changes in the uptake sequence.

The increased atmospheric deposition of sulphur is responsible for the historical acidification of soils and surface waters in Europe and North America. In the Mediterranean region, most SO$_4^{2-}$ is matched by base cations (Rodà et al. 1993) and the precipitation is alkaline (Table 21.1). However, at Montseny, the SO$_2$ dry deposition steeply reduces the alkalinity (to 7 µeq L^{-1}) in the total (wet plus dry) deposition flux.

Three scenarios of atmospheric deposition were envisaged. In scenario a, deposition was kept at background (pre-industrial) levels throughout the simulation. In scenario b, deposition matched the evolution of SO_2 emissions in Europe over the past 140 years until 1985 followed by a 60% reduction over the next 20 years. In scenario c, deposition was similar to b until the present day, but constant at the 1985 levels thereafter. All ions in atmospheric deposition were considered to vary in parallel to SO_2, except for Na^+, which remained at the pre-industrial level. Dry deposition was considered constant in all forest management scenarios, increasing the wet deposition by a factor of 1.8 for chloride, 1.9 for sulphate and 1.5 for the rest of the ions (Table 21.1).

The evolution of two variables (soil base saturation and streamwater alkalinity) which represent the general response of soil and streamwater chemistry to the different scenarios is shown in Figs. 21.2 and 21.3. The effect of atmospheric deposition was similar in all forestry simulations: there was a decrease in soil base saturation and streamwater alkalinity in the simulations of the historical buildup of sulphur (scenarios b and c) with respect to the pre-industrial scenario (scenario a), being most marked in the case where deposition remained at 1985 levels (scenario c). These downward trends resulting from atmospheric pollution were similar for the forestry scenarios of continuous harvest and periodic harvest (Figs. 21.2a, b and 21.3a, b). However, the simulation of periodical thinning/harvest showed a superimposed sinusoidal pattern. The pronounced dip in the streamwater alkalinity (Fig. 21.3b, c) reflected the dilution of streamwater due to reduced evapotranspiration after tree felling. The sinusoidal waves indicated the cycles of replenishment and depletion of soil base cations during phases of low nutrient uptake and higher uptake according to the phase in the management cycle. This pattern also occurred in all cation concentrations, but will not be shown here. Net uptake was set to zero at the beginning of regrowth after forest felling because the reduced biomass could only account for small nutrient uptake rates which were supplied by slash decomposition. The phase of highest uptake was between 20 and 32 years after felling, when the soil base saturation decreased because the base cation inputs in atmospheric deposition and weathering were not high enough to compensate for the output in vegetation uptake and runoff. From 32 to 41 years after felling, net uptake decreased linearly because of lower requirements in a phase of forest maturity: soil base saturation increased. At year 41, the forest was harvested again. If the thinning/harvest practice stopped, base cations accumulated in the soil as a result of

→

Fig. 21.2. Soil base cation saturation over time for the Montseny catchment, under different forestry practices (a–c) and the following atmospheric deposition scenarios: a continued preindustrial deposition; b and c increasing atmospheric deposition over the last 140 years in line with trends in European sulphur emissions up to 1985. For b 1985 levels are reduced by 60% over the next 20 years and kept constant thereafter; for c, there is no reduction: 1985 levels are maintained

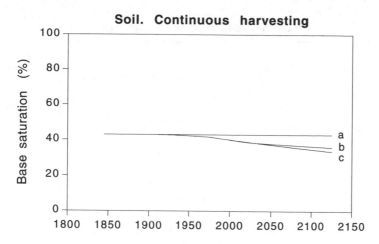

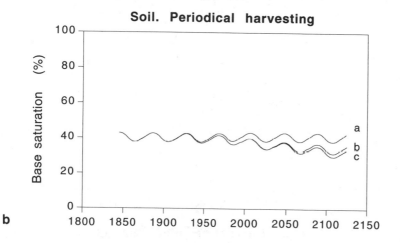

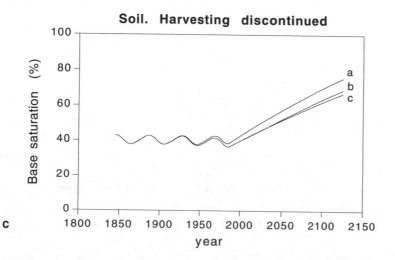

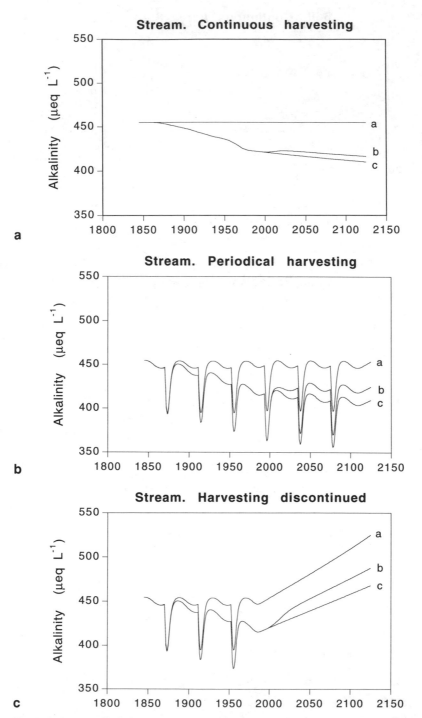

Fig. 21.3. Stream alkalinity over time for the Montseny catchment, under different forestry practices (a–c) and the same three atmospheric deposition scenarios (*a–c*) defined in Fig. 21.2

the continuous production of cations by weathering and from the inputs in precipitation, and the lack of a sink in the harvested biomass. This eventually resulted in an increase in stream alkalinity (Fig. 21.3c).

In short, atmospheric deposition led to a moderate decrease in base cations and alkalinity and had a stronger acidification effect than that caused by a periodical forest management consisting of a mild harvest (removal of 33% of the stand biomass). The combined impact of forest management and atmospheric deposition on streamwater quality was very small, unlike results observed and modelled for more acidic systems in northern Europe, where forest growth alone or in combination with atmospheric deposition had a significant acidifying effect on soils and streamwaters (Jenkins et al. 1990; Jenkins and Wright 1992).

21.3.2 Climate Change Scenarios

MAGIC was also used in Montseny to simulate the effects of climatic change on the streamwater chemistry. To model climatic change, a temperature increase of 4 °C was considered, but, because of greater uncertainty regarding future precipitation (Giorgi et al. 1992), two possible scenarios were devised: a 10% precipitation increase and a 10% precipitation decrease, both with a 4 °C temperature increase. With these changed conditions of temperature and precipitation, a different ratio between precipitation and runoff is to be expected, affecting the streamwater chemistry through simple dilution or concentration mechanisms – the hydrological effect. However, in addition, these climatic changes will lead to the modification of certain soil physical variables and may well affect the weathering rates, which are highly dependent on soil temperature and soil water content (Sverdrup and Warfvinge 1993).

Here, the changes in streamwater chemistry were analyzed for hydrological effects alone and in combination with changes in the weathering rates. The simulations were run with the MAGIC calibration in Montseny of steady state for forest biomass and atmospheric deposition following historical sulphur emissions in Europe until 1985 and constant thereafter (Fig. 21.3a, scenario c).

The model's predictions for the streamwater Ca^{2+} concentrations and alkalinity are shown in Figs. 21.4 and 21.5. The three scenarios are: b, the baseline scenario, with present-day temperature and precipitation; h, the scenario accounting solely for the effects of the hydrological partition; and w, the scenario accounting for both the hydrological effects and the changes in weathering rates. The changes in the weathering rates as a consequence of changes in the climatic conditions were obtained by applying PROFILE, a model which calculates weathering rates at field conditions based on the rock and soil mineralogy (Warfvinge and Sverdrup 1992; Sverdrup and Warfvinge 1993).

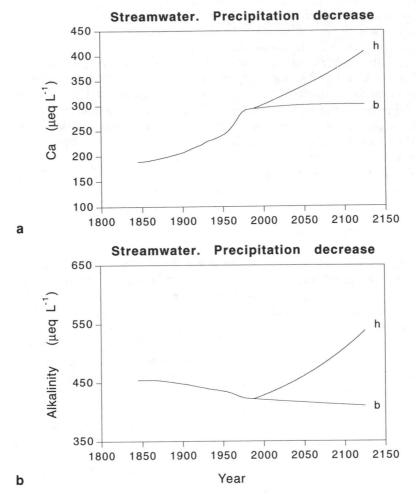

Fig. 21.4. a Ca^{2+} and **b** Alkalinity concentrations in streamwaters of the Montseny catchment over time, in a simulation of a 4 °C temperature increase and 10% precipitation decrease. *b* Represents baseline scenario which accounts for increased atmospheric deposition over the past 140 years in line with trends in European sulphur emissions up to 1985, and constant at 1985 levels thereafter. *h* Represents change upon baseline scenario owing to changed hydrology with climate change

With precipitation decrease and temperature increase, runoff was estimated to fall from 305 to 211 mm year^{-1} (Piñol et al. 1995), thus concentrating solutes in the streamwater (Fig. 21.4). PROFILE revealed no significant changes in the weathering rates since increased weathering related to the temperature increase was virtually cancelled out by a decrease of weathering due to reduced soil moisture.

With precipitation increase and temperature increase, runoff was estimated to increase from 305 to 320 mm year^{-1} (Piñol et al. 1995), thus moder-

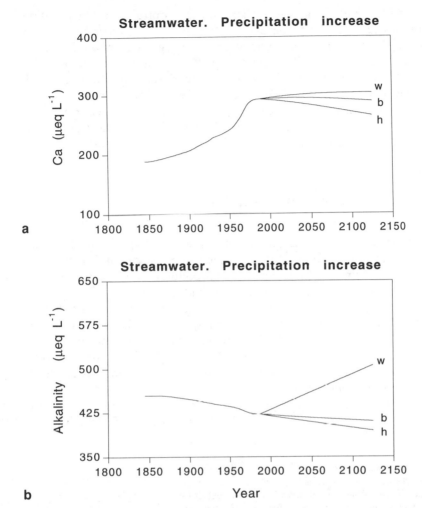

Fig. 21.5. a Ca^{2+} and **b** Alkalinity concentrations in the streamwaters of the Montseny catchment over time, in a simulation of a 4 °C temperature increase and 10% precipitation increase. *b* Represents baseline scenario which accounts for increased atmospheric deposition over the past 140 years in line with trends in European sulphur emissions up to 1985, and constant at 1985 levels thereafter. *h* Represents change upon baseline scenario owing to changed hydrology with climate change. *w* Represents change upon baseline scenario owing to changed hydrology and weathering rates with climate change

ately diluting streamwater concentrations (Fig. 21.5, scenario h). However, when the climatic effects on weathering rates were also considered (scenario w), higher streamwater cation concentrations (here represented by calcium) and alkalinity were predicted by PROFILE since weathering rates increased in relation to higher temperature and higher soil water content. These simulations not only showed the low risk of streamwater acidification for this catchment with climatic change, but also, on the contrary, that base cation concentrations and alkalinity tended to increase. In Mediterranean areas drier than the TM9 catchment, but having similar vegetation and lithology, it is relatively common for calcite to be precipitated along the riverbank under certain conditions, even in catchments entirely underlain by silicate bedrock such as Montseny and Prades (Piñol and Àvila 1992). The increased concentrations simulated for a warmer scenario, with either more or less rain, were not high enough to either pose serious problems for the water quality or reach the calcite saturation equilibria for the Montseny catchment.

21.4 Application of MAGIC to Prades

MAGIC was applied to the Avic catchment of Prades (Chap. 2) with the principal aim of testing whether MAGIC, created for catchments in northern latitudes with a regime of high rainfall, high pollutant loads, acid rain, and soil and water acidification, was able to reproduce the hydrochemical behaviour for Prades, a semi-arid environment, receiving non-acidic deposition, with high soil base saturation and with highly mineralized streamwaters (Bellot et al. 1994). While no acidification was expected for the Avic catchment, the interest was on the calibration procedure itself and the reproduction of historical sequences accounting for the changes in atmospheric deposition and forestry practices between 1845 and 1985.

The parameters used in the calibration are shown in Table 21.2, and the fitted streamwater chemistry is shown in Table 21.4. Data for atmospheric

Table 21.4. Observed (7-year means) and modelled fits for streamwater chemistry at Avic (Prades). Ion concentrations in $\mu eq\ L^{-1}$. The difference between observation and model fit for Ca^{2+} is justified by the fact that some calcite precipitates when water passes from soil to stream

	Observed	Model fit
Ca^{2+}	2574	3569
Mg^{2+}	564	564
Na^+	509	509
K^+	44.4	44.7
NH_4^+	10	10
SO_4^{2-}	658	657
Cl^-	281	282
NO_3^-	1.6	1.6
pH	7.59	7.7
Alkalinity	3745	3746
SBC	3872 .	4687
SAA	953	953

SBC, Sum of base cations; SAA, sum of strong acid anions.

input and vegetation uptake were obtained from Bellot (1989) and Lledó (1990) respectively. Net uptake at this site was calculated to be (in meq m^{-2} year^{-1}) 47 for Ca^{2+}, 5 for Mg^{2+} and 18 for K$^+$. The main calibration problem was in fitting the streamwater Ca^{2+} concentrations because some calcite precipitates along the stream when the water flows from the soil to the channel (with decreasing pCO$_2$) in an environment of very high alkalinity and Ca^{2+} concentrations (Piñol and Àvila 1992). The simulated value of 3569 µeq Ca^{2+} L^{-1} was proposed to account for the calcium in dissolved and precipitated forms (Bellot et al. 1994). A very good fit was obtained for the rest of the variables (Table 21.4).

21.4.1 Historical Sequences

The wet deposition sequence for SO$_4$$^{2-}$, also used for NO$_3$$^-$ and NH$_4$$^+$, was based on the evolution of sulphur emissions in Europe. Note, however, that increasing acid anions in the rainwater from 1845 to 1985 did not result in acidic rain: rainfall alkalinity was 19.6 µeq L^{-1} in 1985, corresponding to an air-equilibrated pH of 6.6 (Table 21.1).

Typical forest use in the Prades mountains between 1845 and 1940 was moderate felling for charcoal production . However, between 1940 and 1950 the forest was totally clear-cut except for a few seed-producing trees. Afterwards the forest in the Prades catchment was protected, producing a dense forest of resprouts (9000 stems ha^{-1} in the tree plus shrub layers). These vegetation changes were simulated in MAGIC with an uptake sequence varying from 3% of present-day uptake between 1845 and 1940, 200% of present-day uptake between 1940 and 1950, and a linear decrease from 200 to 100% between 1950 and 1985.

The results for the historical changes of the base saturation in the soil indicated that from 1845 to 1985 tree felling reduced the base saturation from 84 to 82%, a very small change without acidification consequences. No changes in response to the atmospheric deposition were found.

21.4.2 Forest Management and Atmospheric Deposition Scenarios

MAGIC was also applied to the Avic catchment for different scenarios of forest management and atmospheric deposition. Five scenarios combining different intensities of atmospheric inputs and forest management were devised, in an attempt to forecast the soil and streamwater response to pollution and forestry (Bellot et al. 1995). Scenario 1 was a projection into the future of present-day atmospheric input and forest growth. Scenario 2 considered present-day atmospheric inputs and forest clear-cuts every 60 years. Scenario 3 considered an increasing trend of atmospheric inputs reaching a 30% increment by the year 2045, with the same clear-cuts as in scenario 2. Scenario 4 considered a reduction of atmospheric inputs down by 30% by the year 2045, as well as forest clear-cuts every 60 years. Finally, scenario 5 considered a 30% increase in atmospheric input without forest clear-cutting.

All simulations predicted a decrease in the soil base saturation: the input of base cations in atmospheric deposition was not enough to compensate for the depletion of the soil base cations due to forest growth in scenarios either with or without forest clear-cut. When forest felling was simulated, the decrease in base saturation was steeper, an indication that plant regrowth was consuming base cations at a faster rate than when the forest was left undisturbed. However, even with forest felling the base saturation in year 2045 was estimated to be around 75%, a value too high to pose problems of acidification.

Streamwater response under the different future scenarios is not shown here: until now, no good results were obtained at Prades because of the need to modify the model in order to account for the precipitation–dissolution of calcite. The calcite precipitated in the streambank in conditions of high alkalinity and Ca^{2+} concentration and decreasing pCO_2 may well dissolve when base cation concentrations and alkalinity decrease steeply in scenarios of pollution or forest management. Therefore, the calcite solubility product has to be introduced in the model for it to be able to reproduce accurately the hydrochemical dynamics of the Prades streamwaters.

21.5 Concluding Remarks

The application of MAGIC to Montseny and Prades has shown that high atmospheric pollutant sulphur loadings have not resulted in significant changes in the quality of the streamwater at both sites. At Montseny, atmospheric deposition was responsible of a slight decrease in alkalinity and base cations. Forest management produced a wave-like pattern in streamwater chemistry, reflecting the phase in the forestry cycle, with no consistent trend toward acidification. This simulation confirmed the sustainability of the actual regime of forest harvest currently practised at Montseny.

At Prades, MAGIC was successfully fitted to a catchment of very high soil base saturation and very mineralized streamwaters. However, the solubility product of calcite needs to be included in the model to account for the calcite precipitation/dissolution in these highly mineralized streamwaters. Historical and future changes in soil base saturation showed that vegetation uptake was the main process of depletion of the base saturation for the Prades soils.

References

Àvila A, Bonilla D, Rodà F, Piñol J, Neal C (1995) Soilwater chemistry in a holm oak (*Quercus ilex*) forest: inferences on biogeochemical processes for a montane Mediterranean area. J Hydrol 166:15–35

Àvila A, Neal C, Terradas J (1996) Climate change implications for streamflow and streamwater chemistry in a Mediterranean catchment. J Hydrol 177:99–116

Bellot J (1989) Análisis de los flujos de deposición global, trascolación, escorrentía cortical y deposición seca en el encinar mediterráneo de l'Avic (Sierra de Prades, Tarragona). PhD Thesis, University of Alicante, Alicante

Bellot J, Melià N, Tello E (1994) Calibración y aplicación del modelo MAGIC a las cuencas de encinar mediterráneo de Prades (Tarragona). Stud Oecol 10/11:51–61

Bellot J, Escarré A, Neal C, Wright RF (1995) Effects of atmospheric inputs and forest felling on soil chemistry: MAGIC forecast relating to the Avic catchment in Prades (Spain). In: Jenkins A, Ferrier RC, Kirby C (eds) Ecosystem manipulation experiments. Commission of the European Communities, Brussels, pp 366–369

Cosby BJ, Hornberger GM, Galloway JN, Wright RF (1985a) Modelling the effects of acid deposition: assessment of a lumped parameter model of soil water and streamwater chemistry. Water Resour Res 21:51–63

Cosby BJ, Wright RF, Hornberger GM, Galloway JN (1985b) Modelling the effects of acid deposition: estimation of long-term water quality responses in a small forested catchment. Water Resour Res 21:1591–1601

Cosby BJ, Hornberger GM, Rastetter EB, Galloway JN, Wright RF (1986) Estimating catchment water quality response to acid deposition using mathematical models of soil ion exchange processes. Geoderma 38:77–95

Durand P, Neal C, Lelong F, Didon-Lescot JF (1992) Effects of land-use and atmospheric input on stream and soil chemistry: field results and long-term simulation at Mont Lozère (Cevennes National Park, southern France). Sci Total Environ 119:191–209

Ferrés L (1984) Biomasa, producción y mineralomasas del encinar montano de La Castanya (Montseny, Barcelona). PhD Thesis, Autonomous University of Barcelona, Bellaterra

Giorgi F, Marinucci MR, Visconti G (1992) A 2xCO$_2$ climate change scenario over Europe generated using a limited area model nested in a general circulation model. 2. Climate change scenario. J Geophys Res 97:10011–10028

Hornberger GM, Cosby BJ, Galloway JN (1986) Modelling the effects of acid deposition: control of long-term sulfate dynamics by soil sulfate adsorption. Water Resour Res 21:1591–1601

Jenkins A, Wright RF (1992) ENCORE: application of the MAGIC model to catchments in Norway and the UK. Ecosystem Research Report. Commission of the European Communities, Brussels

Jenkins A, Cosby JB, Ferrier RC, Walker TAB, Miller JD (1990) Modelling stream acidification in afforested catchments: an assessment of the relative effect of acid deposition and afforestation. J Hydrol 120:163–181

Kämäri J, Brakke DF, Jenkins A, Norton SA, Wright RF (1989) Regional acidification models. Springer, Berlin

Lledó MJ (1990) Compartimentos y flujos biogeoquímicos en una cuenca de encinar del Monte Poblet. PhD Thesis, University of Alicante, Alicante

Mason BJ (ed) (1990) Surface Waters Acidification Programme. Cambridge University Press, Cambridge

Neal C (1992) Describing anthropogenic impacts on stream water quality: the problem of integrating soil water chemistry variability. Sci Total Environ 115:207–218

Neal C, Whitehead PG, Neale R, Cosby BJ (1986) Modelling the effects of acidic deposition and conifer afforestation on stream acidity in the British uplands. J Hydrol 86:15–26

Neal C, Àvila A, Rodà F (1995) Modelling the long term impact of atmospheric pollution deposition and repeated forestry cycles on stream water chemistry for a holm oak forest in northeastern Spain. J Hydrol 168:51–71

Piñol J, Àvila A (1992) Stream water pH, alkalinity, pCO₂ and discharge relationships in some forested Mediterranean catchments. J Hydrol 131:205–225

Piñol J, Terradas J, Àvila A, Rodà F (1995) Using catchments of contrasting hydrological conditions to explore climate change effects on water and nutrient flows in Mediterranean forests. In: Moreno JM, Oechel WC (eds) Global change and Mediterranean-type ecosystems. Springer, Berlin, pp 251–264

Rodà F, Àvila A, Bonilla D (1990) Precipitation, throughfall, soil solution and streamwater chemistry in a holm-oak (*Quercus ilex*) forest. J Hydrol 116:167–183

Rodà F, Bellot J, Àvila A, Escarré A, Piñol J, Terradas J (1993) Saharan dust and the atmospheric inputs of elements and alkalinity to Mediterranean ecosystems. Water Air Soil Pollut 66: 277–288

Sverdrup H, Warfvinge P (1993) Calculating field weathering rates using a mechanistic geochemical model PROFILE. Appl Geochem 8:273–283

Warfvinge P, Sverdrup H (1992) Calculating critical loads of acid deposition with PROFILE – a steady-state soil chemistry model. Water Air Soil Pollut 63:119–143

Wright R, Cosby BJ (1987) Use of a process-oriented model to predict acidification at manipulated catchments in Norway. Atmos Environ 21:727–730

Part 6   Responses to Disturbances

22 Soil Fertility After Fire and Clear-Cutting

Isabel Serrasolses and V. Ramon Vallejo

22.1 Introduction

Holm oak (*Quercus ilex* L.) forests have been disturbed for centuries by coppicing for firewood and charcoal production, grazing and wildfires. Though at present many holm oak forests of southern Europe remain unexploited, most holm oak forests in Catalonia (NE Spain) are managed for firewood through selection thinning (Chaps. 3 and 5). Substantial amounts of slash are left to decompose on site after each cutting cycle.

Slash management can have a considerable impact on forest productivity (Fahey et al. 1991). Slash removal gives rise to a net nutrient loss whereas laying slash down protects the soil from erosion and incorporates nutrients into the soil through decomposition. Slash-burning produces a loss of C and certain nutrients, especially N and S, depending on the intensity of the fire (Raison et al. 1985b; Gillon and Rapp 1989). In spite of this nutrient loss, the post-fire soil is enriched by ashes, thereby increasing nutrient availability (Woodmansee and Wallach 1981; Khanna et al. 1994). The question remains though as to what extent the losses produced by volatilization and convection, and eventually by ash leaching or washout, are reversible and affect forest regrowth either in the short or long term.

Clear-cutting and slash-burning treatments were conducted in the holm oak coppices of the Prades experimental area: (1) to analyze the short-term forest response in the case of an intense wildfire; and (2) to assay the effects of two techniques for the management of forest slash: slash-burning and laying down the slash to decompose on the forest floor.

22.2 Experimental Design

The experiment was conducted on the SE-facing slope of Torners (Chap. 2), at an altitude of 900 m a.s.l. Three 40 × 20-m plots were used, one of them being kept undisturbed as a control (Serrasolsas 1994). The trees of the other

Ecological Studies, Vol. 137
Ferran Rodà et al. (eds) Ecology of Mediterranean
Evergreen Oak Forests
© Springer-Verlag, Berlin Heidelberg 1999

two plots were cut in August 1988, and the logs and branches with diameter greater than 2 cm removed. The slash was homogeneously distributed over the plots, though the plot to be burnt received supplemental slash in order to simulate an intense surface fire. Cut plants resprouted after a month. One of the clearcut plots was burned (October 1988) and the other was kept as clearcut treatment. After the fire, resprouting occurred in January 1989.

The slash left on the plots amounted to 51 Mg ha^{-1} in the clearcut plot and 74 Mg ha^{-1} in the burnt plot, corresponding to 50 and 70% of the above-ground forest biomass, respectively. The experimental fire consumed all the slash and part of the forest floor. L and F horizons reached similar tempera-tures at all points of the plot, greater than 400 °C, with a maximum of 750 °C. The H horizon reached temperatures greater than 215 °C in 70% of the points measured, with a maximum of 370 °C. At 2.5-cm depth, heterogeneity increased. The maximum temperature was 240 °C and only 40% of the cases were greater than 110 °C. At 7.5-cm soil depth, the maximum temperature re-corded was 60 °C. According to the fire front advance power recorded, i.e. 8400 cal cm^{-1} s^{-1}, the fire was of moderate–high intensity (Trabaud 1979).

22.3 Direct Effects of the Fire

22.3.1 Nutrient Losses

The fire produced the loss of 79 Mg dry matter ha^{-1}, that is 70% of the dry weight of slash and forest floor (Table 22.1). Known quantities of the differ-ent forest floor layers placed separately on trays located in the plot prior to burning (as described by Raison et al. 1985a) experienced losses for L, F and H layers of, respectively, 70, 45 and 29% of their initial weight.

The fire produced high losses of N (64% of the pre-burn N contained in the slash plus forest floor), S (52%) and K (48%), intermediate losses of P (37%) and Na (33%), and very low or undetectable losses of Mg and Ca (Table 22.1). Similar results were obtained by Raison et al. (1985a,b). Calcium showed a relative accumulation with respect to the rest of nutrients. This phenomenon might have been enhanced by the history of charcoal produc-tion in these forests.

The total N loss was around 800 kg N ha^{-1}, which is comparable to high intensity fires (Grier 1975) or to slash-burning in tropical or temperate oce-anic forests (Kauffman et al. 1992). The overall N loss was proportional to the weight loss of the total fuel, corresponding to 10.2 kg N per Mg of fuel burned. For each forest floor layer, N losses were 83% in L, 74% in F and 44% in H.

Table 22.1. Weight and nutrient amounts of forest floor and slash on the experimental plots before burning (August 1988) and after burning (November 1988). n = 20 for forest floor and n = 10 for slash

Fraction		Weight	C	N	P	S	Ca	Mg	K	Na
						Before burning (kg ha^{-1})				
Slash										
Leaves		12309	6191	160.0	12.3	13.5	118.2	12.3	89.9	2.7
Stems		61805	30161	556.2	80.3	74.2	908.5	74.2	451.2	6.2
Total	Mean	74114	36352	716.3	92.7	87.7	1026.7	86.5	541.0	8.9
	SD	(11730)	(5724)	(105.6)	(15.2)	(14.1)	(172.4)	(14.1)	(85.6)	(1.2)
Forest floor										
L		4500	2173	50.5	3.5	5.0	70.4	5.4	13.6	1.0
F		9820	4384	109.1	10.9	10.8	145.6	21.5	41.0	3.6
H		24960	5776	371.3	27.0	35.4	488.9	127.5	89.7	7.0
Total	Mean	39150	12333	530.9	41.4	51.2	704.9	154.5	144.3	11.6
	SD	(16530)	(4129)	(246.0)	(17.4)	(23.1)	(328.3)	(81.4)	(58.8)	(4.6)
Total before burning										
Total	Mean	113264	48685	1247.1	134.1	138.9	1731.6	240.9	685.3	20.4
	SD	(20269)	(7058)	(267.7)	(23.1)	(27.0)	(370.9)	(82.6)	(103.9)	(4.8)
						After burning (kg ha^{-1})				
Forest floor										
Ashbed		15825	4448	175.0	57.7	35.6	1219.6	101.5	258.5	8.1
H		18640	4397	269.4	26.5	31.0	603.5	97.2	94.6	5.5
Total	Mean	34449	8845	444.4	84.2	66.5	1823.2	198.8	353.0	13.6
	SD	(23086)	(4786)	(285.6)	(35.8)	(34.8)	(840.7)	(106.9)	(143.8)	(6.5)
						Losses during the fire (kg ha^{-1})				
Total	Mean	78815	39840	802.7	49.9	72.4	-91.6	42.2	332.3	6.8
	SD	(30721)	(8527)	(391.4)	(42.6)	(44.0)	(918.8)	(135.1)	(177.4)	(8.1)
Percentage[a]		70	82	64	37	52	-5	18	48	33

[a] Element loss as percentage of pre-fire amount.

22.3.2 The Ashbed Layer After the Fire

C and N concentrations in the ashbed layer were lower than those before burning, but only slightly lower in the case of N. N losses are proportional to the quantity of fuel consumed (Raison et al. 1985a; Little and Ohmann 1988). The concentrations of the other nutrients studied increased in the ashbed as is generally reported. The increases in Ca, K, Mg and P were between 3.5 and 5.8 times the concentrations in the original L+F layer. In the H layer only P, S, Ca and K increased a little.

The pH measured 5 days after the fire was 9.5 in the ashbed and 7.3 in the remaining H layer. The latter had a value of 6.4 before the disturbance. Electrical conductivity (EC) in the H layer showed a two-fold increase due to the ash input. These data follow the common trends found in forest fires (Woodmansee and Wallach 1981; Binkley 1986; Ferran et al. 1992).

22.4 Changes in Forest Floor Weight

The three plots did not show significant differences in their forest floor weight before treatment. Variability increased from L to H layers. Disturbance produced an increased heterogeneity within the plot. In the control plot, the L and F layers showed significant seasonal fluctuations (Fig. 22.1). Maximum L accumulation was recorded just after litterfall, from May to August. The average dry weight for L was 4.7 Mg ha^{-1}, which is similar to the amount of annual litterfall at Torners (V. Diego, unpubl. data; Chap. 3).

The clearcut plot showed significant differences with time in all forest floor fractions (Fig. 22.1). The L layer increased in weight during autumn and winter 1988–1989 from the cut branch litterfall; during spring and summer, the dry weight remained virtually constant suggesting balanced litterfall and decay/transfer; the following autumn, litterfall ceased and a significant L weight loss was observed. The F layer increased in weight after the disturbance as inputs from the L layer were greater than decay. The increments in F weight were of the same order of magnitude as those observed in L, suggesting that most of the litter loss from the L layer was transferred to the F layer. The H layer showed a large spatial variability, which increased considerably in the autumn of 1989, immediately following a period of intense rainfall.

The burned plot completely lost its L and F layers during the fire (Fig. 22.1), and an ashbed layer 2 cm thick was formed. During the 2 years following the fire, ash was incorporated in the underlying layers, the ashbed eventually being mixed with the residual H layer and eroded mineral sediment (Fig. 22.1). In the autumn of 1989, a dramatic decrease in the ashbed and H layer was evident, and the H layer of the clearcut plot decreased as well. Four intense rain events of 7.75 to 19.25 mm 30 min^{-1} in September–October were responsible for this. The L and F layers showed high minerali-

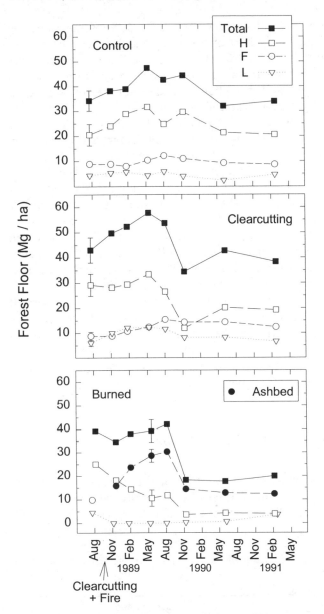

Fig. 22.1. Variation in oven-dry weight of different horizons of the forest floor of studied plots during 2.5 years of experiment. Mean and standard error

zation rates in this period of rains after the summer drought, as corroborated by litter bags results (data not shown). The ashbed plus H layer lost about 50% of its weight after the heavy rains in autumn 1989, that is a mean weight loss of 19.9 ± 6.3 (SE) Mg ha^{-1}. The heavy rains redistributed the forest floor material: in the first post-fire sampling, 80% of the sampling points showed ashbed+H layer, whereas after the heavy rains only 15% of the sampling points conserved these materials. Part of the ashes and H layer were mixed with the top mineral soil. Assuming that all the organic C of these layers was incorporated into the 0–5 cm mineral soil, the resulting enrichment of organic C in the upper mineral soil should be around 1% of the soil weight, i.e. undetectable given the high spatial variability of this layer.

The soil losses collected in Gerlach channels located at the bottom of the plots (Soler et al. 1994) were much higher during the heavy rain events than they had been in the previous period. However, during the first post-fire year the eroded sediment totalled only 363 kg ha^{-1} year^{-1} when referred to the total area of the strip of burnt plot upslope of each Gerlach channel. The sediments collected in the Gerlach channels came, though, from upslope distances of less than 6 m. If the collected sediments are thus referred to the area enclosed by this length, soil losses would amount to 2.4 Mg ha^{-1} year^{-1}. Both results fall within the low range for erosion rates observed in the literature (DeBano et al. 1979; Soto 1993). These soils present low erodibility, in agreement with their high infiltration capacity and structural stability (Josa et al. 1994).

The L layer in the burned plot began to be reconstructed in the second year after the fire. After 2.3 years, the L weight was similar to the pre-fire weight, but was now more greatly enriched with *Cistus albidus* litter in addition to holm oak. Ferran and Vallejo (1992) also found that the weight of L required 2 years to recover in burned holm oak forests, and estimated 23 years for the reconstruction of the F layer weight.

22.5 Nutrient Dynamics in the Forest Floor

In the clearcut plot, the nutrient concentration of the new L layer, coming from the slash, was higher than the L litter of the undisturbed plot for most of the nutrients (N, P, Mg and K) except for Ca (Table 22.2), owing to the lack

Table 22.2. Nutrient concentrations in leaves of the L layer from control and clearcut plots. Mean, maximum and minimum values for the 2.5 years of study, with $n = 32$, and P value of one-way ANOVA

Element	Control plot			Clear-cut plot			ANOVA
(mg g^{-1})	Mean	Maximum	Minimum	Mean	Maximum	Minimum	P
C	511	517	505	511	519	506	0.152
N	13.94	16.40	12.30	17.76	20.90	14.45	0.000
P	0.65	0.82	0.55	0.81	0.98	0.64	0.000
S	1.15	1.38	1.03	1.27	1.59	1.02	0.071
Ca	18.69	20.80	16.15	15.28	18.34	11.40	0.000
Mg	1.22	1.08	1.40	1.47	1.76	1.25	0.000
K	2.29	3.35	1.54	3.41	5.07	2.31	0.000
Na	0.14	0.17	0.09	0.15	0.22	0.12	0.438

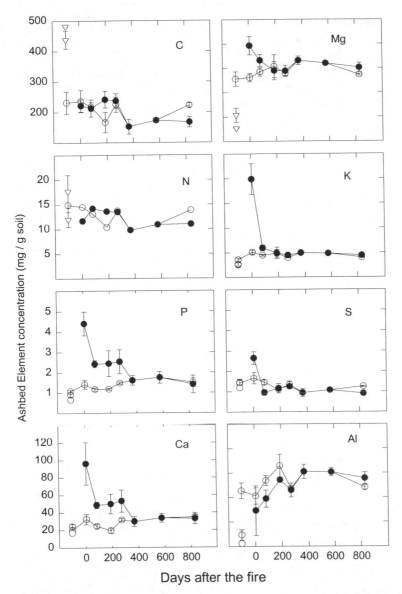

Fig. 22.2. Variation of element concentration in the ashbed layer (*solid circles*) and H layer (*open circles*) of the burned plot with time. Points before fire correspond to L, F and H layers of the undisturbed plot. Mean and standard deviation of four bulk samples

of retranslocation in the green slash (Sabaté 1993). N content in the L layer from the clearcut plot increased with time (and the C:N ratio decreased), due to a rapid N immobilization, to reach an almost steady value some 22 months after the experiment began. This C:N ratio was very similar to that of the fine F fraction in the control plot, suggesting that the clearcut plot had a shorter period of N immobilization, mostly occurring in the L layer, than the control plot, where N increased in both L and F layers. Sauras et al. (1991) found a similar trend when they incubated green and senesced holm oak leaves in the forest of Prades: higher mineralization of N and greater weight loss rates were found for the green litter. In our study, P and Mg concentrations increased with time in the slash L, whereas K decreased rapidly through leaching and accumulated in the F layer.

The burned plot (Fig. 22.2) initially showed high nutrient concentrations in the ashbed. These progressively decreased and were incorporated into the H or deeper layers, becoming stabilized during the second post-fire year. K, and to a lesser extent Ca, showed a rapid washout from the ashbed. This was also observed in the drainage water (Poch 1992; Gázquez 1994). P, S and Mg presented a more gradual decrease. N concentration remained stable until the heavy rains in autumn 1989 when it decreased (the same was found to occur for C). These rains led to the mixing of the ashbed and H layers with the mineral soil, thereby reducing C and N concentrations by dilution, as detected by the increase in Al content, and probably enhanced organic matter and N mineralization.

22.6 Phosphorus and Nitrogen Dynamics

22.6.1 Soluble Phosphorus

Soluble P (mostly inorganic) in the H layer was sensitive to the disturbances applied (Fig. 22.3). In the control plot, both total and organic soluble P fluctuated without a clear trend. In the clearcut plot, total soluble P slowly increased in the first year after clearing and then decreased to reach initial levels in just a few months (Fig. 22.3). In the burnt plot, the ashbed and the H layer showed similar soluble P contents to the control H layer, just after burning. However, 2 months later the ashbed showed a three-fold increase in its total and organic soluble P content, probably because of the increase of P solubility due to the pH raise. Organic soluble P was rapidly mineralized while total soluble P remained constant until the heavy rains in autumn 1989. These rains washed out the soluble P from the ashbed, and both the ashbed and the H layer achieved lower P concentrations than the control H.

22.6.2 Mineral Nitrogen

In the H layer of the control plot (Fig. 22.4), NO_3^--N content was correlated positively with the rainfall in the 5 days prior to sampling ($r^2 = 0.60$, $P < 0.05$). Therefore, nitrification was especially active in spring and autumn when the nitrate concentration was higher than that of ammonium (Chap. 16). In the

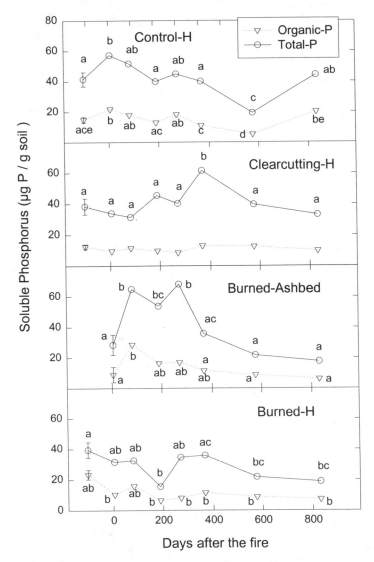

Fig. 22.3. Variation of soluble phosphorus (total and organic) with time. Mean of four bulk samples and standard deviation. Different *letters* within each P fraction show significant (*P*<0.05) differences between sampling times

top mineral layers of soil, most of the mineral N was ammonium. This showed small fluctuations during the 2 years of sampling.

In the clearcut plot, 6 months after clear-cutting, mineral N concentration increased dramatically, but in different forms and proportions depending on the layer considered: in the H layer, mineral N was mostly nitrate and showed the highest increase recorded; in the 0–5 cm mineral soil, both mineral forms increased with an initial peak of nitrate; in the 5–10 cm mineral soil, ammo-

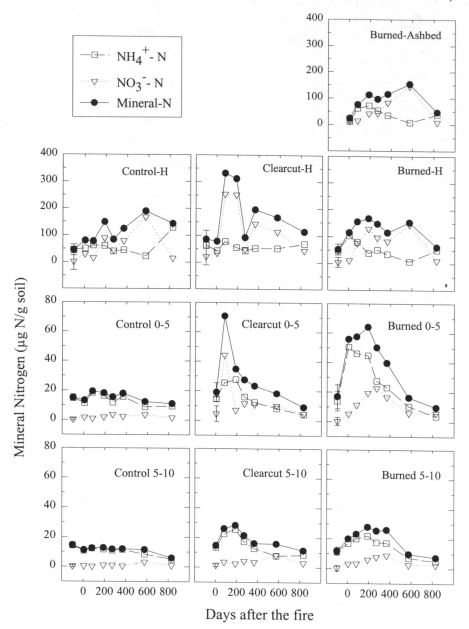

Fig. 22.4. Mineral-N (nitrate and ammonium) over time in forest floor and mineral soil in control, clearcut and burned plots. Mean of four bulk samples and standard deviation

nium predominated and showed the lowest increase registered (Fig. 22.4). The increase in mineral N was observed for 1.5 years and then it decreased to the pre-treatment values. The effect of clear-cutting in increasing N mineralization has been widely observed (Vitousek 1981; Smethurst and Nambiar 1990).

Immediately after the fire, the ashbed presented very low mineral N concentration (Fig. 22.4). The ash usually contains low quantities of mineral N due to N volatilization during the fire (Raison et al. 1985b; Marion et al. 1991). Nevertheless, the H layer and top mineral soil increased their ammonium content as a direct consequence of mild heating during the fire which promoted N mineralization (Kutiel and Shaviv 1989). No nitrate increase was observed in the initial sampling after the fire. The ashbed increased the ammonium content during the following 6 months, but it then decreased. For all the layers sampled, the peak in ammonium was followed by a peak in nitrate. This behaviour was most clearly seen in the H and ashbed layer, to a lesser extent in the 0–5 cm mineral soil and was only slightly detectable in 5–10 cm soil (Fig. 22.4). This delay in nitrification has been attributed to the sterilization of nitrifiers (Dunn et al. 1979) and their relatively slow recovery. The peak in nitrification was reached faster in the clearcut than in the burned plot. This was probably due to the improved moisture conditions and the lack of sterilization in the clearcut plot. The second year after burning, the burned and control plots showed very similar behaviours for total mineral nitrogen and its forms at each of the layers sampled.

22.6.3 Mineral Nitrogen Mobility

Mobile mineral nitrogen was trapped in ion exchange resins (Hübner et al. 1991) incubated in the field seasonally. Resins in the burned plot trapped more NO_3^--N during the year than the other two (Fig. 22.5), according to its higher biological activity measured by CO_2 production (G. Fusté, pers. comm.). The forest floor from the clearcut plot produced higher NO_3^--N than that of the control because of the higher litter content of the former. With the exception of the initial period after the fire, ammonium concentration was lower than that of nitrate and it fluctuated less. The ammonium collected under the H layer was higher in the burned plot than in the control and

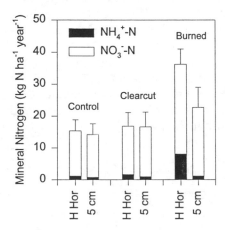

Fig. 22.5. Leaching fluxes of mineral N (nitrate and ammonium) collected during 20 months with resin bags below the organic horizons and 5-cm soil depth

clearcut during the first year after the fire, but the differences were reduced with time and disappeared in autumn 1989. The mineral nitrogen exported from the forest floor layers was higher in the burned plot (35 kg N ha^{-1} year^{-1}) than in the clearcut and control plots (16 kg N ha^{-1} year^{-1}). Under the 5-cm mineral soil the differences between plots were not significant and the losses were around 18 kg N ha^{-1} year^{-1} (Fig. 22.5).

From the data obtained in zero-tension lysimeters in the studied plots (Poch 1992; Gázquez 1994), the burned plot lost more nitrate than the control and clearcut plots. The increase in nitrogen availability in the soil solution of the burned plot exceeded the plant uptake, leading to leaching from the soil.

22.7 Overall Effects of Clear-Cutting and Slash-Burning on Soil Fertility

Slash-laying-down on the forest floor contributed to an increase in the content of nutrients in the forest floor, especially in the L layer during the first years after clear-cutting. This treatment increased N availability in the year and a half following the disturbance, producing a net loss of nitrate from the forest floor.

The effects of fire were dramatically expressed in the direct volatilization of a major proportion of the labile N pool in the ecosystem. In the post-fire period, the residual forest floor layers suffered a reorganization by means of the progressive incorporation of ash into the H layer and upper mineral soil, increasing the spatial heterogeneity in the burned plot. The fire led to high nutrient availability during the first post-fire year. This availability was in excess to plant uptake, and resulted in leaching losses for most of the basic cations and nitrate. The heavy rainfalls 1 year after burning led to detectable losses of soluble P and mineral N. Later on, the burned plot showed lower contents of these nutrients in the remaining H layer. However, the amount of nutrient loss by erosion and leaching was not relevant with respect to the ecosystem pools.

The disturbances studied, clear-cutting and slash-burning, increased soil fertility in the short term. In the medium term, the nutrient content returned to the undisturbed levels by means of different mechanisms (immobilization, leaching). Clear-cutting and slash-laying-down presented lower leaching and erosion losses than were found with slash-burning.

References

Binkley D (1986) Soil acidity in loblolly pine stands with interval burning. Soil Sci Soc Am J 50: 1590–1594

DeBano LF, Rice RM, Conrad CE (1979) Soil heating in chaparral fires: effects on soil properties, plant nutrients, erosion, and runoff. USDA For Serv, Pacific Southwest For Range Exp Stn, Res Pap PSW-145, Berkeley, California

Dunn PH, DeBano LF, Eberlein GE (1979) Effects of burning on chaparral soils: II. Soil microbes and nitrogen mineralization. Soil Sci Soc Am J 43:509–514

Dyrness CT, Van Cleve K, Levison JD (1989) The effect of wildfire on soil chemistry in four forest types in interior Alaska. Can J For Res 19:1389–1396

Fahey TJ, Hill MO, Stevens PA, Hornung M, Rowland P (1991) Nutrient accumulation in vegetation following conventional and whole-tree harvest of Sitka spruce plantations in North Wales. Forestry 64:271–227

Ferran A, Vallejo VR (1992) Litter dynamics in post-fire successional forests of *Quercus ilex*. Vegetatio 99/100:239–246

Ferran A, Serrasolsas I, Vallejo VR (1992) Soil evolution after fire in *Quercus ilex* and *Pinus halepensis* forests. In: Teller A, Mathy P, Jeffers JNR (eds) Responses of forest ecosystems to environmental changes. Elsevier, London, pp 397–404

Gázquez R (1994) Seguimiento lisimétrico del efecto del fuego y de la tala en suelos forestales de la Serra de Prades, Tarragona. MSc Thesis, Escola Superior d'Agricultura de Barcelona, Barcelona

Gillon D, Rapp M (1989) Nutrient losses during a winter low-intensity prescribed fire in a Mediterranean forest. Plant Soil 120:69–77

Grier CC (1975) Wildfire effects on nutrient distribution and leaching in a coniferous ecosystem. Can J For Res 5:599–607

Hübner C, Redl G, Wurst F (1991) In situ methodology for studying N-mineralization in soils using anion exchange resins. Soil Biol Biochem 23:701–702

Josa R, Arias X, Solé A (1994) Effects of slashburning on some soil physical properties in a holm-oak coppice In: Sala M, Rubio JL (eds) Soil erosion and degradation as a consequence of forest fires. Geoforma Ediciones, Logroño, pp 29–42

Kauffman JB, Till KM, Shea RW (1992) Biogeochemistry of deforestation and biomass burning. In: Dunnette DA, O'Brien RJ (eds) The science of global change. The impact of human activities on the environment. American Chemical Society, Washington, DC, pp 426–456

Khanna PK, Raison RJ, Falkiner RA (1994) Chemical properties of eucalyptus litter ash and effects of ash on forest soils. For Ecol Manage 66:107–125

Kutiel P, Shaviv A (1989) Effect of simulated forest fire on the availability of N and P in Mediterranean soils. Plant Soil 120:57–63

Little SN, Ohmann JL (1988) Estimating nitrogen lost from forest floor during prescribed fires in Douglas-fir Western hemlock clearcuts. For Sci 34:152–164

Marion GM, Moreno JM, Oechel WC (1991) Fire severity, ash deposition, and clipping effects on soil nutrients in chaparral. Soil Sci Soc Am J 55:235–240

Poch R (1992) Seguiment lisimètric en sòls d'alzinars de Prades afectats per foc i tala. MSc Thesis, Escola Superior d'Agricultura de Barcelona, Barcelona

Raison RJ, Khanna PK, Woods PV (1985a) Mechanisms of element transfer to the atmosphere during vegetation fires. Can J For Res 15:132–140

Raison RJ, Khanna PK, Woods PV (1985b) Transfer of elements to the atmosphere during low intensity prescribed fires in three Australian subalpine eucalypt forests. Can J For Res 15: 657–664

Sabaté S (1993) Canopy structure and nutrient content in a *Quercus ilex* L. forest of Prades mountains: effects of natural and experimental manipulation of growth conditions. PhD Thesis, University of Barcelona, Barcelona

Sauras T, Roca MC, Vallejo VR (1991) Estudio comparativo de la descomposición de hojas verdes y senescentes de encina. III Jornadas de Ecología Terrestre, León

Serrasolsas I (1994) Fertilitat de sòls forestals afectats pel foc. Dinàmica del nitrogen i del fòsfor. PhD Thesis, University of Barcelona, Barcelona

Smethurst PJ, Nambiar EKS (1990) Distribution of carbon and nutrients and fluxes of mineral nitrogen after clearfelling a *Pinus radiata* plantation. Can J For Res 20:1490–1497

Soler M, Sala M, Gallart F (1994) Post fire evolution of runoff and erosion during an eighteen month period. In: Sala M, Rubio JL (eds) Soil erosion and degradation as consequence of forest fires. Geoforma Ediciones, Logroño, pp 149–161

Soto B (1993) Influencia de los incendios forestales en la fertilidad y erosionabilidad de los suelos de Galicia. PhD Thesis, University of Santiago, Santiago de Compostela

Trabaud L (1979) Etude du comportement du feu dans la garrigue de Chêne kermes à partir des températures et des vitesses de propagation. Ann Sci For 36:13–38

Vitousek PM (1981) Clearcutting and the nitrogen cycle. In: Clark FE, Rosswall TH (eds) Nitrogen cycling in terrestrial ecosystems. Ecol Bull (Stockholm) 33:631–642

Woodmansee RG, Wallach LS (1981) Effects of fire regimes on biogeochemical cycles. In: Clark FE, Rosswall TH (eds) Nitrogen cycling in terrestrial ecosystems. Ecol Bull (Stockholm) 33: 649–669

23 Functional Responses to Thinning

Carlos A. Gracia, Santiago Sabaté, Juan Manuel Martínez and Eva Albeza

23.1 Introduction

Holm oak (*Quercus ilex* L.) has adaptive traits to survive or easily regenerate after disturbances. Its high resprouting capacity (Chap. 5) can explain the structure of most holm oak forests with a very high tree density, which determines a strong competition for the resources and a very low growth rate (Chap. 3). Although self-thinning is the natural mechanism of reduction in tree density as the size of trees increase, the low growth rate leads these forests to an almost permanent state of stagnation in which most of the gross primary production (GPP) is invested in respiration (Chap. 12) leading to a very low net primary production (NPP). This low NPP and the high tree density produce a very slow diameter growth, which often is much less than 1 mm year^{-1} (Gracia et al. 1996; Chap. 3).

Man-induced thinning can be used as a management tool to avoid stagnation and to accelerate the natural processes of forest structure recovery (Aussenac and Granier 1988; Ducrey and Toth 1992). To analyze how holm oak responds to thinning, four different thinning intensities were applied to a dense stand of resprout origin in the Prades experimental area (NE Spain). Stem growth increase, leaf population dynamics, resprouting capacity and water use have been analyzed and the most significant results are presented in this chapter.

23.2 Experimental Design

The experiment was conduced on the north-facing slope of Torners (Chap. 2). Four different thinning intensities – 0 (control), 55, 72 and 79% of basal area reduction – were applied in 1992. Each treatment was replicated in three randomly-assigned circular plots of 0.5 ha. Circular plots were used in order to minimize edge effects. The relatively large area of each plot was decided

Ecological Studies, Vol. 137
Ferran Rodà et al. (eds) Ecology of Mediterranean
Evergreen Oak Forests
© Springer-Verlag, Berlin Heidelberg 1999

Table 23.1. Tree density after thinning, basal area (*BA*) before and after thinning and basal area removed in each treatment. Data are the mean and standard deviation of the three 0.5-ha plots used in each treatment

Treatment	Density (trees ha^{-1})	BA before thinning[a] (m^2 ha^{-1})	BA after thinning[a] (m^2 ha^{-1})	BA removed (%)
Control	12629 ± 1401	37.1 ± 3.1	37.1 ± 3.1	0
Minimum thinning	2432 ± 176	45.9 ± 5.9	16.5 ± 0.8	55 ± 2
Medium thinning	1405 ± 215	42.2 ± 5.3	10.5 ± 1.0	72 ± 3
Maximum thinning	1447 ± 255	44.8 ± 3.5	7.9 ± 1.4	79 ± 4

[a] Measured at 50 cm from the ground.

upon to reduce or eliminate interferences among the different experiments to be carried out in these plots. The thinning intensities were the result of felling all the stems of each stool but the two largest [55% of basal area (BA) reduced], all but the largest one (72% of BA reduced) or all but the second largest in diameter (79% of BA reduced). The structure of the forest before and after thinning is summarized in Table 23.1.

23.3 Stem Growth Increase

The mean annual diameter increment (measured over bark at 50 cm from the ground) during the first 3 years of the experiment (1992–1995) was only 0.029 cm in the control plots. Trees from the heavily thinned plots grew on average 0.419 cm year^{-1}, i.e. 14 times faster than the control trees. The diameter of the trees in the medium and minimum thinning intensities increased 3.2 and 10.6 times more than did the control trees (Table 23.2). The modelled gross primary production of the forest remained constant despite the treatments (data not shown). Nevertheless, the reduction in respiration due to the elimination of part of the aboveground biomass and the redistribution of the net primary production among a lower number of trees in the thinned plots explain the great differences observed in tree growth rates between the different treatments.

Table 23.2. Mean annual increment of tree diameter over bark[a] during first 3 years after thinning (1992–1995)

Treatment	Mean annual diameter increment (cm)	Growth relative to control
Control	0.029	100
Minimum thinning	0.093	320
Medium thinning	0.308	1062
Maximum thinning	0.419	1440

[a] Measured at 50 cm from the ground.

23.4 Leaf Population Dynamics After Thinning

Leaves and fine roots are the most active parts of trees. As a consequence, it can be expected that the changes induced by thinning that can affect the physiology of the tree will be quickly reflected in the structure and dynamics of the leaves (Castell et al. 1994). Evergreen trees such as holm oak have different cohorts of leaves with different morphological and physiological traits (Chap. 9). This heterogeneity arises because these leaf cohorts are produced in different years under different environmental conditions, and also because the position of a leaf within the canopy determines its structural and physiological characteristics (Harper 1989). The analysis of leaf population dynamics is of the utmost importance to understand the recover capacity of holm oak after disturbances.

On a representative sample of branches from trees selected at random (Gracia et al. 1996) leaf population dynamics was analyzed following the thinning treatments. Results obtained during the first year after thinning are summarized in Table 23.3. Thinning reduced the leaf area index of remaining trees by 54, 70 and 78%, depending on thinning intensity. Leaf shedding during the first year ranged between 36 and 61% of the initial number of leaves in the control and maximum thinning, respectively. Leaf mortality was higher in the more severe treatments. New leaves produced during the first year amounted only to 22% of the leaves initially present in the control plots, but this value increased with the intensity of the thinning until 217% in the maximum thinning plots. The morphology of the new leaves differed among treatments reflecting the environmental conditions under which they were produced. Leaf specific mass was higher and mean leaf area lower in the more intense treatments where the leaves were more exposed to solar radiation.

Leaf turnover increased with thinning intensity. Trees therefore allocated more carbon to the formation and maintenance of the leaves. The production of new leaves was much higher than leaf shedding during the years

Table 23.3. Leaf population dynamics during first year after thinning: leaf area index (LAI) of canopy and resprouts, leaf mortality and new leaf formation (as percentage of initial number of leaves in canopy), leaf specific mass and mean leaf area 1 year after thinning. Mean ± SD; $n = 3$

| Variable | Thinning intensity | | | |
	Control	Minimum	Medium	Maximum
LAI of remaining stems	3.63	1.66	1.07	0.78
Leaf mortality (%)	36 ± 7	47 ± 6	50 ± 6	61 ± 12
Leaf formation (%)	22 ± 14	133 ± 46	156 ± 15	217 ± 24
Leaf specific mass (mg cm^{-2})	14.8 ± 1.2	17.0 ± 1.0	17.8 ± 0.8	17.5 ± 0.3
Mean leaf area (cm^2)	5.9 ± 0.3	5.3 ± 0.4	4.5 ± 0.7	4.6 ± 0.3
LAI of new resprouts of holm oak	0	0.3	–	1.1
LAI of new resprouts of other species	0	0.51	–	1.5
Biomass of holm oak resprouts (kg ha^{-1})	0	800	–	3400
Biomass of other species resprouts (kg ha^{-1})	0	1700	–	5800

following thinning and the leaf area index of the forest thus increased, recovering the pre-treatment values in about 6 years. Competition among trees was reduced since there were more water, nutrients and light available for the remaining trees. These results agree with those obtained by Abrams and Mostoller (1995) who found a reduction in the mean leaf lifespan as water or nutrient availability increased. Brix and Mitchell (1986), Sabaté and Gracia (1994) and Sheriff (1996) found analogous results analyzing the effects of irrigation or fertilization on different tree species.

23.5 Leaf Water Potential

Soil water availability in Prades is closely related to the course of precipitation throughout the year. During the experiment, a period of severe drought occurred. From September 1993 to September 1994 rainfall was less than 100 mm. Predawn leaf water potential (Ψ_{pd}) was measured at three different levels of the vertical profile: the upper canopy, involving sun-exposed leaves, lower canopy, involving shade leaves of the canopy, and leaves of the new resprouts. The results obtained for 1 year, from June 1994 to March 1995, are plotted in Fig. 23.1. In June 1994, after several months of severe drought, Ψ_{pd} of mature trees in control plots was about -2 MPa, i.e. much lower than in the thinned plots. In the treatment of maximum thinning the leaf water potential at that time was about -0.5 MPa, and slightly less than -1 MPa in the medium thinning treatment. These results show the improved water status of the trees induced by thinning compared with the trees in control plots. During August 1994, Ψ_{pd} in the control treatment was lower than -4.2 MPa - the maximum measuring capacity of our pressure gauge - while the trees in the maximum thinned plots reached values around -3 MPa. In control plots, visual symptoms of foliar damage were evident at this time with the browning of the leaves of more than 50% of the trees. No brown leaves were observed in the trees of the thinned plots, which remained green and healthy-looking all summer. During the drought period, the higher competition for water among trees and the higher rainfall interception in control plots led to a decrease in the soil water content compared with the thinning treatments, in agreement with results of Bréda et al. (1995).

Following the severe drought period, more than 450 mm of rain fell in only 2 days in October 1994. Ψ_{pd} in December 1994 and spring 1995 was about -0.5 MPa (Fig. 23.1). There were no statistical differences between the upper and lower canopy, either in the control treatment or the maximum thinning. However, resprouts in the maximum thinning treatment showed a slightly better water status (higher Ψ_{pd}) than the canopy leaves. Differences between resprouts were not always significant, but at most times Ψ_{pd} was less negative in the maximum thinning treatment than in the minimum thinning. So during March 1995, Ψ_{pd} was higher ($P < 0.04$) in the maximum thinning,

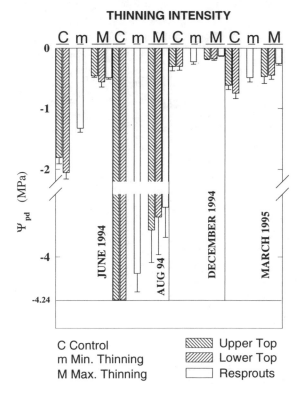

THINNING INTENSITY

Fig. 23.1. Seasonal course of predawn xylem water potential in three of the treatments and in three levels along the vertical canopy profile; –4.24 MPa was the lowest potential that could be measured by the used pressure gauge

C Control
m Min. Thinning
M Max. Thinning

Upper Top
Lower Top
Resprouts

but the difference was only marginally significant in August ($P < 0.08$) and December ($P < 0.11$).

23.6 Effect of Treatments on Gas Exchange

23.6.1 Photosynthesis and Leaf Conductance

Seasonal and daily courses of net photosynthesis (A) and leaf conductance (g_s) were monitored in the three different levels of the canopy in control and maximum thinning plots. In December 1994 and March 1995, when the water availability was not limiting, both photosynthesis and leaf conductance in the upper canopy were similar in both treatments (Figs. 23.2 and 23.3). However, during the dry season (June and August 1994), when ecophysiological activity was largely water-limited, A and g_s were higher in the thinned than in the control plots. These higher rates of gas exchange can be explained by the higher amount of water available per unit of leaf area in the thinned plots. Bréda et al. (1995) found similar results for thinning treatments in *Quercus petraea* forests. Analogous responses of leaf activity have been re-

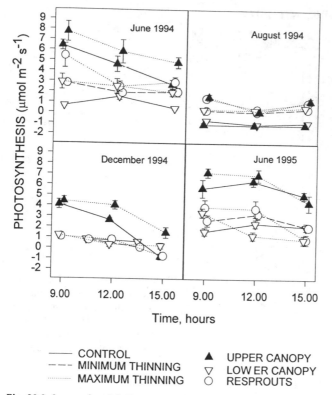

Fig. 23.2. Seasonal and daily courses of leaf photosynthetic activity. Levels along vertical canopy profile are shown for each treatment

ported by Tenhunen et al. (1990), Epron and Dreyer (1993) and Valentini et al. (1995) in plants living under dry conditions.

The lowest values of A and g_s were recorded in August when the very high air temperature (near 40 °C at midday) and elevated vapour pressure deficit induced the stomatal closure. In August 1994, the leaves of control plots were below the photosynthetic compensation point, with a net release of CO_2, and close to zero in the thinned plot (Fig. 23.2). Similar patterns have been observed by Méthy and Trabaud (1993) in holm oak leaves, and by Damesin and Rambal (1995), Trabaud and Méthy (1992) and Goulden (1996) in several *Quercus* species.

23.6.2 Differences of A and g_s Along the Vertical Canopy Profile

Both control and thinned plots showed higher A and g_s rates in the upper canopy when compared with the lower canopy and with resprouts. The reduction of A and g_s in the lower levels was due to the lower irradiance resulting from the interception of light by upper canopy leaves. After thinning,

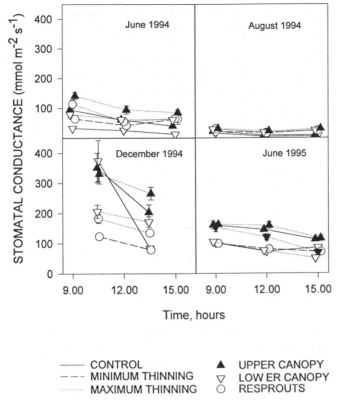

Fig. 23.3. Seasonal and daily courses of leaf conductance. Levels along the vertical canopy profile are shown for each treatment

irradiance on leaves of the lower layers increased, leading to an increase in A and g_s in these leaves. In the maximum thinning treatment, differences in A and g_s between the lower canopy and the resprouts were usually non-significant, and when they were, resprouts showed higher rates.

23.6.3 Differences Between Resprouts

Both A and g_s rates were higher for the resprouts in the maximum thinning than in the minimum thinning plots, although these differences were not always significant. Three years after thinning, the biomass and LAI of resprouts in the maximum thinning treatment were higher than in the minimum thinning treatment. Resprouts in the maximum thinning plots probably transpired collectively more water than resprouts in the minimum thinning plots, since the former had a much higher leaf area (Table 23.3). Competition for water between resprouts and trees was thus more intense in the maximum thinning treatment.

Table 23.4. Instantaneous water use efficiency (*WUE*) of sun and shade leaves, carbon uptake on a leaf and a ground area basis, and water transpired by holm oak canopies in control and heavily thinned plots on four different dates. Summer observations (June and August 1994) were made during a severe drought

Variable	Treatment	June 1994	August 1994	December 1994	March 1995
WUE sun leaves	Control	1.86	−1.44	0.75	2.29
(mmol CO_2 mol^{-1} H_2O)	Thinning	1.92	0.71	1.05	2.70
WUE shade leaves	Control	0.92	−1.67	24	1.34
(mmol CO_2 mol^{-1} H_2O)	Thinning	1.20	0.36	32	1.56
Carbon uptake	Control	2.61	−0.96	1.02	2.66
(g C m^{-2} leaf day^{-1})	Thinning	5.75	1.06	1.44	4.02
Carbon uptake	Control	4.88	−1.74	1.31	3.23
(g C m^{-2} ground day^{-1})	Thinning	5.39	0.94	1.19	3.14
Transpiration	Control	254	94	225	142
(mmol H_2O m^{-2} ground day^{-1})	Thinning	279	125	148	127

23.6.4 Instantaneous Water Use Efficiency and Productivity

Estimations of the instantaneous water use efficiency (WUE) in control and thinned plots were made in the upper and lower canopy (Table 23.4). Results showed no significant differences between treatments, except in August 1994, when WUE in control plots was negative in both sun and shaded leaves while leaves in thinned plots had a positive WUE.

Carbon uptake per unit of leaf area was always higher in the thinned plots. In August 1994, the leaves in the control plots had a net loss of carbon of − 0.96 g m^{-2} day^{-1} while the leaves in the thinned plots still have a low but positive carbon gain of 1.06 g m^{-2} day^{-1}. To best understand the meaning of these results, the difference in leaf area index in the canopies between control and thinned plots must be taken into account. Integrating the leaf gas exchange for the whole canopy, the carbon uptake per unit of ground area was higher in June. No significant differences between control and thinned plots were found in December and March.

23.7 Conclusions

The thinning treatments reduce the aboveground biomass, and hence the maintenance respiration of the trees. The soil water available to the remaining trees increases as do the transpiration and carbon uptake per unit of leaf area. All these changes lead to a significant improvement in the growth rates of the trees. Leaf turnover increases with thinning intensity. This increase in leaf turnover is the result of the balance between leaf shedding and new leaf production. Both increase with the intensity of thinning, but leaf formation increases faster than leaf shedding. The net result is a fast recovery of the canopy leaf area, which can reach pre-thinning values in about 6 years.

Thinning improved the relative water availability and light intensity on leaves, which promoted higher net photosynthesis and leaf conductance than in control plots. Integrating these results at the canopy level, these differences took place during the dry season, specially during August 1994, and disappeared in winter and spring when water was not a limiting factor. The increment of water availability, detected as increased predawn Ψ_{pd} in the thinning treatments, was mainly due to the redistribution of water between the remaining trees, and to the reduction of canopy interception. This increment promoted higher GPP and NPP in the thinning treatments than in the control, which meant a fast growth of the remaining trees and recovery of the tree canopy. However, water use efficiency did not increase because the increase in the photosynthetic rate was proportional to the increase in the transpiration rate.

The water available to adult trees was possibly reduced through competition with the new resprouts after thinning, which, moreover, received high irradiation through gaps in the canopy. This fact was specially important in the maximum thinning plot, where the high radiation allowed high photosynthetic rates, even 3 years after thinning.

References

Abrams MD, Mostoller SA (1995) Gas exchange, leaf structure and nitrogen in contrasting successional tree species growing in open and understory sites during a drought. Tree Physiol 15:361–370

Aussenac G, Granier A (1988) Effects of thinning on water stress and growth in Douglas-fir. Can J For Res 18:100–105

Bréda N, Granier A, Aussenac G (1995) Effects of thinning on soil and tree water relations, transpiration and growth in an oak forest (Quercus petraea (Matt.) Liebl.). Tree Physiol 15: 295–306

Brix H, Mitchell AK (1986) Thinning and nitrogen fertilisation effects on soil and tree water stress in a Douglas-fir stand. Can J For Res 16:1334–1338

Castell C, Terradas J, Tenhunen JD (1994) Water relations, gas exchange, and growth of resprouts and mature plant shoots of Arbutus unedo L. and Quercus ilex L. Oecologia 98:201–211

Damesin C, Rambal S (1995) Field study of leaf photosynthetic performance by a Mediterranean deciduous oak tree (Quercus pubescens) during a severe summer drought. New Phytol 131: 159–167

Ducrey M, Toth J (1992) Effects of clearing and thinning on height growth and girth increment in holm oak coppices (Quercus ilex L.). Vegetatio 99/100:365–376

Epron D, Dreyer E (1993) Long-term effects of drought on photosynthesis of adult oak trees [Quercus petrea (Matt.) Liebl.) and Quercus robur L.] in a natural stand. New Phytol 125: 381–389

Goulden ML (1996) Carbon assimilation and water-use efficiency by neighboring Mediterranean-climate oaks that differ in water access. Tree Physiol 16:417–424

Gracia C, Bellot J, Sabaté S, Albeza E, Djema A, León B, Martínez J M, Ruiz I, Tello E (1996) Respuesta del encinar a tratamientos de aclareo selectivo. In: Vallejo R (ed) La restauración de la cubierta vegetal en la Comunidad Valenciana. Fundación CEAM, Valencia, pp 547–601

Harper JL (1989) Canopies as populations. In: Russell G, Marshall B, Jarvis PG (eds) Plant canopies: their growth, form and function. Cambridge University Press, Cambridge, pp 105–128

Méthy M, Trabaud L (1993) Seasonal courses of photosynthetic activity and sublethal temperature tolerance of Quercus ilex leaves. For Ecol Manage 61:339–348

Sabaté S, Gracia C (1994) Canopy nutrient content of a Quercus ilex L. forest: fertilization and irrigation effects. For Ecol Manage 68:31–37

Sala A, Tenhunen JD (1994) Site-specific water relations and stomatal response of Quercus ilex in a Mediterranean watershed. Tree Physiol 14:601–617

Sheriff DW (1996) Response of carbon gain and growth of Pinus radiata stands to thinning and fertilizing. Tree Physiol 16:527–536

Tenhunen JD, Sala A, Harley PC, Dougherty RL, Reynolds JF (1990) Factors influencing carbon fixation and water use by Mediterranean sclerophyll shrubs during summer drought. Oecologia 82:381–393

Trabaud L, Méthy M (1992) Effects de températures sub-létales sur l'appareil photosynthétique du chêne vert (Quercus ilex L.). Ann Sci For 49:637–649

Valentini R, Epron D, De Angelis P, Matteucci G, Dreyer E (1995) In situ estimation of net CO_2 assimilation, photosynthetic electron flow and photorespiration in Turkey oak (Q. cerris L.) leaves: diurnal cycles under different levels of water supply. Plant Cell Environ 18:631–640

Part 7 Animals in the Forest: Ecology of Two Main Groups

24 Soil Arthropods

Pilar Andrés, Eduardo Mateos and Carlos Ascaso

24.1 Introduction

The Mediterranean soil fauna is mainly composed of mites (Acari) and springtails (Collembola) (Lossaint and Rapp 1978), with mites being usually more abundant than springtails (Poinsot-Balaguer 1988). The greatest arthropod diversity is often found in the litter layer, where the arthropods closely depend on climatic factors, with Collembola being strongly influenced by water deficits (Poinsot 1971). Most of the hypogeal groups inhabiting the Mediterranean soil face the seasonal climatic pattern by adjusting the timing of their life cycle or by migrating through the soil layers (Sgardelis et al. 1993). Life in the depths of the soil habitat has to be considered as a mechanism of avoidance related to the fluctuations of the Mediterranean climate, particularly to the summer drought (Di Castri 1973).

Information about the soil fauna of holm oak (*Quercus ilex* L.) forests is scant, and differences in sampling systems or in taxonomic levels of approach impede a consistent interpretation of the available data. Results offered in this chapter come from our studies (Ascaso 1986; Andrés 1990; Espuny 1992; Mateos 1992) conducted in the experimental areas of Montseny and Prades (NE Spain; Chap. 2). We deal successively with (1) the characteristics of the hypogeal arthropod mesofauna of recently undisturbed holm oak forests, (2) the effects of an experimental fire on this fauna, and (3) the epigeal soil arthropods.

24.2 Hypogeal Arthropods

24.2.1 Density, Abundance and Trophic Characteristics

Soil mesofauna (body width < 2 mm) was collected by Berlese-Tullgren funnels. The mean density of the hypogeal arthropods in the upper 15 cm of the soil in Prades and in Montseny (Table 24.1) is within the normal range in

Ecological Studies, Vol. 137
Ferran Rodà et al. (eds) Ecology of Mediterranean
Evergreen Oak Forests
© Springer-Verlag, Berlin Heidelberg 1999

European temperate forests (from 50 10³ to 300 10³ individuals m⁻²; Lebrun 1971). As expected, Acari and Collembola are the most abundant taxa in our plots. The Acari/Collembola ratio (6.1 in Montseny and 15.3 in Prades) is especially great in relation to other Mediterranean holm oak forests. According to Poinsot-Balaguer (1988), it can be explained either by the lower anthropic pressure they endure compared with the French Mediterranean area, or by the more xeric conditions of our forests. Collembola, whose activity depends strongly on the soil water availability, have a lower relative abundance in our plots (13% of the sampled individuals in Montseny and 6% in Prades) than in other Mediterranean forests. The Diplopoda/Chilopoda ratio, which is higher when the turnover of organic matter is faster (Demange 1979), is 0.74 in Prades and 1.07 in Montseny. The mean arthropod biomass is 11.3 g dry wt. m⁻² in Prades and 5.2 g dry wt. m⁻² in Montseny. In both forests, Acari account for most of the mesofaunal dry weight (46% in Montseny and 86% in Prades).

Acari were studied in detail in Montseny. Most of the individuals (73%) and of the biomass (76%) belong to Cryptostigmata (oribatid mites). Mesostigmata (17% of the Acari and 22% of their biomass) is the second most abundant group of Acari, and the family Uropodidae, characteristic of mature soils (Athias-Binche 1987), is notable among them. Prostigmata (7% of the Acari and 2% of their biomass) are less abundant in Montseny than in

Table 24.1. Mean density (individuals m⁻²) of hypogeal arthropods of some Mediterranean holm oak forests. A, Sainte-Baume (Lions 1972); B, Mediterranean sclerophyllous forest (Di Castri 1973); C, Le Rouquet (Lossaint and Rapp 1978); D, Prades (Mateos 1992); E, Montseny (Andrés 1990).

Taxa	A	B	C	D	E
Pseudoescorpionida	n.a.	n.a.	n.a.	28	103
Mesostigmata	2567	9410	3189	n.a.	10099
Prostigmata + Astigmata	7100	51790	n.a.	n.a.	5397
Cryptostigmata	7396	43290	5586		24913
Total acarina	17063	104490	n.a.	213261	46398
Diplopoda	59	540	n.a.	68	271
Pauropoda	87	220	n.a.	67	485
Chylopoda	56	130	n.a.	92	256
Symphyla	62	280	n.a.	294	831
Total myriapods	264	1170	n.a.	521	1843
Protura	70	470	n.a.	20	173
Diplura	n.a.	n.a.	n.a.	34	366
Collembola	6413	19000	7460	13939	7529
Psocoptera	n.a.	n.a.	n.a.	1017	308
Diptera+Coleoptera (larvae)	n.a.	n.a.	n.a.	1040	1312
Pterigota	1629	8900	n.a.	n.a.	n.a.
Total	18431	136070	n.a.	229860	58033

n.a., No available data.

other Mediterranean holm oak forests. According to Di Castri and Vitali di Castri (1981), the Prostigmata density increases towards the xeric parts of the environmental gradient.

In the soil arthropod trophic pyramid of Montseny, predators account for 18% of the individuals and 37% of the live weight. Decomposers (i.e. microphytophagous, mycophagous, and detritivorous arthropods)[1] account for 82% of the individuals and 63% of the dry weight. The Mesostigmatic families Rhodacaridae and Parasitidae are the most abundant predators. Despite their abundance, these small individuals (58 µg individual^{-1} in Rhodacaridae) contribute less biomass to the trophic web than the large predators like Chilopoda (4026 µg individual^{-1}). The trophic pyramid changes along the soil profile. In Montseny, the biomass ratio of decomposers to predators is 3 in the litter layer (L), 2 in the humus layers (F and H; these two layers were intermixed in the Montseny site and were sampled as a unit) and 1 in the upper mineral soil (S). This ratio is higher when applied to the number of individuals: 11 in L, 6 in F+H and 3 in S.

24.2.2 Spatial Structure

Fifty per cent of the soil arthropods in Montseny and 46% in Prades inhabit the humus layers The litter layer is less suitable for animal life, and shelters only 18% of the soil mesofauna in Montseny and 16% in Prades. The harshness of the edaphic microclimate and the low trophic quality of the fresh holm oak leaves can explain the poor nature of this layer. The conditions of the underlying mineral soil are intermediate between the humus and the litter layers.

Taxa exhibiting a great surface to volume ratio or soft teguments (such as Pauropoda, Symphyla, Protura and Diplura) seem unable to endure drought and do not colonize the L layer of Prades. In Montseny, under a smaller water deficit (Chap. 2), they sporadically appear. The same reasons can explain the scarcity of Collembola in the L layers of our two forests. On the opposite side, Psocoptera are the only arthropods showing preference for the L layer. According to García Aldrete (1991), this behaviour is characteristic of forest litter-dwelling insects, whose presence in the underlying layers is merely sporadic. Nearly all of the Acari families occupy mainly the humus layer, with the exception of Rhodacaridae and Rhaphignatidae, two families that share characteristics of small size, light teguments and predatory behaviour.

In each layer, arthropods usually occur in tridimensional aggregates whose form, size and distribution are related to factors such as water availability, soil temperature, animal behaviour, seasonal and daily specific rhythms, and patterns of food and plant distribution (Butcher et al. 1971). In Montseny, the Lexis Index of aggregation (Greig-Smith 1983) is lowest in the

---

[1] Editors' Note: although in ecology "decomposer" is usually applied to microorganisms, we have respected the authors' meaning, which is widespread among soil zoologists.

L layer (16 in L; 29 in both F+H and S). In the L layer, soil arthropods tend to be randomly distributed. On the one hand, the relatively large gaps between fresh-litter leaves facilitate animal movements. On the other, the seasonal pattern of leaf-fall and the horizontal transport of litter by wind promote the distribution at random of the microhabitats able to be colonized by soil fauna (Garay 1985).

24.2.3 Seasonal Pattern

Mediterranean soil climate includes summers with water deficit, wet and cold winters, and wet and warm autumns and springs. Litter endures strong daily and seasonal climatic variations. Temperature of the L layer in Montseny ranges from 28 °C in August to 0 °C in January. After summer rains and under an intense sunshine, its water content can drop in a few hours from saturation to 2%. The maximum fall of holm oak leaves, rich in aliphatic and aromatic compounds unpalatable for the mesofauna, occurred in our study from June to September. In autumn and winter, the plant litter is leached by rain and loses tannins and phenols. The following summer, humification starts and the litter, previously attacked by fungi and bacteria, becomes palatable to the arthropods.

The litter acts as a buffer to the underlying soil layers. The humus layers are relatively protected from the daily climatic variations, but undergo a seasonal fluctuation in water content. In spring and autumn, water availability and warm temperatures enhance the growth of fungi that mycophagous arthropods use as an essential part of their diet. Less-fluctuant environmental conditions prevail in the mineral soil, in which the biological activity is low and the water content is moderate and much less temporally variable than in the forest floor.

In Mediterranean soils, water is the most limiting environmental factor. At Montseny, population parameters (such as density, relative abundance, and indices of spatial distribution) of most mesofaunal species tend to vary in parallel with soil water availability, though correlations with soil water content were significant only for some species (Table 24.2). Water surplus is less

Table 24.2. Montseny. Mesofaunal species, all of them oribatid mites, having significant linear correlation coefficients (r) between water content in the litter layer and species density in this layer, aggregation Index of Lexis (AI), and Usher Index of vertical distribution (DV). Maximum $n = 182$ (26 sampling dates x seven spatial replicates)

Species	Density	Aggregation (AI)	Vertical distribution (DV)
Eueremaeus granulatus	0.77*	0.52	0.75*
Xenillus tegeocranus	0.84*	0.44	0.73*
Tectocepheus velatus	0.79*	0.76*	0.62
Chamobates cuspidatus	0.79*	0.75*	0.77*
Genus Oppia	0.70*	0.71*	0.32

* $P < 0.001$.

limiting than water deficit. Above a critical threshold characteristic of each species (pF = 4.7–5 in most Collembola or pF = 5–6 in oribatid mites; Vannier 1971), survival depends on the ability of the animals either to endure the dry periods in an inactive state (by delaying egg hatching or by living in a state of diapause or dormancy), to aggregate in moist microhabitats or to migrate to the deeper soil layers (Poinsot-Balaguer 1988).

Summer drought gives rise to the minimum annual arthropod density (10 000 individuals m^{-2} in Montseny; 98 580 individuals m^{-2} in Prades). At this time, oribatid mites account for 82% of the arthropods, and most of the taxa, mainly Prostigmata and myriapods (except Diplopoda, whose density remains similar to the annual mean), disappear from the soil. This dramatic density fall is partly explained by mortality but is also overestimated by the abundance of inactive forms that are not captured by Berlese-Tullgren funnels.

There is a positive relation between drought harshness and the depth at which most of the animals are collected (Fig. 24.1). The ability of some taxa to remain active in the soil throughout the year is accomplished by different strategies. The most drought-sensitive taxa, like Symphyla, remain in the mineral layer all the time and move to the soil surface only in spring, probably attracted by the food supply. Some arthropods having strong and well sclerosed exoskeletons (such as oribatid mites, Uropodidae or Zerconidae)

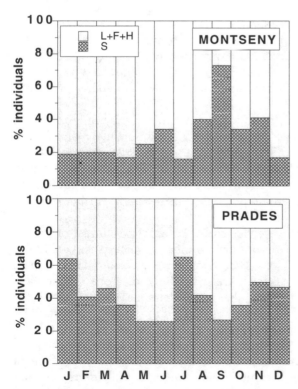

Fig. 24.1. Percentage of arthropods inhabiting the forest floor (L+F+H) and upper mineral soil (S) throughout the year in Montseny and Prades

spend most of the year in the organic layers, but move to the S layer at the end of the summer. For the small Prostigmata, which appear sporadically in the soil, survival of the population is accomplished by massive egg hatching coinciding with the wet period. The Astigmata can undergo drought as hypopus. Collembola are strictly dependent on the water availability and exhibit a large range of drought resistance strategies (Poinsot-Balaguer 1988). In spite of these mechanisms, summer drought causes a decrease in their density both in Montseny (down to 585 individuals m^{-2}) and in Prades (down to 735 individuals m^{-2}).

Food supply also influences the mesofaunal dynamics, but as omnivory is a common trend belowground (Wallwork 1983; Moore et al. 1988), many arthropods do not depend on a single food-type, and the availability or scarcity of a given trophic resource must not be regarded as a determinant factor.

24.2.4 Main Hypogeal Decomposers: Oribatid Mites and Collembola

In Montseny, oribatids account for 58% of the hypogeal arthropods. Mean annual density is 34 017 individuals m^{-2}, mean annual biomass 1633 mg m^{-2} (26% of the arthropod biomass), and cumulative annual specific diversity (Shannon index) is 2.46 bits individual^{-1}. Most of the oribatid mites (51%) live in the F+H layers, 27% in S and 22% in L. The larger species live near the ground surface, and, as a consequence, mean individual biomass decreases from the shallower to the deeper soil layers (0.03 µg in L, 0.01 in F and 0.007 in S). Specific diversity is lower in S (1.6 bits individual^{-1}) than in L (2.5) or F (2.4). This low diversity of the mineral layer results from its low number of species (18 in S, 34 in L and 35 in F) and from its low equitability (59% in S, 67% in L and 66% in F).

The maximum oribatid density occurs in autumn and the minimum at the end of summer. Specific diversity also follows a seasonal pattern and falls to 1.27 in the dry season. Most animals move to the deeper layers at the end of summer, so that in September 86% inhabit S and the litter layer is empty. In the litter layer, where moisture is highly variable, there is a significant correlation between water content and population densities and distributional parameters for some species (Table 24.2).

The 35 species of Collembola collected in the holm oak forest of Prades account for 6% of the soil arthropods, with a mean annual density of 13 939 individual m^{-2}, mean annual biomass of 395 mg m^{-2}, and cumulative annual diversity of 3.35 bits individual^{-1}. The relative abundance of Collembola is highest in the humus layer (45% of the soil Collembola) and lowest in the litter layer (10%). The mineral layer can be divided into two 5-cm parts: 32% of the soil Collembola live in the upper one, and 10% in the deeper one.

There is a specific vertical stratification of Collembola at Prades. The most abundant species can be ranked according to their depth preference. Some of them inhabit exclusively the organic layers (*Odontella vallvidrerensis, Heteromorus major, Pseudochorutella catalonica, Bilobella aurantiaca, Spaerida*

Table 24.3. Spearman rank correlation coefficient (rs) between daily rainfall and density and number of species of Collembola in the forest floor (L, F and H) and upper mineral soil (A_{11} and A_{12}) in Prades

Variable	Soil layer				
	L	F	H	A_{11}	A_{12}
Density	0.80 **	0.55 *	0.70 **	−0.06 ns	−0.20 ns
Species number	0.72 **	0.82 **	0.77 **	0.07 ns	−0.28 ns

ns, not significant; * $P < 0.05$; ** $P < 0.01$.

pumilis) or exhibit a clear preference for the mineral layers (*Onychurus minutus, Ptotaphorura prolata, Mesaphorura italica, M. critica*), whereas others occupy all soil layers. Mean annual specific diversity varies with depth. In the forest floor as a whole, diversity is higher (2.04) than in the mineral horizons (1.9 in A_{11} and 1.39 in A_{12}). In the organic layers, the density and the specific diversity are highest in spring and autumn and lowest in winter and summer. In the mineral layers, the seasonal pattern is less clear, but minimum density and specific diversity occur in summer. Summer drought is limiting for Collembola, so rainfall is positively correlated with density and with number of species in the organic layers (Table 24.3). Hyper-dominance of some species occurs in summer, so, in June, *Xenylla schillei* accounts for 64% of the Collembola. The active species are then scarce (*Mesaphorura macrochaeta, Entomobrya nivalis, Entomobrya strigata*) and move to the deeper layers where moisture is higher. Most species withstand the summer as eggs or in a state of dormancy or diapause to become active again in autumn, after the first rains.

24.3 Effects of Fire on Soil Arthropods

Man and fire have coexisted for a long time in the Mediterranean Basin. Fire usually destroys the litter layer which constitutes both a trophic resource for soil organisms and a protective layer regulating the soil climate (Prodon et al. 1987), and it induces an immediate loss of animals, more or less intense depending on the type of animal (Athias-Binche et al. 1987).

Our data on the effects of fire on hypogeal arthropods come from an experimental fire in the holm oak forest of Prades (Chap. 22). The fire lasted for 1 hour and temperature reached maxima of 750 °C in litter, 370 °C in the humus layer and 60 °C 7.5 cm below the ground surface. The L and F layers completely disappeared. The humic layer was partially replaced by an ash-bed, 2 cm thick, composed of ash and charred plant materials mixed with the remains of the H layer. In the mineral soil layer, the water storage capacity and the size of soil aggregates decreased. Erosion increased after fire, and the ashbed was in part exported out of the site and in part remobilized and mixed with the mineral soil. One year after the fire, plant cover recovered to 40%, but during that year soil surface temperature in the burnt plot was higher than in the control, and reached 60 °C in June. As a consequence of the lack of plant protection and of the degradation of the soil structure, soil water content was extremely variable.

In comparison with the pre-fire situation, the mean arthropod annual density fell to one fifth in the whole soil and to one thirtieth in the upper organic layers, with most of the arthropods living in the mineral soil (Table 24.4). Sensitivity to fire depends on the biological characteristics of the animals. Density of Diplura decreased in the organic layers and remained constant in the mineral soil. Pauropoda and Iulida disappeared from the soil. According to Saulnier and Athias-Binche (1986), Iulida, large epigeal decomposers unable to dig are among the most fire-sensitive soil taxa. Protura and Psocoptera disappeared from the organic layers, but their density increased in the mineral soil (547 and 110% respectively). As the migratory ability of these animals is not enough to explain their population increase, there is probably a reproductive stimulation due to the new favourable post-fire characteristics of the upper mineral soil.

To sum up, Iulida and Pauropoda were the most damaged taxa. On the contrary, Proturan densities increased after fire. Population sensitivity depends to a great extent on the date of the fire. As the deeper layers are less injured by fire than the upper ones, dry season fires, when most of the fauna inhabit the upper mineral soil, allow the survival of a stock of animals from which the organic layers will be colonized (Majer 1984). Fire eliminates the strictly litter-dwelling arthropods, and restoration will depend then on the

Table 24.4. Fire effects on hypogeal arthropods in Prades. Density changes 1 year after fire, expressed as percentage increment (+) or reduction (−) with respect to pre-burn densities, and changes in vertical distribution of fauna in soil

Taxa	Density changes (%)	Pre-fire vertical distribution (%)			Post-fire vertical distribution (%)		
		L	F + H	S	Ash.	H	S
Pseudoscorpionida	−90	11	61	28	0	100	0
Acarina	−80	16	46	38	2	8	90
Chilopoda	−79	8	55	37	0	55	45
Diplopoda	−82	0	75	25	0	27	73
Pauropoda	−100	0	0	100	0	0	0
Symphyla	−60	0	19	81	0	0	100
Protura	+82	0	15	85	0	0	100
Diplura	−41	0	50	50	0	15	85
Collembola	−87	10	48	42	2	8	90
Psocoptera	−23	40	27	33	4	5	91
Pterigota (larvae)	− 86	10	49	41	7	22	71
Total	−79	16	46	38	2	8	90

L, litter layer; F and H, humic layers; S, upper mineral soil; Ash., ashbed.

colonization from neighbouring ecosystems. The ability for self-movement or for using transport vectors (as phoretic Acari or Pseudoscorpionida do) is then of great importance.

The effect of fire on Collembola was studied in detail at Prades. Immediately after fire, Collembola disappeared from the ashbed. Density decreased 99% in the organic layers and 26% in the upper mineral soil, but increased 60% in the deeper mineral layer. Eighty-six per cent of the species in the organic layers disappeared, but the number of species doubled in the deeper mineral layer. The vertical pattern of specific diversity was similar.

One year after the fire, the depth-integrated density of Collembola was still 87% lower than in the unburnt plots, and the number of species was 17% lower. Most Collembola (80%) were then collected in the mineral soil. The litter-dwelling species were absent or were collected in a very low density. Some of the species characteristic of forest ecosystems, such as *Isotomiella minor* and *Protaphorura prolata*, disappeared from the burnt plot, whereas the density of some species that characterize open ecosystems, such as *Metaphorura affinis* and *Onychiurus minutus*, increased. Some pioneer species specialized in burnt soils, such as *Willemia scandinavica*, *Paratullbergia callipygos* and *Entomobrya quinquelineata*, appeared during this first year. Two years after fire the density and number of species of Collembola remained lower than in the unburnt plots, but the vertical distribution of animals was similar to the pre-fire situation.

There is no agreement about the short- and long-term effects of fire on soil fauna (Huhta et al. 1967; Majer 1984; Saulnier and Athias-Binche 1986; Neumann and Tolhurst 1991). Estimates of the post-fire populations' recovery time range from 2 to 7 years, but other authors state that the hypogeal fauna of sclerophyllous forest soils remains unaltered when fires are neither frequent nor intense, and even that the variations caused by fire are smaller than the spatial and temporal natural fluctuations of the soil communities. In Prades, 2 years was probably not enough to complete the faunal recovery in the soil, despite the occurrence of some factors favouring the recovery, such as the small plot area and the proximity of an unburnt holm oak forest.

24.4 Epigeal Arthropods

24.4.1 Density, Abundance and Trophic Characteristics

Data on the epigeal soil fauna come from pit-fall traps in the permanent plot located in the holm oak forest at La Castanya (Montseny; Chap. 2). Half (49%) of the collected arthropods (8% of the dry weight biomass) are mainly hypogeal, but they are collected in pit-fall traps when they emerge to the surface under favourable climatic and trophic conditions (Table 24.5).

Table 24.5. Epigeal arthropods collected in holm oak forest at La Castanya (Montseny)

Taxa	N	Ra (%)	B	Rb (%)
Chilopoda[a]	8.6	1.5	83.4	1.4
Diplopoda[a]	12.6	2.2	359.7	6.2
Pseudoscorpionida[a]	5.0	0.9	0.5	0.01
Phalangida	64.4	11.4	153.2	2.6
Araneae	48.6	8.6	424.4	7.3
Acarina[a]	72.4	12.8	11.5	0.2
Isopoda	12.4	2.2	74.0	1.3
Collembola[a]	153.0	27.1	3.2	0.1
Protura + Diplura[a]	22.6	4.0	0.3	0.01
Tysanura	5.6	1.0	16.6	0.3
Orthoptera	0.4	0.1	12.8	0.2
Lepidoptera	2.4	0.4	5.4	0.1
Diptera	46.0	8.1	13.6	0.2
Hymenoptera	13.0	2.3	2.1	0.04
Psocoptera	4.2	0.7	1.5	0.02
Thysanoptera	2.4	0.4	4.4	0.1
Homoptera	0.6	0.1	5.0	0.1
Coleoptera	83.2	14.7	4588.1	79.2
Insecta varia	7.0	1.2	9.6	0.2
Total	564.4	100.0	5789.8	100.0

N, individuals trap^{-1} year^{-1}; Ra, relative abundance of taxa; B, biomass (mg dry wt. trap^{-1} year^{-1}); Rb, relative biomass of taxa.

[a] Mainly hypogeal taxa.

Averaged throughout the year, beetles are the most abundant epigeal arthropods (15% of the individuals and 79% of the biomass; Table 24.5), with Lathridiidae being the most abundant Coleopteran family (7% of the arthropods). Carabidae is the most important family in relative weight (76% of the arthropod biomass). Predators (Araneae, Phalangidae, Chilopoda, Pseusoscorpionida, Carabidae, Lampyridae, Staphylinidae, Scydmaenidae) account for 90% of the biomass and for 27% of the individuals feeding on the ground surface, and decomposers account for 9% of the biomass and for 71% of the individuals. Phytophagous arthropods are scarcely represented.

24.4.2 Seasonal Pattern

The seasonal pattern of the number of individuals captured in pit-fall traps is more related to fluctuations in their activity than in their population density (Briggs 1961). Data have then to be expressed as "activity density" or number of individuals captured by trap and per day or year (Tretzel 1955; Heydemann 1956).

In La Castanya, arthropods are active all year long, but their activity density fluctuates seasonally. During the study year, 50% of the annual captures occurred from July to September, and only 5% from December to February. A relative minimum of activity density was observed in June, which was a very dry month. Many species show summer diapause, timing and duration of which depend on the onset of soil water stress. As a consequence, vernal populations must either have a very short biological cycle or exhibit summer dormancy.

Eighty-five per cent of the annual biomass was collected from June to August whereas only 1% was collected in winter. This seasonality reflects largely the activity patterns of the large predators and, specially, of the family Carabidae which accounts for 90% of the summer biomass. In winter and spring, the active predators are smaller and the effect of individual size disappears. Collected biomass is then proportional to the number of arthropods and, specially, to the saprophagous and mycophagous activity density.

24.4.3 Epigeal Predators

Hypogeal communities are dominated by decomposers, but predators are the most important epigeal arthropods. Most abundant taxa are Chilopoda, Phalangida, Araneae and Carabidae (Table 24.5). Mean annual specific diversity is 3.6 bits individual^{-1}, maximum diversity 5.0, and equitability 70%. These values reflect the codominance of some species, whose activity varies in space according to the plant cover phenology, and in time according to the seasonal pattern. Climate is warm enough to allow some faunal activity in all seasons. However, there are two well-defined subcommunities of predators active in spring and in winter, respectively.

Spiders are the most abundant taxa throughout the year except in summer, when they are substituted by Phalangida. The summer abundance of Phalangida has to be explained by the coincidence on the ground surface of the adult stages of both litter-dwelling and shrub-dwelling species. Phalangida inhabit mainly the litter, where they prey on small insects and mites (Rueda and Lobo 1984). Two genera (*Troqulus* and *Nemastoma*) belong to the superfamily Troquloidea, associated with the epigeal environment. The other species belong to the superfamily Phalangioidea, which inhabits the litter sporadically and mainly in the adult stage. However, Phalangioidea includes *Nelima silvatica* which accounts for 79% of the Phalangida.

Twenty-nine species of spiders were captured in the experimental plot at La Castanya, and *Tegenaria fuesslini* and *Lepthyphantes flavipes* were the most abundant. A great part of these species are not characteristic of holm oak forests and can be found in the neighbouring herbaceous or bushy environments also. Most of the species strictly depending on the forest environment are not restricted to the soil surface and move along the vertical plant strata, from soil to bushes and trees.

Two species of Carabidae were collected: *Carabus (Chrysotribas) rutilans aragonensis* and *Ceuthosphorus (Actenipus) oblongus orientalis*. The specific diversity of the Carabidae in our forest is very low. On the one hand, summer drought is not favourable for the species characteristic of the Pyrenees. On the other, the shady conditions of the plot are not suitable for the Mediterranean species. The two species found at La Castanya are "spring carabids" (Loreau 1985). The hibernating adult stage appears in autumn, and the reproductive period occurs in spring and continues until the summer, as long as the soil temperature is higher than 5 °C.

24.5 The "Soil Arthropod Community"

The expression "soil arthropod community" is surely a misleading conceptual simplification in a spatially and temporally variable ecosystem in which seasonal patterns are strongly defined by water availability. Even though some taxa belong strictly to a given part of the vertical forest structure, a great number of arthropods move across the different soil layers and even across the epigeal and hypogeal environments. The litter marks the border between the above- and belowground subsystems, and, here, the composition of the fauna changes continuously according to a seasonal pattern. When moisture conditions are advantageous, the litter is inhabited by arthropods characteristic of the deeper soil layers. On the other hand, arthropods that inhabit the aboveground parts of plants or temporary habitats (some Hemiptera, larvae of Coleoptera or Diptera, and some Psocoptera species), such as dung or decaying stumps, are mainly active on the ground surface from July to September. Reproductive rhythms and food search define the epigeal activity pattern.

Belowground, mycophages and detritivores account for most of the arthropods, and the humus layers are the focus from which colonization extends towards the upper and lower horizons. Here, the seasonal pattern is strongly related to the organic matter decomposition rhythm and to the evolution of the fungal colonies. For most of the mesofauna, the upper mineral soil is only an emergency shelter, to be occupied during summer drought.

References

Andrés P (1990) Descomposición de la materia orgánica en dos ecosistemas forestales del Macizo del Montseny (Barcelona): Papel de los ácaros oribátidos (Acarina, Oribatei). PhD Thesis, Autonomous University of Barcelona, Bellaterra

Ascaso C (1986) Análisis cuantitativo de poblaciones a partir de muestreos indirectos: aplicación a comunidades de artrópodos en dos bosques del Montseny. PhD Thesis, Autonomous University of Barcelona, Bellaterra

Athias-Binche F (1987) Modalités de la cicatrisation des écosystèmes méditerranèens après incendie: cas de certains arthropodes du sol. 3. Les acariens Uropodides. Vie Milieu 37:39–52

Athias-Binche F, Briard J, Fons R, Sommer F (1987) Study of the ecological influence of fire on fauna in Mediterranean ecosystems (soil and above-ground layer). Patterns of post-fire recovery. Ecol Mediterr 13:135–153

Athias-Henriot C, Cancela da Fonseca JP (1976) Microarthropodes édaphiques de la Tillaie (Forêt de Fontainebleau). Composition et distribution spatio-temporelle d'un peuplement en placette de litière de hêtre pure (Acariens et Collemboles). Rev Écol Biol Sol 13: 315–329

Briggs JB (1961) A comparison of pit-fall trapping and soil sampling in assessing populations of two species of ground beetles (Col. Carabidae). Rep E Malling Res Stn, pp 108–112

Butcher JW, Snider R, Snider RJ (1971) Bioecology of the edaphic Collembola and Acarina. Annu Rev Entomol 16:249–288

Demange JM (1979) Les mille pattes. Boubée, Paris

Di Castri F (1973) Animal biogeography and ecological niche. In: Di Castri F, Mooney HA (eds) Mediterranean-type ecosystems, origins and structure. Springer Berlin, pp 279–283

Di Castri F, Vitali di Castri V (1981) Soil fauna of Mediterranean-climate regions. In: Di Castri F, Goodall DW, Specht RL (eds) Mediterranean-type shrublands. Elsevier, Amsterdam, pp 444–478

Espuny A (1992) Distribució vertical i altres dades autoecològiques dels aranèids d'un alzinar muntanyenc del Montseny. PhD Thesis, Autonomous University of Barcelona, Bellaterra

Garay I (1985) Hétérogeneité spatiale et fonctionnelle dans la décomposition de la litière. CNRS PIREN-ATP. Matières organiques dans les sols. Internal Report CNRS

García Aldrete AN (1992) Insecta: Psocoptera. In: Dindal DL (ed) Soil biology guide. John Wiley, New York, pp 1033–1052

Greig-Smith P (1983) Quantitative plant ecology. Blackwell, London

Heydemann B (1956) Über die Bedeutung der "Formalinfallen" für die zoologische Landesforschung. Faun Mitt Norddtsch 6:19–24

Huhta V, Karppinen E, Nurminen M, Valpas A (1967) Effect of silvicultural practices upon arthropods, annelid and nematode populations in coniferous forest soils. Ann Zool Fenn 4:87–135

Lebrun P (1971) Écologie et biocénotique de quelques peuplements d'arthropodes édaphiques. Mem InsT R Sci Nat Belg 165

Lions JC (1972) Écologie des Oribates (Acariens) de la Sainte Baume (Var). Thèse doct sci nat, Université de Provence. AO 7248 cent doc CNRS 81972-05-29 vol 2:1–549

Loreau M (1985) Annual activity and life cycles of carabid beetles in two forest communities. Holarct Ecol 8:228–235

Lossaint P, Rapp M (1978) La forêt Méditerranéenne de chênes verts. In: Lamotte M, Bourliere F (eds) Problèmes d'écologie. Structure et fonctionnement des écosystèmes terrestres. Masson, Paris

Majer JD (1984) Short-term responses of soil and litter invertebrates to a cool autumn burn in Jarrah (*Eucalyptus marginata*) forest in Western Australia. Pedobiologia 26: 229–246

Mateos E (1992) Colémbolos (Collembola, Insecta) edáficos de los encinares de la Serra de l'Obac y la Serra de Prades (Sierra prelitoral catalana). Efectos de los incendios forestales sobre estos artrópodos. PhD Thesis, University of Barcelona, Barcelona

Moore JC, Walter DE, Hunt HW (1988) Arthropod regulation of micro- and mesobiota in belowground detrital food-webs. Annu Rev Entomol 33:419–439

Neumann FG, Tolhurst K (1991) Effects of fuel reduction burning on epigeal arthropods and earthworms in dry sclerophyll eucalypt forest of west-central Victoria. Austr J Ecol 16:315–330

Poinsot N (1971) Éthologie de quelques espèces de Collemboles Isotomides de Provence. Ann Fac Sci Marseille 44: 33–53

Poinsot-Balaguer N (1988) Stratégies adaptatives des arthropodes du sol en région méditerranèenne. In: Di Castri F, Floret Ch, Rambal S, Roy J (eds) Time scales and water stress. Proc 5th Int Conf on Mediterranean ecosystems (MEDECOS). IUBS, Paris

Prodon R, Fons R, Athias-Binche F (1987) The impact of fire on animal communities in Mediterranean area. In: Trabaud L (ed) The role of fire in ecological systems. SPB Academic Publishing, The Hague, pp 121–157

Rueda F, Lobo JM (1984) La vida en el suelo. Ediciones Penthatlon, Madrid

Saulnier L, Athias-Binche F (1986) Modalités de la cicatrisation des écosystemes méditerranéens après incendie: cas de certains Arthropodes du sol. 2. Les Myriapodes édaphiques. Vie Milieu 36:191–204

Sgardelis SP, Sarkar S, Asikidis MD, Cancela da Fonseca JP, Stamou GP (1993) Phenological patterns of soil microarthropods from three climate regions. Eur J Soil Biol 29:49–57

Tretzel E (1955) Technik und Bedeutung des Fallenfanges für ökologische Untersuchungen. Zool Anz 155: 276–287

Vannier G (1971) Signification de la persistance de la pédofaune après le point de flétrissement permanent dans les sols. Rev Écol Biol Sol 8:343–365

Wallwork JA (1983) Oribatids in forest ecosystems. Annu Rev Entomol 28:109–130

25 Composition and Dynamics of the Bird Community

Germán López-Iborra and José A. Gil-Delgado

25.1 Introduction

Birds have been a favourite subject of study in forest ecosystems. They are easily detectable in comparison with other vertebrate groups, and their reproductive and feeding ecology are critically determined by forest characteristics, such as the composition and structure of both the tree canopy and understory. Forest birds have been a focus of theoretical interest in competition, regulation of populations and life history strategies, and include one of the most studied groups of birds – the Tits (Paridae). They have received much attention also from an applied point of view, since conservation of their populations is in many cases linked to the conservation and appropriate management of their forest habitat. Thus, much information is available especially for temperate forests (Keast 1990).

Within the Palaearctic, bird communitites of deciduous and northern coniferous forests are best known, while studies on Mediterranean forest bird communities started in the 1970s, mainly with the works of J. Blondel and C.M. Herrera. Although composition of forest bird communities is broadly similar throughout European forests (Blondel and Farré 1988), some specific characteristics of Mediterranean ecosystems could determine important differences in bird community dynamics with respect to their more northern counterparts. The mild Mediterranean climate and the evergreen condition of most forests in this area make it possible for an important percentage of birds to remain in the forests during winter. On the contrary, the hot and dry summer may impose new limiting conditions for birds in some Mediterranean forests. Moreover, the geographical position of these forests on the main migratory pathways allows for the overlap of different populations of the same species in Mediterranean habitats. Thus, seasonal dynamics of bird communities may be more complex in Mediterranean than in more northern forests. This aspect needs to receive more attention, since only a small fraction of studies published have followed the bird community during a period longer than 1 year; thus, intra-annual and inter-annual fluctuations in Mediterranean bird communities remain poorly known.

Ecological Studies, Vol. 137
Ferran Rodà et al. (eds) Ecology of Mediterranean
Evergreen Oak Forests
© Springer-Verlag, Berlin Heidelberg 1999

In this chapter we describe the seasonal dynamics of the bird community in the holm oak (*Quercus ilex* L.) forest of the Prades experimental area (NE Spain), specifically in the Avic catchment. In particular, the seasonal variations of density and diversity are analyzed, and the relative abundance of summer migrants and the seasonal pattern of density variation are compared between several Iberian forest types.

25.2 Census Methods

We estimated the abundance of bird species using the mapping method. Although this method was originally designed to be used in the breeding season, it has been also employed to study the dynamics of bird communities throughout the year. It was used, for instance, in the study of the role of birds in the nutrient cycles in the Hubbard Brook Experimental Forest (Sturges et al. 1974; Holmes and Sturges 1975).

We used two plots of 5 ha (250 × 200 m) within the Avic catchment (Chap. 2). The area included in both plots represents 19.4% of the catchment area. One plot (plot 1) was located at the bottom of the catchment, between 740 and 800 m a.s.l., while the other (plot 2) was located in the mid–upper catchment, between 850 and 950 m a.s.l. In this way we may better cover the range of variation of vegetation structure within the catchment, since holm oaks had greater diameters and were taller at the bottom of the catchment while their density was greater at the top. Diversity of the tree layer also varied within the catchment, being greater at the bottom (López 1991).

A grid of 25 × 25 m was marked within each plot to locate the contacts with birds. Each plot was censused in two different days in the central week of each month. Censuses were performed in the morning and lasted for about 2 h. The plots were walked following the lines of the grid parallel to the longest side and the contacts and displacements of birds were registered on a map of the plot. Singing males were counted as a pair. In order to reduce the problem of high border–surface ratio in small plots, individuals or pairs located less than 50 m from the plot border have been counted as half an individual or half a pair since part of their home range probably lies out of the plot. Test censuses were performed in March of 1985 in order to become familiarized with the plots, and then real censuses were performed monthly from April 1985 to April 1987, always by the same person (G.L.I.). The data of the two census days of each month were combined to obtain a density estimate per month in each plot.

25.3 Seasonal Variation in Bird Density and Diversity

Both plots show a similar pattern of bird density fluctuation (Fig. 25.1A), with maximum densities in summer (June in plot 1, August in plot 2) and minimum in winter (December to February). The density of the bird community is lower in plot 2 during most of the year (paired samples t-test, $t = 6.06$, $df = 23$, $P < 0.001$), and only by the end of summer are densities in both plots similar or greater in plot 2. The difference between both plots is greatest in winter, showing that in this season the abundance of birds decreased more in the upper part of the catchment.

Considering both plots together, bird density in the breeding season (April–May) was 56.5 birds 10 ha^{-1} in 1985 and 58.3 birds 10 ha^{-1} in 1986. These values are in the lower third of the range for Fagaceae forests shown in the review by Potti (1986) and close to the maximum densities found for shrubby habitats. Other studies on holm oak woods have found slightly lower values, such as 41.9 birds 10 ha^{-1} in Herrera (1980) and 41.1 birds 10 ha^{-1} in

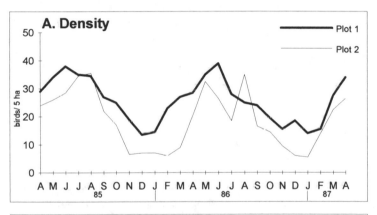

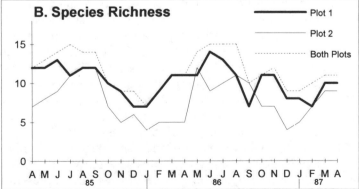

Fig. 25.1. A Seasonal variation in bird density (birds 5 ha^{-1}). B Seasonal variation in bird species richness in each plot studied and in both plots pooled

Tellería and Garza (1983). However, Zamora and Camacho (1984a) found greater values (100 birds 10 ha^{-1}) in a holm oak forest of just 28 ha, but this was probably due to the presence of birds foraging in habitats surrounding the forest.

The pattern of seasonal variation of species richness (S) was similar in both plots ($r = 0.51$, df = 22, $P < 0.02$), reaching maximum values in summer and minima in winter (Fig. 25.1B). Species richness tended to be higher in plot 1 than in plot 2, although the difference was less marked in spring-summer (April to September: paired samples t-tests, $t = 2.38$, df = 12, $P < 0.05$) than in autumn–winter (October to March: $t = 7.78$, df = 10, $P < 0.001$). The values of S for both plots pooled were very similar to their values for plot 1 alone, showing that very few species were detected in plot 2 that were lacking in plot 1. The greatest differences between total S and S for plot 1 occurred in late spring and summer when some irregular species in our study area (such as Hoopoe, Cuckoo, Green Woodpecker and Nightjar) were detected only in plot 2.

López (1992) compared the number of bird species expected in 10 ha for several types of forests in the breeding season. Bird species richness in the Avic catchment (Fig. 25.1B) lies in the middle of the range found for Iberian holm oak forests (12–17 species 10 ha^{-1}) and close to the average for this kind of forest (14.8 species 10 ha^{-1}, SD = 1.7, $n = 6$). Holm oak forests are less diverse than Iberian deciduous oak forests (22 species 10 ha^{-1}, SD = 6.2, $n = 4$) and mixed forests (26 species 10 ha^{-1}, SD = 7.8, $n = 3$), but present similar values to pine forests (12.4 species 10 ha^{-1}, SD = 4.3, $n = 17$).

25.4 Phenological Status of Birds

Owing to their capacity to fly over long distances, birds are potentially able to exploit resources appearing in a wide geographical range. Thus, most bird communities include species that exploit local resources throughout the year and species that just use seasonal bursts of food, whether in spring–summer or in winter in mild climates. The relative importance of both kinds of species conditions the dynamics of any bird community. To study this aspect, the species detected in the census plots have been classified (Table 25.1) into three categories: "constant species", including resident species present in the plots at least 50% of the study months; "summer species", including summer migrants present in the plots regularly in the breeding seasons of both years; and "irregular species", including all the rest of the species.

Figure 25.2 shows the seasonal variation of the percentage of individuals included in the above phenological categories. The community is dominated in both plots by constant species, since on average 91.6% of birds present in plot 1 and 89.2% of birds present in plot 2 belong to this group. The average percentage of the community included in the summer species group during

Table 25.1. Bird community composition in the Avic catchment. Average density (birds 5 ha^{-1}) is shown for each plot and two periods of the year: April–September ('Spring') and October–March ('Winter'). Species found present only on one or two occasions in each season are marked with a cross. Phenological status of birds (Phen., see text) is shown by: C (constant species), S (summer visitor), and I (irregular species). Paired samples t-test compares abundance between plots for each season. Correlation coefficient (r) between abundance of each species in both plots is also shown

Species	Scientific name	Phen.	Plot 1		Plot 2		Paired samples t-test		
			Spring	Winter	Spring	Winter	Spring	Winter	r
Hoopoe	Upupa epops	I							
Wood Pigeon	Columba palumbus	C	0.96	0.33	0.27	0.32	4.18**	0.29	0.078
Cuckoo	Cuculus canorus	I	+		+				
Nightjar	Caprimulgus europaeus	I			+				
Green Woodpecker	Picus viridis	I	+	+	+				
Great Spotted Woodpecker	Dendrocopos major	I	+						
Wren	Troglodytes troglodytes	C	0.31	0.50	0.04	0.00	1.53	2.39	-0.132
Robin	Erithacus rubecula	C	6.81	4.58	7.96	1.18	1.08	5.41***	0.443*
Redstart	Phoenicurus phoenicurus	I			+				
Blackbird	Turdus merula	C	2.31	3.42	1.61	0.68	2.00	4.98***	0.105
Subalpine Warbler	Sylvia cantillans	S	1.62	–	0.15	–	4.16**	–	0.294
Garden Warbler	Sylvia borin	S	0.58	–	0.27	–	1.43	–	0.532**
Sardinian Warbler	Sylvia melanocephala	I	+						
Blackcap	Sylvia atricapilla	S	0.28	–	0.04	–	1.39	–	-0.072
Chiffchaff	Phylloscopus collybita	I		+					
Bonelli's Warbler	Phylloscopus bonelli	S	0.62	–	0.69	–	0.23	–	0.486*
Firecrest	Regulus ignicapillus	C	3.92	2.96	4.11	3.77	0.43	1.99	0.443*
Long-tailed Tit	Aegithalos caudatus	C	1.15	1.79	0.77	0.64	0.75	2.42*	-0.159
Crested Tit	Parus cristatus	I	0.19	0.00	0.38	0.09	0.79	1.49	-0.174
Blue Tit	Parus caeruleus	C	3.54	1.88	2.65	1.14	2.14	2.41*	0.665***
Great Tit	Parus major	C	3.81	1.58	2.58	0.86	2.31*	1.72	0.508*
Short-toed Treecreeper	Certhia brachydactyla	C	1.00	0.79	0.85	0.68	0.65	0.43	0.306
Jay	Garrulus glandarius	C	0.54	0.38	0.96	0.45	1.50	0.27	0.294
Chaffinch	Fringilla coelebs	C	3.73	0.75	3.11	0.23	1.92	2.95*	0.918***
Rock Bunting	Emberiza cia	I		+					

* $P<0.05$, ** $P<0.01$, *** $P<0.001$.

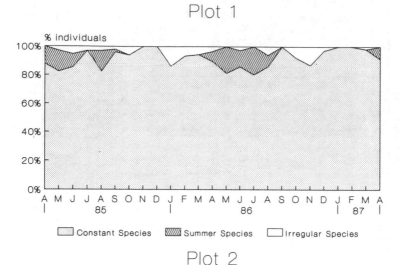

Plot 1

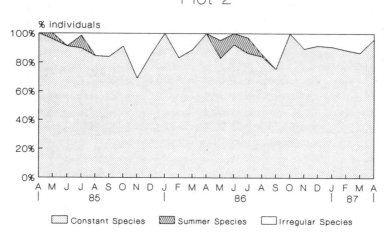

Plot 2

Fig. 25.2. Seasonal variation of percentage of individuals belonging to the phenological categories defined in the text: constant species, summer migrants, and irregular species

the breeding season is higher in plot 1 (9.7%) than in plot 2 (3.4%), the difference being significant (paired t-test, t = 5.06, df = 12, P < 0.001). The high dominance of constant species in this community is due both to the low number of summer visitor species that breed in the area, limited to the Bonelli's Warbler, the Subalpine Warbler, the Garden Warbler and the Blackcap, and to the low abundance of these species, well under the abundance of most resident species (Table 25.1).

The high percentage of birds belonging to constant species is characteristic of Iberian forests. Herrera (1978) analyzed the percentage of long-distance migrants in breeding bird communities in Europe and found a clear latitudinal gradient. No Iberian community in his data set had more than

20% of their individuals belonging to these species, while in localities north of the Pyrenees higher percentages were frequent. This gradient could be explained by regression of percentage of long-distance migrants on latitude or on mean temperature of the coldest month with almost the same determination coefficient. Herrera (1978) then suggested that the percentage of migrants depended both on resource abundance during the breeding season and on the severity of the winter. The lower the level of resources in winter the lower would be the population of resident birds that survive and then greater the relative availability of resources to summer migrants.

If this hypothesis is true we could expect that, within the Iberian forests, the percentage of summer migrants would be smaller in evergreen than in deciduous forests. The reason is that in evergreen forests just a small fraction of leaves is renewed each year, and only the young leaves may be eaten by caterpillars which are the main food for breeding birds (Iglesias 1996). In deciduous forests, on the contrary, all leaves are newly produced each spring, thus supplying a burst of food for insects and therefore for birds. Evergreen forests have a reduced level of summer resources compared to deciduous forests, but they offer more cover and food (insects, fruits) in winter. This results in a lower difference between summer and winter resource availability for birds and a lower percentage of summer migrants in evergreen forests.

To test this prediction, we have plotted in Fig. 25.3 the percentage of summer migrants found in several studies of Iberian bird communities. About 75% of the studied pine and evergreen oak forests presented percentages of summer migrant individuals lower than 15% of the breeding density. On the contrary, all studies in deciduous oaks presented percentages between 15 and 29% of summer migrants, thus seemingly supporting the hypothesis. How

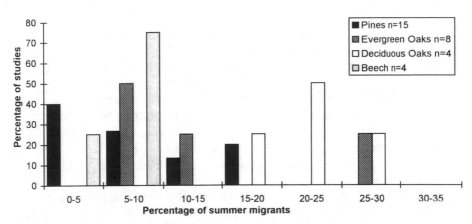

Fig. 25.3. Percentage of summer migrant birds in the breeding community of several Iberian forests. Data from: Purroy (1974, 1975a,b), Pedrocchi (1975), Herrera (1980), Santos and Suárez (1983), Tellería and Garza (1983), Costa (1984, 1993), Zamora and Camacho (1984a,b), Torres and León (1985), Carrascal (1986), Fernández and Galarza (1986), Obeso (1987), Zamora (1987), Maicas (1989), Carrascal and Tellería (1990), Santos and Alvarez (1990), Tellería and Galarza (1990), López and Gil-Delgado (this Chap., both plots pooled)

ever, beech (*Fagus sylvatica*) forests do not fit the predicted pattern, since they show low percentages of summer migrants (range 4.8–6.7%). Thus, it appears that other factors besides the deciduous/evergreen dichotomy are involved. One possibility is that the structural complexity of vegetation could also play a role, and forests with more diversified vertical structure could support a greater number of individuals of summer visitor species. This could explain why beech forests, which usually lack a developed understory, present so few summer migrant birds.

25.5 Dynamics of Selected Species

Table 25.1 shows the average density of the most abundant species in both plots and the correlation coefficient between density in both plots for each species. The seasonal pattern of variation for summer migrants is similar in all species detected, since they arrive in May and remain in the forest until August. However, the low densities that these species attain, especially in plot 2, prevent most correlations between their abundances in the study plots from being significant (Table 25.1).

The seasonal dynamics of constant species are more varied. Although all these species are present in the forest for most of the year, their response to the changing conditions in the upper and lower parts of the catchment can be different. This situation can be illustrated by some selected constant species (Fig. 25.4). The Robin and the Blackbird are species whose abundance is similar in spring in both plots (Table 25.1), but that present greater density in plot 1 in winter. With the Robin, the pattern of fluctuation is relatively similar in both plots, with abundance maxima in summer and a significant correlation between plots (Table 25.1), although its density decreases in winter much more in plot 1 than in plot 2. In contrast, the correlation between plots is not significant for the Blackbird, indicating a different pattern of fluctuation in both plots. Blackbird abundance shows a summer maximum and a winter minimum in plot 2, but in plot 1 it shows a relative maximum in summer and then an absolute maximum in winter, in both years. Thus, it appears that in these two species, part of the population living at higher altitudes concentrates in winter at the bottom of the valleys, where the vegetation is more diverse and shrubs producing berries are more abundant. The available data suggest that most of the Blackbirds remain in the forest in winter, since its density in the central part of the winter in plot 1 is close to the density in spring in both plots added together. On the contrary, a greater proportion of Robins would leave the forest studied to live probably in man-made habitats at lower altitudes.

The tendency for bird density to be greater in plot 1 than in plot 2 during winter is also observed in other species, such as the Blue Tit and the Chaffinch. These species presented similar abundance fluctuations in both plots

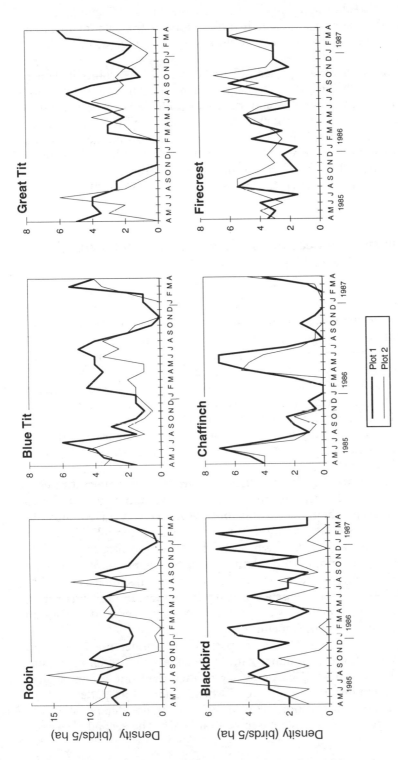

Fig. 25.4. Seasonal variation in density (birds 5 ha⁻¹) of some selected constant species in the two plots studied

(Fig. 25.4; see also the high correlations between plots in Table 25.1), but their density in winter was greater in the bottom of the catchment (Table 25.1), although the difference between plots was not as high as in the Robin and Blackbird.

The abundance of other species did not differ between plots in winter. This is the case with the Firecrest and the Great Tit (Fig. 25.4; Table 25.1). The Firecrest presented similar abundance in both plots in both seasons, with a tendency to be even more abundant in winter in plot 2 (paired t-test, $t = 1.99$, $df = 10$, $P = 0.07$). This is the species with least pronounced seasonal variations in abundance, being able to maintain relatively high densities in winter in both plots. The abundance of the Great Tit decreased to low values in winter in both plots, but in spring tended to be greater in plot 1.

25.6 Comparison of Seasonal Patterns of Density Fluctuation in Iberian Forest Types

The seasonal pattern of fluctuation of total bird abundance in a community should reflect the changes in resource availability throughout the year (Herrera 1980; Potti and Garrido 1986), which in turn will be affected by the climate and the specific composition of vegetation. The increase in the number of studies on the composition of Iberian bird communities that present data for a period of at least 12 months allows the comparative study of changes in bird density throughout the year. In this way, we could determine whether the pattern obtained in this study (summer maximum and winter minimum of abundance), which is very often found in studies of temperate bird communities, can be extended to other forest types in the Iberian Peninsula, or if other possibilities exist.

For this analysis, we transformed the total bird density obtained in each month into a relative density with respect to the maximum density obtained in each annual cycle. Then, the data for both years were averaged, thus obtaining a profile of density variation for each plot. We compiled literature data and analyzed them in the same way, thus transforming the absolute density values of each month into relative values to the month of maximum abundance, which received a value of 100. A matrix of similarity was obtained for the bird community data set using the Spearman rank correlation index, and then this matrix was analyzed to construct a dendrogram (Fig. 25.5) using the UPGMA algorithm.

This analysis shows that the simple pattern of summer maximum and winter minimum of bird abundance cannot be generalized to all bird communities of Iberian forests. The dendrogram identified one group of bird communities, including our study plots, which follows this pattern. It included not only holm oak woods, but also pine forests, deciduous oak forest, and one non-forest habitat, a steppe in central Spain. The second identified

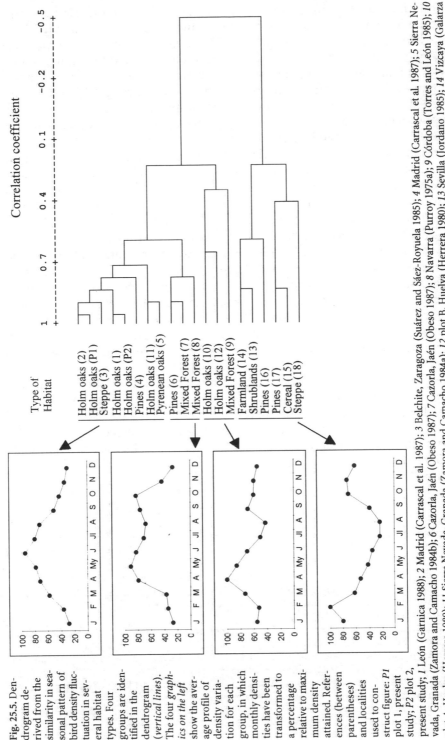

Fig. 25.5. Dendrogram derived from the similarity in seasonal pattern of bird density fluctuation in several habitat types. Four groups are identified in the dendrogram (vertical lines). The four graphics on the left show the average profile of density variation for each group, in which monthly densities have been transformed to a percentage relative to maximum density attained. References (between parentheses) and localities used to construct figure: P1 present plot 1, present study; P2 plot 2, present study; 1 León (Garnica 1988); 2 Madrid (Carrascal et al. 1987); 3 Belchite, Zaragoza (Suárez and Sáez-Royuela 1985); 4 Madrid (Carrascal et al. 1987); 5 Sierra Nevada, Granada (Zamora and Camacho 1984b); 6 Cazorla, Jaén (Obeso 1987); 7 Cazorla, Jaén (Purroy 1975a); 9 Córdoba (Torres and León 1985); 10 vada, Granada (Zamora and Camacho 1984b); 6 Cazorla, Jaén (Obeso 1987); 7 Cazorla, Jaén (Purroy 1975a); 8 Navarra (Purroy 1975a); 9 Córdoba (Torres and León 1985); 10 plot A, Huelva (Herrera 1980); 11 Sierra Nevada, Granada (Zamora and Camacho 1984a); 12 plot B, Huelva (Herrera 1980); 13 Sevilla (Jordano 1985); 14 Vizcaya (Galarza 1987); 15 Cuenca (Potti and Garrido 1986); 16 Doñana, Huelva (Costa 1984); 17 Doñana, Huelva (Costa 1984); 18 Alcañiz, Teruel (Suárez and Sáez-Royuela 1985).

group included two mixed forests and one pine forest, and it also presented the typical pattern of summer maximum and winter minimum, but with a marked peak at the end of summer and beginning of autumn corresponding to the post-breeding migration. The third group included bird communities with abundance maximum in spring, and a pronounced decrease in summer, due probably to the high temperatures and summer drought. This group included two holm oak woods and one mixed forest that was also dominated by holm oak, all of them located in southern Spain (Andalusia). Finally, the fourth group included bird communities with abundance maximum in winter and minimum in summer, a pattern opposite to that found in our study area. This group is heterogeneous with respect to habitat types: it included Mediterranean shrublands, cultivated fields, steppes, and coastal pine woods. The only two forest habitats included are coastal pine woods located in southernmost Spain (Doñana).

All the holm oak forests reviewed here exhibited maximum bird abundance in spring or summer, with minimum values in winter, but two groups are clearly identifiable. Holm oak forests in the northern half of Spain reached the maximum abundance in mid-summer, while holm oak forests located in the southern half of the Peninsula presented the maximum abundance earlier, at the end of spring. In southern forests, most of the birds disappeared from the area at the beginning of summer, producing a summer minimum. After this season, bird abundance recovered and decreased again in winter. Thus, it appears that in holm oak forests of southern Spain, the summer conditions impose more important limitations on the birds than in more northern areas, where winter would be the critical season. However, altitude seems to play an important role in determining the seasonal pattern of bird abundance fluctuation, as the holm oak forest studied by Zamora and Camacho (1984b), placed at about 1000 m a.s.l., suggests. Its bird community is included within the first group in Fig. 25.5, despite being located in southern Spain.

References

Blondel J, Farré H (1988) The convergent trajectories of bird communities along ecological successions in European forests. Oecologia 75:83–93

Carrascal LM (1986) Estructura de las comunidades de aves de las repoblaciones de *Pinus radiata* del País Vasco. Munibe 38:3–8

Carrascal LM, Tellería JL (1990) Impacto de las repoblaciones de *Pinus radiata* sobre la avifauna forestal del norte de España. Ardeola 37:247–266

Carrascal LM, Potti J, Sánchez-Aguado FJ (1987) Spatio-temporal organization of the bird communities in two Mediterranean montane forests. Holarct Ecol 10:185–192

Costa L (1984) Composición de la comunidad de aves en pinares del Parque Nacional de Doñana (suroeste de España). Doñana Acta Vertebr 11:151–183

Costa L (1993) Evolución estacional de la avifauna en hayedos de la montaña cantábrica. Ardeola 40: 1–11

Fernández A, Galarza A (1986) Estructura y estacionalidad de las comunidades de aves en distintos medios del tramo costero del País Vasco. Bol Estac Centr Ecol 29:59–66

Galarza A (1987) Descripción estacional de las comunidades de passeriformes en una campiña costera del País Vasco. Munibe 39:3–8

Garnica R (1988) Ciclo anual de la ornitocenosis del encinar de llanura en la provincia de León (estudio zoogeográfico y aspectos tróficos). Stud Oecol 5:191–204

Herrera CM (1978) On the breeding distribution pattern of European migrant birds: MacArthur's theme reexamined. Auk 95: 496–509

Herrera CM (1980) Composición y estructura de dos comunidades mediterráneas de passeriformes. Doñana Acta Vertebr 7(Esp):11–340

Holmes RT, Sturges FW (1975) Bird community dynamics and energetics in a northern hardwoods ecosystem. J Anim Ecol 44:175–200

Iglesias DJ (1996) Efecto de la abundancia de alimento sobre los parámetros reproductores de los Páridos en un encinar mediterráneo. PhD Thesis, University of Valencia, Valencia

Jordano J (1985) El ciclo anual de los passeriformes frugívoros en el matorral mediterráneo del sur de España: importancia de su invernada y variaciones interanuales. Ardeola 32:69–94

Keast A (ed) (1990) Biogeography and ecology of forest bird communities. SPB Academic Publishing, The Hague

López G (1991) Intervención de la avifauna en el ciclo de nutrientes de un encinar mediterráneo. PhD Thesis, University of Alicante, Alicante

López G (1992) A comparative study about bird diversity of different forest types in the Iberian Peninsula and central Europe. In: Thanos CA (ed) Plant-animal interactions in Mediterranean type ecosystems. MEDECOS VI, pp 165–170

Maicas R (1989) Comunidades de aves en matorrales y ecosistemas forestales de la Sierra de Filabres (Almería). IARA Junta de Andalucía, Sevilla

Obeso JR (1987) Comunidades de passeriformes en bosques mixtos de altitudes medias de la Sierra de Cazorla. Ardeola 34:37–59

Pedrocchi C (1975) Efecto topoclimático en la densidad de nidificación de aves. Publ Centr Pir Biol Exp 7:163–167

Potti J (1986) Densidad y riqueza de aves en comunidades nidificantes de la Península Ibérica. Misc Zool 10:267–276

Potti J, Garrido G (1986) Dinámica estacional de una ornitocenosis agrícola en el centro de España. Alytes RECN 4:29–48

Purroy FJ (1974) Contribución al conocimiento ornitológico de los pinares pirenaicos. Ardeola 20:245–261

Purroy FJ (1975a) Evolución anual de la avifauna de un bosque mixto de coníferas y frondosas en Navarra. Ardeola 21:669–697

Purroy FJ (1975b) Avifauna nidificante e invernante del robledal atlántico de Quercus sessiliflora. Ardeola 22:85–95

Santos T, Alvarez G (1990) Efectos de las repoblaciones con eucaliptos sobre las comunidades de aves forestales en un maquis mediterráneo (Montes de Toledo). Ardeola 37:319–324

Santos T, Suárez F (1983) Comparative study of the results obtained from the use of three different methods in a beech forest (Fagus sylvatica L.) of the Cordillera Cantábrica. In: Purroy FJ (ed) Bird census and Mediterranean landscape. Proc VII Int Congr Bird Census IBCC/V Meeting EOAC, Universidad de León, León, pp 96–99

Sturges FW, Holmes RT, Likens GE (1974) The role of birds in nutrient cycling in a northern hardwoods ecosystem. Ecology 55:149–155

Suárez FS, Sáez-Royuela C (1985) Variación estacional de la estructura y demanda energética de dos comunidades de passeriformes de zonas semiáridas. Stud Oecol 6:181–203

Tellería JL, Garza V (1983) Methodological features in the study of a Mediterranean forest bird community. In: Purroy FJ (ed) Bird census and Mediterranean landscape. Proc VII Int Congr Bird Census IBCC/V Meeting EOAC, Universidad de León, León, pp 89–92

Tellería JL, Galarza A (1990) Avifauna y paisaje en el norte de España: efecto de las repoblaciones con árboles exóticos. Ardeola 37:229–245

Torres JA, León A (1985) Estudio de la comunidad de passeriformes del bosque mixto mediterráneo de la Sierra de Hornachuelos (Córdoba-España). Servicio de Publicaciones de la Universidad de Córdoba, Córdoba

Zamora R (1987) Variaciones altitudinales en la composición de las comunidades nidificantes de aves de Sierra Nevada (S. de España). Doñana Acta Vertebr 14:83–106

Zamora R, Camacho I (1984a) Evolución estacional de la comunidad de aves en un encinar de Sierra Nevada. Doñana Acta Vertebr 11:25–43

Zamora R, Camacho I (1984b) Evolución estacional de la comunidad de aves en un robledal de Sierra Nevada. Doñana Acta Vertebr 11:129–150

Subject Index

Ecological Studies
Volumes published since 1992

Ecological Studies
Volumes published since 1992

Springer
and the
environment

At Springer we firmly believe that an international science publisher has a special obligation to the environment, and our corporate policies consistently reflect this conviction.
We also expect our business partners – paper mills, printers, packaging manufacturers, etc. – to commit themselves to using materials and production processes that do not harm the environment. The paper in this book is made from low- or no-chlorine pulp and is acid free, in conformance with international standards for paper permanency.

Springer

Printing: Saladruck, Berlin
Binding: Buchbinderei Lüderitz & Bauer, Berlin